AF510435

Solid State Physics: Essential Concepts

Solid State Physics: Essential Concepts

Augusta Lawrence

STATES
ACADEMIC PRESS
www.statesacademicpress.com

Solid State Physics: Essential Concepts
Augusta Lawrence
ISBN: 978-1-63989-491-8 (Hardback)

STATES
ACADEMIC PRESS

Published by States Academic Press,
109 South 5th Street,
Brooklyn, NY 11249, USA

Cataloging-in-Publication Data

Solid state physics : essential concepts / Augusta Lawrence.
p. cm.
Includes bibliographical references and index.
ISBN 978-1-63989-491-8
1. Solid state physics. 2. Solids. 3. Physics. I. Lawrence, Augusta.
QC176 .I58 2022
530.41--dc23

For more information regarding States Academic Press and its products, please visit the publisher's website www.statesacademicpress.com

Table of Contents

Preface

Solids are made up of densely packed atoms. The interactions between these atoms decide the various mechanical, electrical, thermal, optical and magnetic properties of solids. These solids can be broadly classified into crystalline solids and amorphous solids. Solid state physics is the sub-discipline of condensed matter physics which is concerned with the study of such solids. It focuses on how the large scale properties of matter result from its atomic scale properties. The wide variety of techniques used in solid state physics range from electromagnetism, metallurgy, crystallography and quantum mechanics. Some of the emerging areas of research in this field are quasicrystals, spin glass, nanomaterials, two dimensional materials and superconductivity. The subject of solid state physics finds extensive application in the fields of consumer electronics, fiber optics and silicon-based memory bits. This book attempts to understand the multiple branches that fall under the discipline of solid state physics and how such concepts have practical applications. The topics covered herein deal with the core subjects of solid state physics. This book is an essential guide for both academicians and those who wish to pursue this discipline further.

To facilitate a deeper understanding of the contents of this book a short introduction of every chapter is written below:

Chapter 1- The area of study within condensed matter physics which focuses on rigid matter by using methods from the fields of crystallography, metallurgy and quantum mechanics is known as solid state physics. It studies the relationship between the large-scale properties of solid materials and their atomic-scale properties. This is an introductory chapter which will introduce briefly all the significant aspects of solid state physics.

Chapter 2- The description of the ordered arrangement of ions, molecules or atoms in a crystalline material is known as the crystal structure. The interruptions in the regular pattern of crystalline solids are known as crystal defects. This chapter has been carefully written to provide an easy introduction to the varied facets of crystal structure and defects such as inelastic interaction of X-rays with matter and elastic interaction of X-Rays with matter.

Chapter 3- Some of the different aspects studied in relation to the thermal properties of solids are atomic bonding, heat capacity of solids, lattice vibrations, specific heat capacities and the Dulong-Petit Law and Einstein's and Debye's theory. The diverse applications of these thermal properties of solids have been thoroughly discussed in this chapter.

Chapter 4- The magnetic properties of solids are due to the electrons present in the molecules or atoms. The materials which are used and studied for their magnetic properties are known as magnetic materials. The topics elaborated in this chapter will help in gaining a better perspective about the magnetic properties of solids as well as magnetic materials and theories of magnetism.

Chapter 5- The dielectric properties of solids deal with the storage and dissipation of magnetic and electric energy in materials. Some of the concepts studied in relation to the dielectric properties of solids are polarization density, dielectric constant, dielectric loss and pyroelectricity. All these diverse dielectric properties of solids have been carefully analyzed in this chapter.

Chapter 6- The range of energy levels which might be present within electrons as well as those which might be missing from it is described through the electronic band structure. The energy levels which are missing are known as band gaps. The topics elaborated in this chapter will help in gaining a better perspective about electron band structure, band gap and Fermi liquid theory.

I owe the completion of this book to the never-ending support of my family, who supported me throughout the project.

Augusta Lawrence

1

Understanding Solid State Physics

The area of study within condensed matter physics which focuses on rigid matter by using methods from the fields of crystallography, metallurgy and quantum mechanics is known as solid state physics. It studies the relationship between the large-scale properties of solid materials and their atomic-scale properties. This is an introductory chapter which will introduce briefly all the significant aspects of solid state physics.

The branch of physics that deals with the study of rigid solid matter is called Solid state physics. It is done through a variety of methods including quantum mechanics, electromagnetism, metallurgy and crystallography and forms the theoretical foundation of materials science as a whole. It can be defined as:

"The study of rigid matter, or solids, through methods such as quantum mechanics, crystallography, electromagnetism, and metallurgy."

Function

The basic function of solid-state physics is to study how the atomic properties of a solid material affect its overall properties. It did not emerge as a separate field of study till the 1940s, although the properties of solids have been studied by scientists for centuries.

Every solid material is made up of densely packed atoms, the interaction between which decides the properties of the material. The mechanical, electrical, optical, thermal and magnetic properties of the item are all dependent on the interaction of the atoms making it up. The alignment of the atoms in a regular, geometric pattern or an irregular pattern depends on the materials involved and the conditions under which it was formed.

The example of regular and irregular patterns can be crystalline solids such as metals and amorphous solids such as common window glass respectively.

In its modern form, solid state physics is usually said to have begun around the end of World War II. The development of the transistor, based on theories about the electrical properties of semiconductor solids, was announced in 1948. It replaced the much bulkier vacuum tube in radios, in computers, and in many scientific instruments.

Semiconductor devices similar to transistors have made possible the construction of tiny electronic circuits. These have found application in a wide range of fields—from extremely expensive computers with a very limited market to such low-priced, mass-produced consumer goods as radios, television sets, and telephones.

In 1960 the first laser was developed, based on theories about how solid ruby absorbs and emits light waves. Since then, laser light has been sent to the moon and back, has been used in delicate eye surgery, and has guided the construction of a two-mile-long linear accelerator.

Applications of Solid State Theory

For many centuries, people learned how to use solids by trial and error. Man's first pottery and metallurgical techniques were acquired in this way. For example, early in man's development it was discovered that heating certain kinds of earth—the metal ores—caused the metals in them to separate out.

Then it was learned that metallic mixtures might have desirable properties different from those of their individual metals. But no one knew what internal changes caused these improvements, and there was no way to foresee whether a particular mixture would have the qualities sought. Man depended on accidental discoveries.

The situation is now quite different. New solid state devices are often preceded by theoretical predictions that they will have desirable properties. For example, scientists have applied solid state theory to control the magnetic properties of ferrites. Since ferrites are not electrical conductors, they are used in television tubes and as antenna cores. The magnetic orientation of ferrites is easily and quickly changed by the application of an external magnetic field. This explains their use as the "memory" storage units of computers.

Many solid state applications have developed from the theories of imperfections in solids. Alloys—mixtures of metals—may be stronger than any of their metallic components if the atoms of one of these metals fill microscopic gaps, called edge dislocations, in the crystal structure of another. But sometimes the atoms of a metal in an alloy may act as lubricants rather than cements, and the alloy may then be weaker than any of its component metals.

The functioning of transistors and solar cells depends on the addition of impurity atoms to a semiconductor. When an impurity atom adds extra electrons, a negative semiconductor area is formed. When it provides positions where electrons can settle, a positive semiconductor area is formed. A series of three alternating negative and positive areas can easily conduct an electric current.

Phosphors—solids that give off visible light when excited by radiation—are similarly dependent on the presence of impurities. Phosphors are used to create color television pictures. As new phosphors that emit more desirable colors are developed, the quality of these pictures will be improved. Some impurities delay the action of phosphors by capturing electrons that are involved in the light-emitting process. But other impurities seem to be necessary for light emission to take place.

Concepts of Solid State Physics

In 1912 Max von Laue established that crystals diffract X rays in an orderly manner. X-ray diffraction photographs revealed that a crystal is an ordered arrangement of atoms or molecules in a regular repeating pattern. The particular pattern often helps explain properties of a given crystal. For example, metals that have one kind of crystalline arrangement become brittle at low temperatures, while metals that have another kind remain strong.

While crystallographers were analyzing the patterns of crystal atoms, atomic and quantum theorists tried to determine what forces caused these patterns. Atomic theory postulates that an atom has a central, positively charged nucleus surrounded by outer, negatively charged electrons. Crystalline solids have been classified as ionic, covalent, metallic, or molecular crystals, depending on how the outermost of these electrons interact.

Quantum mechanics is a way of describing the relationships between energy and matter. This discipline is preferred because it has been very successful in explaining otherwise inaccessible atomic phenomena. Its predictions are the most precise and the best checked of any in physics; some of them have even been tested and found accurate to better than one part per billion. For this reason, quantum mechanics has been used to explain mathematically the electrical conductivity of metals, semiconductors, and insulators. In such calculations, three energy regions are significant.

Filled energy bands contain all the electrons that are attached to their atoms. When electrons occupy the conduction band they are completely free of their atoms and available to an electric current. Energies falling between the conduction band and a filled band are said to be in a forbidden region. Quantum calculations predict that other atoms in the crystal reflect electrons having forbidden energies, thereby preventing their movement in an electric current.

In metals, some electrons have enough energy to occupy the partially empty conduction band even at very low temperatures (in other words, they are free from their atoms). In insulators, all the bands are completely filled (electrons are attached to their atoms), and no electrons are left over to join the conduction band. Furthermore, the jump to the empty conduction band is very great—five to seven electron volts—so electrons are not likely to gain enough energy to reach it. Finally, in semiconductors, the energy bands are completely filled, with no electrons left over. But the gap between one filled band and the empty conduction band is very small—about 0.5 electron volts—so a small energy input such as the thermal energy at ordinary temperatures can boost some of the electrons into the conduction band.

A metal can carry an electric current because it always has electrons in the conduction band. Its electrical conductivity decreases as the temperature rises because the more energetic atoms interfere with electron movement. A semiconductor permits electricity to pass through it only if an external source of energy, such as an electric current or heat, is introduced. Its electrical conductivity therefore increases with temperature. Finally, an insulator—for example, a rubber sole—will not ordinarily conduct electricity. But if an enormous electric field is applied—for instance, if an object is struck by a lightning bolt—enough energy is supplied to raise some electrons over the large forbidden region into the conduction band. In such circumstances the insulator will also conduct electricity.

Issues in Solid State Physics

Many solids we interact with—metals, stones, etc.—are random assemblies of grains, held together by strong adhesion forces. Like those of sand, quartz, or salt, those grains are very likely to be themselves crystals (which has said do not imply they are perfect: they may contain lots of impurities and defects). But there are two particular aspects of crystals we are concerned with here. The first is that unlike complex systems, which may display emergent structures at each scale (think, e.g., of mitochondria, cells, tissues, organs, etc.), crystals are very simple: they are huge assemblies of elementary building blocks (be they atoms, molecules, nanoclusters, or whatever). The second is that since the building blocks obey quantum mechanics, crystals inherit the quantum character (despite being themselves macroscopic).

Band Spectra in the Tight-Binding Approach: Effects of the Overlaps Between Neighboring Orbitals

Quantum Mechanics in a Nutshell

- Dynamical states are vectors: One can account for the wavelike behavior of quantum objects (e.g., diffraction of single electrons by two slits) by letting their dynamical state $|\psi\rangle$ belong to a vector space over the complex numbers. In few problems (e.g., addition of angular momenta), this vector space is finite-dimensional. But most problems entail infinite sequences (e.g., energy spectrum of the hydrogen atom) or even a continuum of values (e.g., in the measurement of positions and momenta), so the notion of dimension is replaced by that of completitude (any state can be spanned in suitable "bases"). By assigning a complex number $\langle\varphi|\psi\rangle$ (their "internal product") to every pair of dynamical states $|\psi\rangle$, $|\varphi\rangle$, the complete vector space is made into a Hilbert space.
- Probabilistic interpretation: If $|\psi\rangle=\sum_I \alpha_I |\psi_I\rangle$ (be aware that the index set I may be infinite or may even be a patch of $\mathbb{R}^d$), then $|\alpha_I|^2$ yields the probability to find an outcome represented by $|\psi_I\rangle$, when the system is in state $|\psi\rangle$. This obviously requires normalization: $\langle\psi|\psi\rangle=1$.
- Dynamical magnitudes are linear operators L, which take a vector into another vector. For instance, the projector $P_\varphi=|\varphi\rangle\langle\psi|$ projects state $|\psi\rangle$ onto $|\varphi\rangle$. Measuring a dynamical magnitude thus means finding one of its eigenvalues $L|l_I\rangle=l_I|l_I\rangle$. Also of interest is the mean (or expectation) value $\langle\psi|L|\psi\rangle$ of L in a generic state $|\psi\rangle$. Correspondence with classical physics imposes that those eigenvalues be real, and thus dynamical magnitudes must be self-adjoint (Hermitian) operators (P_φ is thus *not* a dynamical magnitude).
- Unitary evolution: In order to conserve the probabilistic interpretation, the dynamic evolution of the state is accomplished by a unitary operator. Again, correspondence with classical physics (already implicit in Schrödinger's equation) forces this operator to be exp(−iH).
- Wave function: A possible "basis" set is that of eigenstates $(X|x\rangle=x|x\rangle)$ of the position operator, namely, $|\psi\rangle=\int dx\,\psi(x)|x\rangle$. The *wave function* $\psi(x)$ plays here the role of the coefficients α_I. In modern notation, $\psi(x)$ is written as $\langle x|\varphi\rangle$, so one writes I $|\varphi\rangle=\int dx=|x\rangle\langle x|\varphi\rangle$.
- Orthogonality $(\langle\varphi|\psi\rangle=0)$: A given eigenvalue l_1 of a Hermitian operator L may have a single eigenstate $|l_1\rangle$ (the normalized one out of a dimension-1 subspace over the

complex numbers) or more (in this case, it is said to be *degenerate*). Eigenstates corresponding to *different* eigenvalues are *automatically* orthogonal.

Naive Tight-Binding Approach to Band Theory

The starting point of this approach is to express the electron's state vector as a linear combination of atomic orbitals (LCAO) located at the crystal's lattice sites. The eigenvalue problem of the isolated atom centered at x_c is $H_{atom}|\psi_{atom}\rangle = E_{atom}|\psi_{atom}\rangle$, with $H_{atom} = T + V(x - x_c)$. We then place a copy $|i\rangle$ $|\psi_{atom}\rangle$ of centered at each lattice site $i(x_i = i_a)$ and write the electron's state in the crystal as LCAO:

$$|\psi_{crystal}\rangle = \sum_i c_i |i\rangle$$

(clearly, $c_i = \langle i|\psi_{crystal}\rangle$). Now, even though the interatomic distance in the crystal (the "lattice spacing" a) is usually larger than the range x_0 of the atomic orbitals, the atomic cores *do* interact, and one should include at least two effects:

- A *correction* to the isolated atomic level E_{atom} (we shall call αα the corrected level).
- Electron tunneling between neighboring orbitals (let γ be a gauge of the energy involved in such a "hopping" process).

It thus makes sense to write up the lattice Hamiltonian in terms of projection operators as:

$$H_{crystal} = \alpha \sum_i |i\rangle\langle i| - \gamma \sum_i \left(|i+1\rangle\langle i| + |i\rangle\langle i+1|\right)$$

Two timely comments are:

The presence of $|i\rangle\ \langle i+1|$, the adjoint of $|i+1\rangle\ \langle i|$, ensures that Hcrystal be Hermitian.

The minus sign in the second term ensures crystal stability (energy is released by forming a crystal).

Using Eqs. $|\psi_{crystal}\rangle = \sum_i c_i |i\rangle$ and $H_{crystal} = \alpha \sum_i |i\rangle\langle i| - \gamma \sum_i \left(|i+1\rangle\langle i| + |i\rangle\langle i+1|\right)$, the eigenvalue problem $H_{crystal}|\psi_{crystal}\rangle = E_{crystal}|\psi_{crystal}\rangle$ for the electron in the crystal reads:

$$\left[\alpha \sum_i |i\rangle\langle i| - \gamma \sum_i \left(|i+1\rangle\langle i| + |i\rangle\langle i+1|\right)\right] \sum_j c_j |j\rangle = E_{crystal} \sum_j c_j |j\rangle$$

Assuming the states |i⟩ to be *orthogonal* to each other, the left-hand side of previous equation reads $\sum_i c_i |i\rangle - \gamma \sum_i \left(c_i |i+1\rangle + c_{i+1}|i\rangle\right)$. If the number of sites in the crystal is large enough (usually it is ~10^6), one can greatly simplify the problem by assuming periodic boundary conditions (PBC). This allows to rearrange the sums (their indices become dummy), and previous equation reads $\sum_i \left[\left(E_{crystal} - \alpha\right)c_i - \gamma \sum_i \left(c_{i-1} + c_{i+1}\right)\right]|i\rangle = 0$. Clearly, the LCAO assumes that the |i⟩ are *linearly independent* (be they orthogonal or not), so we are left with the system of difference equations:

$$(E_{crystal} - \alpha)c_i - \gamma\sum_i (c_{i-1} + c_{i+1}) = 0, i = 1\ldots N \equiv 0$$

Again invoking PBC, one tries the form $c_j = \exp ijka$ with $-\pi < ka \le \pi$ (Bloch phase factors) and obtains the known cosine spectrum:

$$E_{crystal} = \alpha - 2\gamma\cos ka,\ -\pi < ka \le \pi$$

Tight-Binding Band Calculation: Properly Done

Recognizing that $H_{crystal} = \sum_i \left(H_i^{atom} + W_i\right)$ and using Eq. $\left|\psi_{crystal}\right\rangle = \sum_i c_i \left|i\right\rangle$, $E_{crystal}$ turns out to be:

$$E_{crystal} = E_{atom} + \left[\sum_i \alpha_i \left|c_i\right|^2 + \sum_{ij}\gamma_{ij}c_i^* c_j\right] / \left[\sum_i \left|c_i\right|^2 + \sum_{ij} S_{ij}c_i^* c_j\right]$$

Where,

$$\alpha_i = H_{ii} = \left\langle i\middle|W_i\middle|i\right\rangle, \gamma_{ij} = H_{ij} = \left\langle i\,|\,W_i\,|\,j\right\rangle, S_{ij} = \left\langle i\middle|j\right\rangle, j \ne i$$

The contribution of the S_{ij} (known as overlap integrals) to the band spectrum is the main concern in this section. But not less interesting are that of the α_i terms—which, as argued, shift the electronic energy in an atom from its isolated value E_{atom}, as a collective effect of the other atoms—and that of the γ_{ij}. The latter can be regarded as the sum of two contributions, as $V_j = V(x - x_j)$ can be singled out from W_i. Then whereas the two-center integrals $\gamma_{ij}^{(2)} = \left\langle i\middle|V_j\middle|j\right\rangle$ involve only sites i and j, the three-center integrals $\gamma_{ij}^{(3)}$ also involve the sum $\sum_{l \ne i,j} V(x - x_l)$ of the potentials of the remaining atoms in the solid. Hence, the $\gamma_{ij}^{(3)}$ can be interpreted as the collective effect on the overlap between orbitals i and j.

Variation of Eq. $E_{crystal} = E_{atom} + \left[\sum_i \alpha_i \left|c_i\right|^2 + \sum_{ij}\gamma_{ij}c_i^* c_j\right] / \left[\sum_i \left|c_i\right|^2 + \sum_{ij} S_{ij}c_i^* c_j\right]$ with respect to the LCAO coefficients of Eq. $\left|\psi_{crystal}\right\rangle = \sum_i c_i \left|i\right\rangle$—namely, $\partial E_{crystal} / \partial a_j^* = 0$ —yields $(H_{ij} - E_{crystal} S_{ij})a_j = 0, \forall j$. Assuming PBC, H_{ij} and S_{ij} are functions only of the interatomic distance na, with $n = \left|i - j\right|$. Again using $a_n = \exp inka$ with $-\pi < ka \le \pi$, Eq. $E_{crystal} = E_{atom} + \left[\sum_i \alpha_i \left|c_i\right|^2 + \sum_{ij}\gamma_{ij}c_i^* c_j\right] / \left[\sum_i \left|c_i\right|^2 + \sum_{ij} S_{ij}c_i^* c_j\right]$ yields:

$$E_{crystal} = E_{atom} + \left[\alpha + 2\sum_n H_n \cos nka\right] / \left[1 + 2\sum_n S_n \cos nka\right]$$

Note however that the number of multicenter integrals to be computed is immense! Because of that, most tight-binding calculations plainly ignore almost all the multicenter integrals (keeping only those involving nearest neighbors) and neglect orbital non-orthogonality. This way, the familiar cosine spectrum is obtained. Often, multicenter integrals are just regarded as parameters to fit the results of more sophisticated calculations made by other methods at the highest symmetry points of the Brillouin zone.

In the following, we compute *all* the multicenter integrals *exactly* in the framework of a simple model for the atomic potential. The results help get an intuition on the effect on band spectrum of neglecting overlap integrals and distant-neighbor interactions.

A Simple Model that Yields an Exact Tight-Binding Band Spectrum

We restrict the selves to a 1D monoatomic crystal and assume the interatomic distance a to be larger than the effective range of the screened Coulomb potential representing the atomic core. In such a situation, we can approximate the latter by a Dirac δ -function (complete screening up to the scale of the nucleus):

$$V_{crystal}(x) = -V_0 \sum_n \delta(x - na)$$

The solution to $H_{atom}|\psi_{atom}\rangle = E_{atom}|\psi_{atom}\rangle$, with $H_{atom} = -\frac{\hbar^2}{2m}\frac{d^2}{dx^2} - V_0\delta(x)$ is an exponential function of the form $\psi_{atom}(x) = \langle x|\psi_{atom}\rangle = x_0^{-\frac{1}{2}}\exp\left(-\frac{|x|}{x_0}\right)$. Its range x_0 is related to E_{atom} by $-E_{atom} = \hbar^2/2mx_0^2 = mV_0/2\hbar^2$.

The only two spatial scales involved in this problem are x_0 and the lattice spacing a. The parameter $t = a/x_0$ will thus allow us to follow the formation of energy bands (k-space picture) as atoms get close together (real-space picture). All the multicenter integrals can be computed analytically in terms of t. The results are $S_n = (1+nt)\exp(-t)$, $\alpha = 2E_{atom}\exp(-t)/\sinh t$, and $\gamma_n = 2E_{atom}[n + \exp(-t)/\sinh t]\exp(-nt)$. We thus get the following closed expression for $\lambda = (E_{crystal} - E_{atom})/E_{atom}$:

$$\lambda(k,t) = [A_0(t) + A_1(t)\cos ka]/[1 + S(t)\cos ka], -\pi < ka \le \pi$$

Where,

- $A_0 = \exp(-t)\sinh t/[\sinh t \cosh t - t]$,
- $A_1 = \sinh t/[\sinh t \cosh t - t]$,
- $S = [t\cosh t - \sinh t]/[\sinh t \cosh t - t]$.

Explicit evaluation of Eq. $\lambda(k,t) = [A_0(t) + A_1(t)\cos ka]/[1 + S(t)\cos ka], -\pi < ka \le \pi$ at the bottom (ka = 0) and top (ka = π) of the band shows that for t < 4, the cosine spectrum of Equation

$E_{crystal} = \alpha - 2\gamma \cos ka, -\pi < ka \leq \pi$. underestimates both. Moreover, the multicenter integrals neglected in the cosine spectrum shift unevenly the top and bottom of the exact spectrum. Hence, the approximation performs worse for the top than for the bottom of the band.

Quantum Monte Carlo Method for Systems with Strongly Correlated Fermions

Quantum Statistical Mechanics in a Nutshell

When one deals with statistical ensembles of quantum states, the object of interest is the Hermitian operator $\exp(-\beta H)$ called the density matrix operator (here $\beta = (k_B T)^{-1}$ and $k_B = R/N_A$ are Boltzmann's constant).

What drives the interest in the density matrix—namely, the matrix elements between pure states of $\exp(-\beta H)$ is the fact that it can be used to find the ground state of many-body systems by stochastic methods. For β large enough, $\exp(-\beta H)$ acts effectively as a projector over the lowest-lying energy eigenstate to which the initial (trial) state $|\varphi\rangle$ is not definitely orthogonal. Let E be the corresponding eigenvalue, and consider another trial state $|\chi\rangle$ over which we will project the result. Then we may numerically compute E from:

$$\exp(-\Delta\beta E) = \lim_{\beta\to\infty}\left[\langle\chi|\exp\left[-(\beta+\Delta\beta)H\right]|\varphi\rangle / \langle\chi|\exp(-\Delta\beta H)|\varphi\rangle\right]$$

But what is yet more interesting is that in the process, we find a good estimate of the eigenstate itself, namely, its composition in terms of a known basis.

Monte Carlo Pursuit of the Ground State

The first step in this computation is to divide the interval [o, β] into L "time" slices of width $\Delta\tau = \beta / L$. Some comments are in order:

We take the language from the formal analogy between the density matrix and the evolution operators.

Note that in the case, $\exp(-\beta H)$ is not meant to be traced over as it should be in a thermodynamic calculation: here it must rather be considered as a formal tool to make sense in the limit $\beta \to \infty$.

We may call $U = \exp(-\Delta\tau H)$ the transfer matrix operator.

If we can decompose H into a sum of several terms H_i which (although not commuting among them) are themselves sums of commuting terms, then for L large enough, the error of approximating:

$$U = \exp\left[-\Delta\tau(H_1 + 2)\right] = \exp(-\Delta\tau 1)\exp(-\Delta\tau H_2)\exp\{-(\Delta\tau)^2 [H_1, H_2]\}$$
$$= U_1 U_2 \{1 - (\Delta\tau)^2 [H_1, H_2] + \dots\} \sim U_1 U_2$$

Would be at most of order $(\Delta\tau)^2$. Hence,

$$\langle\chi|\exp(-\beta H)|\phi\rangle \sim \langle\chi|(U_1U_2)^L|\varphi\rangle.$$

In order to evaluate previous expression, we introduce complete sets of states at each time slice.

The clue to quantum Monte Carlo simulation of Equation $\exp(-\Delta\beta E) = \lim_{\beta\to\infty}\left[\langle\chi|\exp\left[-(\beta+\Delta\beta)\right]H\right]|\varphi\rangle / \langle\chi|\exp(-\Delta\beta H)\varphi\rangle\right]$ resides in evaluating the sums over complete states by importance sampling: in order to do that, observe first that we can rather arbitrarily decompose:

$$\langle\psi_j|(U_1U_2)^L|\psi_i\rangle = S_{ij}\ P_{ij}$$

As the product of a probability times a (complex) number which we will call a "score." The probability distribution P_{ij} is at the disposal in order to optimize numerical convergence, minimize statistical error, etc. It can be shown that the way to achieve the last goal is by assigning to every matrix element the same score: that is the basis for the so-called population method. Here the initial (trial) state is represented by a "population" in which there are n_i copies of state $|\psi_i\rangle$. The latter corresponds to a definite assignment of occupation numbers both in coordinate and spin (always belonging to the Hilbert space of the problem, i.e., compatible with the conserved quantum numbers). To each individual in the population, we apply the evolution operator, thus obtaining a new state after one time slice. That particular matrix element can be decomposed as indicated in Eq. ($\Phi(u,v) = \lambda_2^{-1}(\beta u - v)^2 + (\lambda_1\varepsilon)^{-1}\left(\beta u^2 - 2\int_{u_0}^{u} f(x)dx\right)^2$) (but being now $S_{ij} = S = \text{const}$). The way in which we implement the P_{ij} is by making as many copies of that particular resulting state as indicated by $\langle\psi_j|(U_1U_2)^L|\psi_i\rangle/S$. Proceeding this way, we will get a different population after each time slice which we expect to approach successively to one representing the lowest reachable energy eigenstate.

We have not said anything about the way in which we evaluate the alluded matrix elements, besides the fact that we resort to the decomposition ($\Phi(u,v) = \lambda_2^{-1}(\beta u - v)^2 + (\lambda_1\varepsilon)^{-1}\left(\beta u^2 - 2\int_{u_0}^{u} f(x)dx\right)^2$): if, as we have already assumed, the term H_i can itself be decomposed into mutually commuting terms, we need only to focus on the Hilbert space of that (much smaller) system. We can compute exactly the matrix elements of the evolution operator for that cluster, write them as the product of a probability times a score (now we can choose the probability distribution to minimize total computing time), and make transitions among cluster states according to those probabilities, assigning then the corresponding score to the particular transition.

The Case of Fermions

Again within the tight-binding approach to crystalline solids, quantum creation $(c_{is}^{\dagger})$ and annihilation (c_{is}) operators determine the existence of electrons with spin projection σ at site i. For the state vector of the whole set of electrons in the crystal to be totally antisymmetric under exchange, those operators must anticommute with each other, unless they refer to the same site and spin projection. In such a case, there can be at most one electron per site and spin projection, as required by Pauli's principle.

In the case of the 1D Hubbard model, we chose the following decomposition of the Hamiltonian, which allows us to consider clusters of only two sites:

$$H_1 = -t\sum_{oddj}\sum \left(c^{\dagger}_{j+1\sigma}c_{j\sigma} + h.c.\right) = -t\sum_{oddj}\sum_{\sigma} h_{j,j+1}$$

$$H_2 = -t\sum_{evenj}\sum_{\sigma}\left(c^{\dagger}_{j+1\sigma}c_{j\sigma} + h.c.\right) = -t\sum_{evenj}\sum_{\sigma} h_{j,j+1}$$

$$H_3 = -t\sum_{all\,j} n_{j\uparrow}n_{j\downarrow}$$

The corresponding matrix elements are then $\langle \psi_{i+1}|U_3U_2U_1|\psi_i\rangle$ with $U_1 = \prod_{oddj}\exp(-\Delta\tau h_{j,j+1}), U_2 = \prod_{evenj}\exp(-\Delta\tau h_{j,j+1}), U_3 = \prod_j \exp(-\Delta\tau n_{j\uparrow}n_{j\downarrow})$, and

$$\langle 01|\exp(-\Delta\tau h_{j,j+1})|01\rangle = \langle 10|\exp(-\Delta\tau h_{j,j+1})|10\rangle = \cosh\Delta\tau$$

$$\langle 10|\exp(-\Delta\tau h_{j,j+1})|01\rangle = \langle 01|\exp(-\Delta\tau h_{j,j+1})|10\rangle = \sinh\Delta\tau$$

$$\langle 00|\exp(-\Delta\tau h_{j,j+1})|00\rangle = \langle 11|\exp(-\Delta\tau h_{j,j+1})|11\rangle = 1$$

From which we write up the (a priori) transition probabilities. Then, in case there is only one occupied site in the block, we draw a random number r and compare it with the a priori transition probability p for the state to remain the same. In case that r>p, we make a hopping, i.e., exchange empty and occupied states in the block.

The a priori probabilities can be better chosen if we take into account the occupation of those same two sites by electrons with the other spin projection, thus anticipating to the fact that they will penalize doubly occupied sites. This will certainly improve convergence.

Non-Equilibrium Routes to Soft Solids

Up to now, we have dealt with crystalline solids. This means that disregarding the topology of the interaction network, we paid attention to the underlying geometry of the quantum problem. At present, a host of synthetic materials has outperformed metals at their initial tasks. Some of them still display a varying degree of crystalline character, but others are not crystalline at all. Vulcanized rubbers are an example: created by forcing random chemical bonds in a melt (a "spaghetti dish"), they are inhibited to flow, and, thus, they are amorphous solids. But they exhibit a varying degree of viscoelastic behavior. In the last decades, the vast discipline of soft condensed matter has incorporated to mainstream research in solid-state physics, at equal footing with crystalline solids. The scope of soft condensed matter is very wide. In particular, it considers many non-equilibrium routes to self-assembled emergent structures. Of huge interest is the neocortex (not just because understanding the brain's behavior is one of the "Holy Grails" of science, but because in doing it we may achieve to master a computational strategy which is far more efficient than the present one).

References

- Solid-state-physics-277131: britannica.com, Retrieved 21, April 2020
- Issues-in-Solid-State-Physics-67259: intechopen.com, Retrieved 04, August 2020

2

Crystal Structure and Defects

The description of the ordered arrangement of ions, molecules or atoms in a crystalline material is known as the crystal structure. The interruptions in the regular pattern of crystalline solids are known as crystal defects. This chapter has been carefully written to provide an easy introduction to the varied facets of crystal structure and defects such as inelastic interaction of X-rays with matter and elastic interaction of X-Rays with matter.

Crystal Structure

Based on the atomic arrangement in a substance, solids can be broadly classified as either crystalline or non-crystalline. In a crystalline solid, all the atoms are arranged in a periodic manner in all three dimensions where as in a non-crystalline solid the atomic arrangement is random or non-periodic in nature. A crystalline solid can either be a single crystalline or a polycrystalline. In the case of single crystal the entire solid consists of only one crystal and hence, periodic arrangement of atoms continues throughout the entire material. A polycrystalline material is an aggregate of many small crystals separated by well-defined grain boundaries and hence periodic arrangement of atoms is limited to small regions of the material called as grain boundaries. The non-crystalline substances are also called as amorphous substances materials. Single crystalline materials exhibit long range as well as short range periodicities while long range periodicity is absent in case of poly-crystalline materials and non-crystalline materials.

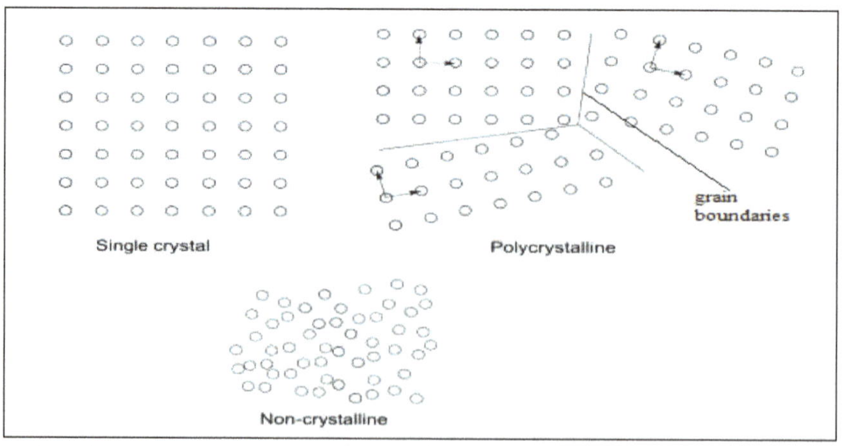

Two dimensional representations of single crystal, polycrystalline and non-crystalline materials.

A crystal has regular shape and when it is broken all broken pieces have the same regular shape. A crystal has sharp melting point. Since the crystals may have different periodic arrangements in all three dimensions, the physical properties vary with direction and therefore they are called anisotropic substances. Crystalline solids may be made up of metallic crystals or non-metallic crystals. Copper, Silver, tungsten and magnesium are examples of metallic crystals while carbon, crystallized polymers and plastics are examples of non-metallic crystals. Let us now discuss the elementary concepts of crystallography.

Table: Distinction between Crystalline and amorphous solids.

Crystalline Solids	Amorphous Solids
The atoms or molecules of the crystalline solids are periodic in space.	The atoms or molecules of the amorphous solids are not periodic in space.
Some crystalline solids are anisotropic i.e., the magnitude of physical properties [such as refractive index, electrical conductivity, thermal conductivity, etc.,] we different along different Erections of the crystal.	Amorphous solids are isotropic i.e., the magnitude of the physical properties are same along all directions of the solid.
Crystalline solids have sharp melting points.	Amorphous solids do not possess sharp melting points.
Breaks are observed in the cooling curve of a crystal solid.	Breaks are not observed in the cooing curve.
A crystal breaks along certain crystallographic planes.	When an amorphous solid breaks, the broken surface is irregular because it has no crystal planes.

Table: Examples for different categories of solid materials.

Type of materials	Examples
Single Crystals	Silicon chips used in integrated circuits, quartz crystals used in watches, ruby rods used in ruby lasers, gallium arsenide crystals used in light-emitting diodes, etc.
Polycrystalline Materials	Metals like copper, steel, aluminium, brass etc., used in ordinary engineering and household appliances. Organic and inorganic materials used in powder form like sugar, salts, etc.
Non-Crystalline Materials	Soda lime glass (ordinary window glass). Polymers (plastics like polyethylene, polyvinylchloride, polytetrafluoroethylene (teflon), plexiglass, polyurethane foam, bakelite, etc.).

Space Lattice or Crystal Lattice

In a solid crystalline material, the atoms or molecules are arranged regularly and periodically in all three dimensions. The atomic arrangement in a crystal is called crystal structure. To explain crystal symmetries easily, it is convenient to represent an atom or a group of atoms that repeats in three dimensions in the crystal as a unit. If each such atom or unit of atoms in a crystal is replaced by a point in space, then the resultant points in space are called space lattice. Each point in a space lattice is called a lattice point and each atom or unit of atoms is called basis or pattern. A space lattice represents the geometrical pattern of crystal in which the surroundings of each lattice point is the same. If the surroundings of each lattice point is same or if the atom or all the atoms at lattice points are identical, then such a lattice is called Bravais lattice. On the other hand, if the atom or the atoms at lattice points are not same, then it is said to be a non-Bravais lattice. Figure below shows a two- dimensional lattice.

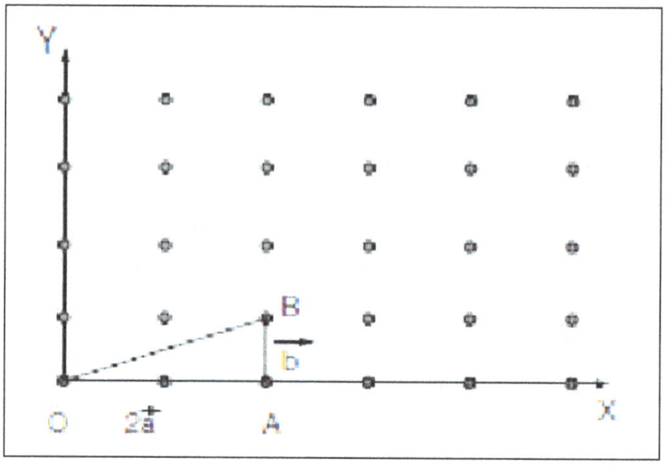

Two-dimensional lattice.

In the same manner, it is very convenient to imagine periodic arrangement of points in space in 3-dimensions about which these atoms are located. "A space lattice or a crystal lattice is defined as a three dimensional infinite array of points in space in which every point has surroundings identical to that of every other point in the array".

To represent translational vectors or basis vectors, considers a coo-ordinate system with its origin at the lattice point 'O'. Let $\overrightarrow{OA} = 2\vec{a}$ and $\overrightarrow{AB} = \vec{b}$, such that $\overrightarrow{OB} = 2\vec{a} + \vec{b}$ where $\vec{a}$ and $\vec{b}$ are called translational or basis vectors along X and Y directions. The position vectors $\vec{R}$ of any lattice point can be represented as $\vec{R} = n_1\vec{a} + n_2\vec{b}$ where n_1 and n_2 are integers; their values depend on the position of the lattice point under consideration with respect to the origin. In three dimensions, the position vector of a point can be expressed as $\vec{R} = n_1\vec{a} + n_2\vec{b} + n_3\vec{c}$, where $\vec{a}, \vec{b}$ and $\vec{c}$ are the translational or basis vectors along X, Y and Z directions, respectively. They are also called translational primitives.

Basis and Crystal Structure

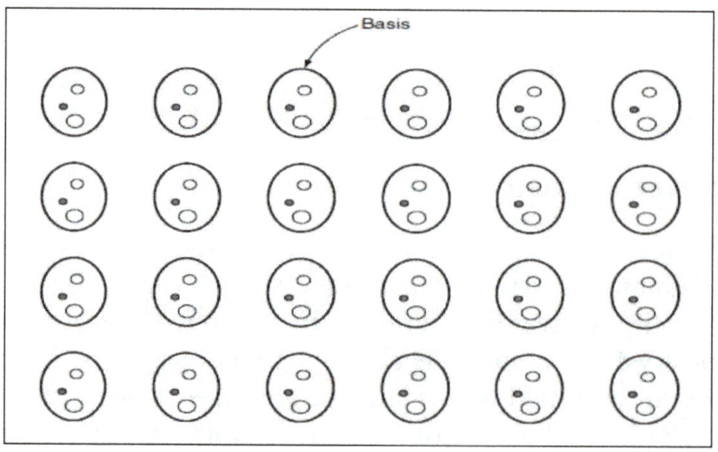

Two-dimensional crystal structure.

The atomic arrangement in a crystal is called crystal structure. The crystal structure is formed by associating every lattice point with an atom or an assembly of atoms or molecules or ions, which are identical in composition, arrangement and orientation, called the basis. i.e., an atom, or a group of atoms or molecules identical in composition is called the basis or the pattern. The basis provides the number of atoms per lattice point, their types, mutual orientations and distances of separation between the atoms. If the basis is substituted for the lattice points, then the resulting structure is called crystal structure as shown in figure. Thus lattice + basis = crystal structure. The basis shown in figure below contains three different atoms. In copper and sodium crystals the basis is single atoms; in NaCl, the basis is diatomic and in CaF_2 the basis is triatomic. A lattice is an imaginary assumption while the crystal structure is a real concept.

Unit Cell and Lattice Parameters

Unit cells for most of the crystals are parallelepipeds or cubes having three sets of parallel faces. A unit cell is the basic structural unit or building block of the crystal. A unit cell is defined as the smallest parallelepiped volume in the crystal, which on repetition along the crystallographic axes gives the actual crystal structure or the smallest geometric figure, which on repetition in three-dimensional space, gives the actual crystal structure is called a unit cell. The choice of a unit cell is not unique but it can be constructed in a number of ways; figure below shows different ways of representing unit cells in a two-dimensional lattice. A unit cell can be represented as ABCD or A′B′C′D′ or A″B″C″D″, etc.

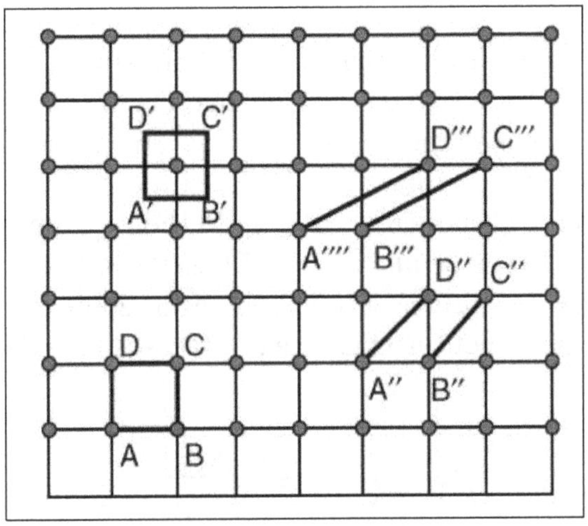

Unit cells in crystal lattice.

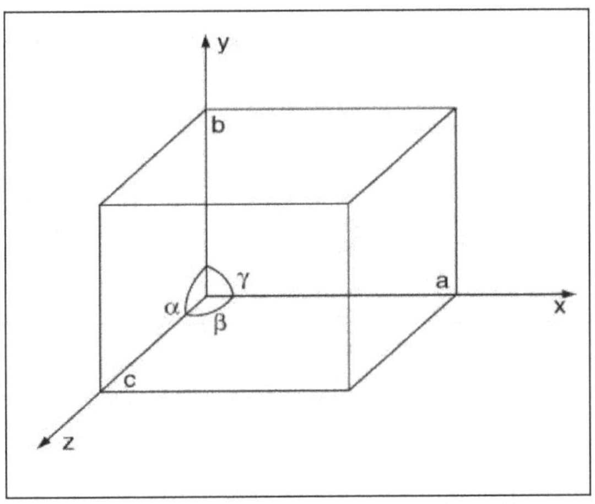

Unit cell parameters.

To define the unit cell parameters or lattice parameters, first we define crystallographic axes. These axes are obtained by the intersection of the three non-coplanar faces of the unit cell. The angle between these faces or crystallographic axes is known as inter-facial or inter-axial angles. The angles between the axes Y and Z is α, between Z and X is β and between X and Y is γ. The translational vectors or primitives a, b, c of a unit cell along X, Y, Z axes and interaxial angles α, β, γ are called cell parameters. These cell parameters are shown in figure. The cell parameters determine the actual size and shape of the unit cell.

Primitive and Non-Primitive Unit Cells

The unit cell formed by primitives is called a simple or a primitive unit cell. A primitive unit cell contains only one lattice point. If a unit cell contains more than one lattice point, then it is called a non-primitive unit cell or a multiple unit cell. Most of the unit cells of various crystal lattices contain two or more lattice points and hence it is not necessary that unit cell should be a primitive unit cell. For example, BCC and FCC are non-primitive unit cells.

Crystal Systems, Bravais Lattices

Seven Crystal Systems

For representing the type of distribution of lattice points in space, seven different coordinate systems are required. These co-ordinate systems are called crystal systems. The crystal systems are named on the basis of geometrical shape and symmetry. The seven crystal systems are: (1) Cubic (2) Tetragonal (3) Orthorhombic (4) Monoclinic (5) Triclinic (6) Trigonal (or Rhombohedral) and (7) Hexagonal. Space lattices are classified according to their symmetry. In 1948, Bravais showed that 14 lattices are sufficient to describe all crystals. These 14 lattices are known as Bravais lattices and are classified into 7 crystal systems based on cell parameters. The Bravais lattices are categorized as primitive lattice (P); body-centered lattice (I); face-centered lattice (F) and base-centered lattice (C).

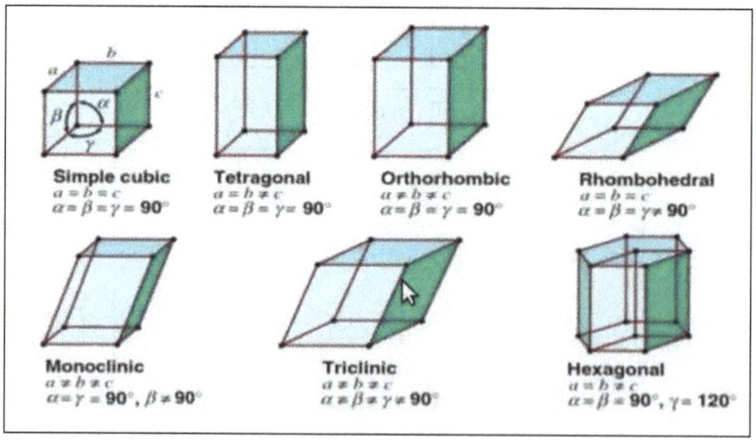

Seven crystal systems.

Bravais Lattices

If environment of each lattice point is same or if the atom or all the atoms at lattice points are identical, then such a lattice is called Bravais lattice. Based on the number of lattice points present per unit cell and stacking sequence of base atoms in a crystal structure, the Bravais lattices are classified into Simple or Primitive (P), Body Centered (I), Face Centered (F) and Base Centered (C) lattices. In simple or primitive lattice, 8 lattice points or 8 atoms are present at the 8 corners of the unit cell, each contributing 1/8th to the unit cell and hence there will be 8 x 1/8 = 1 lattice point per unit cell. In body-centered lattice, in addition to the 8 atoms at 8 corners each contributing 1/8th to the unit cell, there will be one complete atom at the center of the unit cell. Therefore number of atoms or lattice points in a body centered unit cell becomes 8 x 1/8 + 1 x 1 = 2. In case of face-centered lattice, in addition to the 8 atoms at 8 corners each contributing 1/8th to the unit cell, six atoms will be present at the center of six faces of the cell each contributing 1/2nd to the unit cell. Therefore number of atoms or lattice points in a face centered unit cell becomes 8 x 1/8 + 6 x 1/2 = 4. Similarly, in case of base-centered lattice, in addition to the 8 atoms at 8 corners each contributing 1/8th to the unit cell, two atoms will be present at the center of upper and lower faces of the unit cell each contributing 1/2nd to the unit cell and hence the number of atoms or lattice points in a base centered unit cell becomes 8 x 1/8 + 2 x 1/2 = 2.

Bravais showed that 14 lattices are sufficient to describe all crystals. These 14 lattices are known as Bravais lattices and are classified into 7 crystal systems based on cell parameters. The Bravais lattices are categorized as primitive lattice (P); body-centered lattice (I); face-centered lattice (F) and base-centered lattice (C).

1. Cubic crystal system: In this crystal system, all the unit cell edge lengths are equal and are at right angles to one another i.e., $a = b = c$ and $\alpha = \beta = \gamma = 90°$. In cubic system, there are three Bravais lattices; they are simple (primitive); body-centered and face-centered. Examples for cubic system are Au, Cu, Ag, NaCl, diamond, etc. In simple cubic lattice, lattice points or atoms are present at the corners of the cube. In body-centered cube, atoms are present at the corners and one atom is completely present at the center of the cube. In the case of face-centered cube, atoms are present at corners and at the centers of all faces of cube.

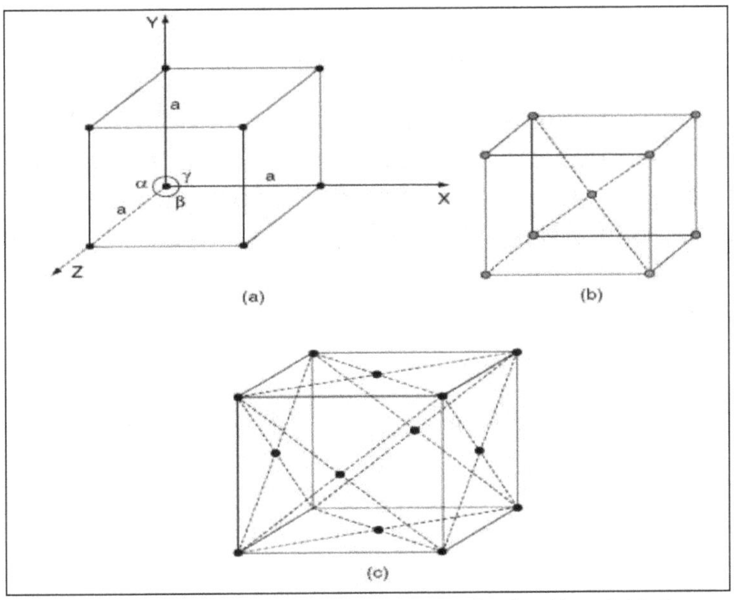

Cubic crystal system: (a) Simple cubic (P); (b) Body-centered cube (I) and (c) Face-centered cube (F).

2. Tetragonal crystal system: In this crystal system, two lengths of the unit cell edges are equal

whereas the third length is different. The three edges are perpendicular to one another i.e., a = b ≠ c and α = β = γ = 90°. In tetragonal system, there are two Bravais lattices; they are simple and body-centered. These are shown in figure below. Examples for tetragonal crystal systems are TiO_2, SnO_2, etc.

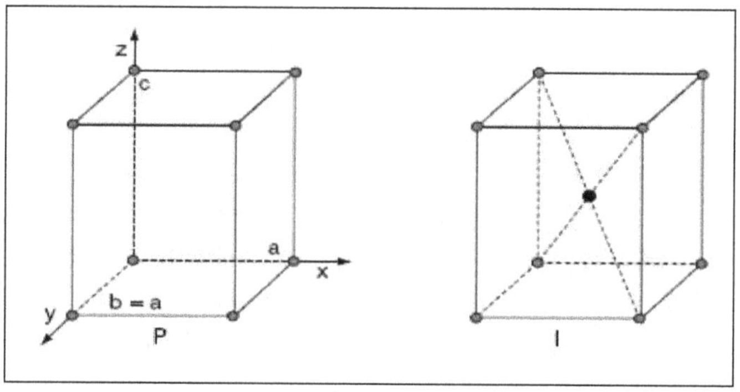

Tetragonal crystal system.

3. Orthorhombic crystal system: In this crystal system, unit cell edge lengths are different and they are perpendicular to one another i.e., a ≠ b ≠ c and α = β = γ = 90°. There are four Bravais lattices in this system. They are simple; face centered, body centered and base centered. These are shown in figure below. Examples for orthorhombic crystal system are $BaSO_4$, K_2SO_4, $SnSO_4$, etc.

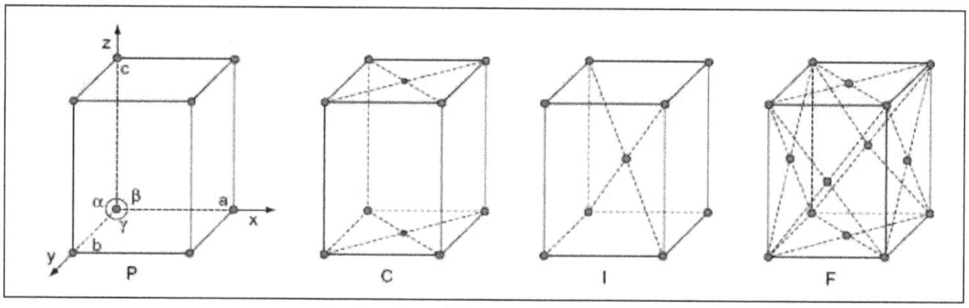

Orthorhombic crystal system.

4. Monoclinic crystal system: In this crystal system, the unit cell edge lengths are different. Two unit cell edges are not perpendicular, but they are perpendicular to the third edge i.e., a ≠ b ≠ c and α = β = γ = 90°. This crystal system has two Bravais lattices; they are simple and base centered. These are shown in figure below. Examples for Monoclinic crystal system are $CaSO_4.2H_2O$ (gypsum), Na_3AlF_6 (cryolite), etc.

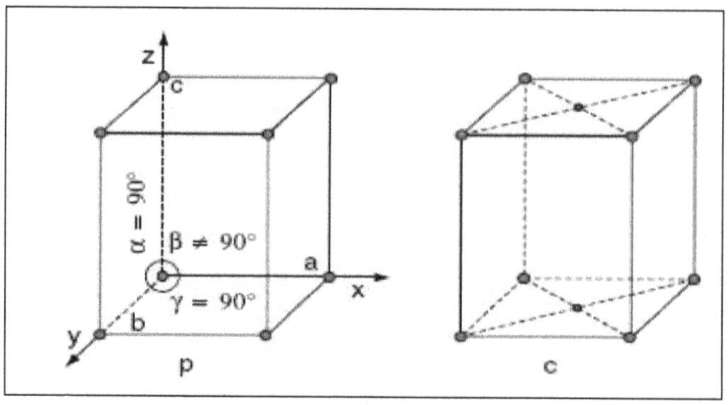

Monoclinic crystal system.

5. Triclinic crystal system: In this crystal system, the unit cell edge lengths are different and are not perpendicular i.e., a ≠ b ≠ c and a ≠ b ≠ c and α ≠ β ≠ γ ≠ 90° and all the angles are different. This crystal exists in primitive cell only. Examples for triclinic crystal system are $K_2Cr_2O_7$, $CuSO_4$. $5H_2O$, etc.

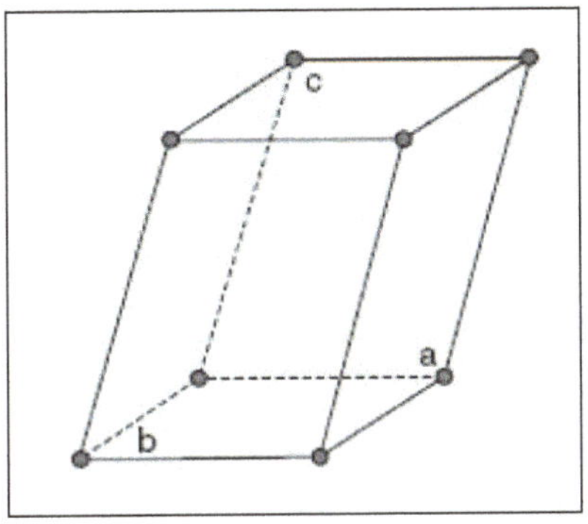

Triclinic crystal system.

6. Trigonal or Rhombohedral crystal system: In this crystal system, all the lengths of unit cell edges are equal. The angles between the axes are equal but other than 90° i.e., a = b = c and α = β = γ ≠ 90°. The Bravais lattice is simple only as shown in figure below. Examples for Rhombohedral crystal system are As, Bi, Sb, etc.

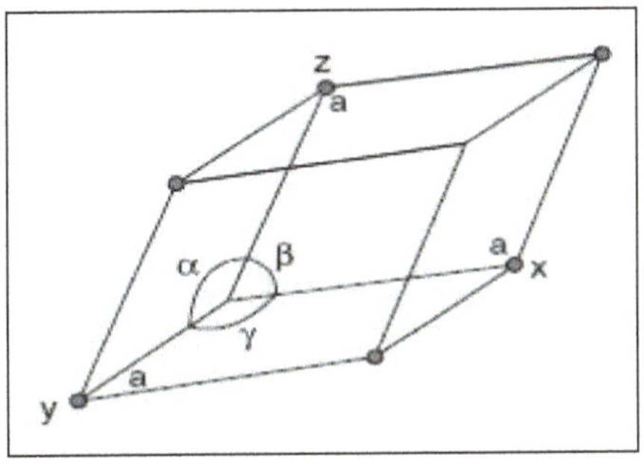

Rhombohedral crystal system.

7. Hexagonal crystal system: In this crystal system, two sides of the unit cell edge lengths are equal and the angle between these edges is 120°. These two edges are perpendicular to the third edge, and not equal in length i.e., a = b ≠ c and α = β = 90°; γ = 120°. The Bravais lattice is primitive only. The atoms in this crystal system are arranged in the form of a hexagonal close pack.

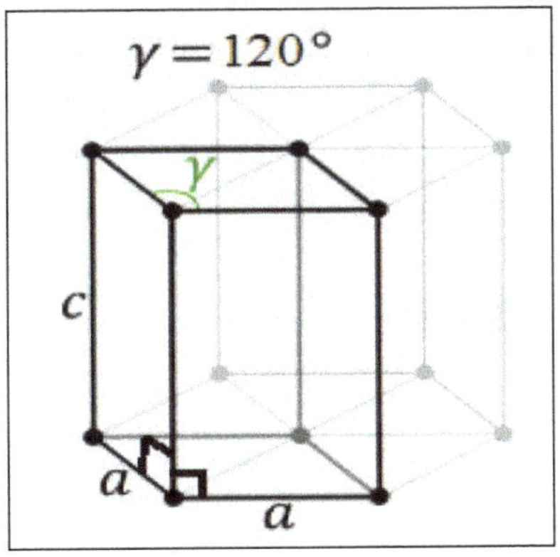

Hexagonal crystal system.

Table: Seven crystal systems and fourteen Bravais lattices.

SI. No	Crystal System	Types of Bravais Lattices	No. of Bravais Lattices	Relation between Lengths and Angles
1	Cubic	PIE	3	$a = b = c$ $\alpha = \beta = \gamma = 90^\circ$
2	Tetragonal	P I	2	$a = b \neq c$ $\alpha = \beta = \gamma = 90^\circ$
3	Orthorhombic	P. I, F, C	4	$a \neq b \neq c$ $\alpha = \beta = \gamma = 90^\circ$
4	Monoclinic	P. C	2	$a \neq b \neq c$ $\alpha = \gamma = 90^\circ \neq \beta$
5	Triclinic	P	1	$a \neq b \neq c$ $\alpha \neq \beta \neq \gamma$
6	Rhombohedral (Trigonal)	P	1	$a = b = c$ $\alpha = \beta = \gamma \neq 90^\circ$
7	Hexagonal	P	1	$a = b \neq c$ $\alpha = \beta = 90^\circ$ $\gamma = 120^\circ$

Structures of SC, BCC and FCC

Structure of SC

There are three different types of cubical crystal lattice systems namely Simple Cubic (SC), Body-Centered Cubic (BCC) and the Face Centered Cubic (FCC). Using parameters such as, co-ordination number, nearest neighbor distance, number of atoms per unit cell and atomic packing fraction we can analyze structure of crystalline solids. Atomic packing factor is defined as the fraction of the space occupied by atoms in the unit cell or it is the ratio of the volume occupied by atoms in the unit cell to the unit cell volume. An example for simple cubic structure is polonium crystal.

Structure of Simple Cubic Crystal

- Let a be the side of unit cell.
- Coordination number: 6.
- Nearest neighbor distance: a units.

- Relationship between atomic radius (R) and side of unit cell: a = 2R.
- Number of atoms per unit cell, n = 1.
- Atomic packing fraction: $APF = \frac{n.(4/3)\pi R^3}{a^3} = \frac{1.(4/3)\pi R^3}{8R^3} = \frac{\pi}{6} = 0.52.$

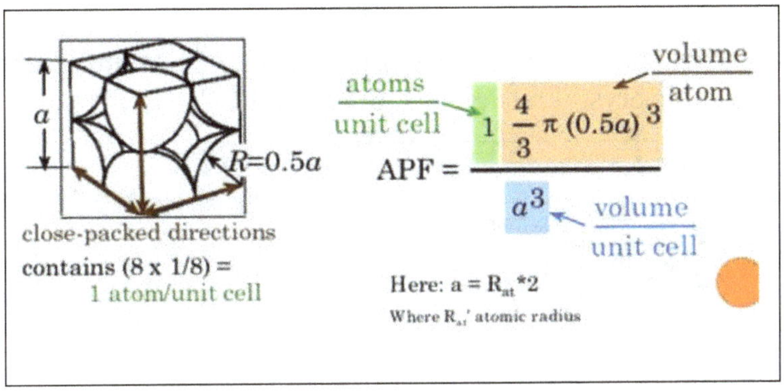

Structure of BCC

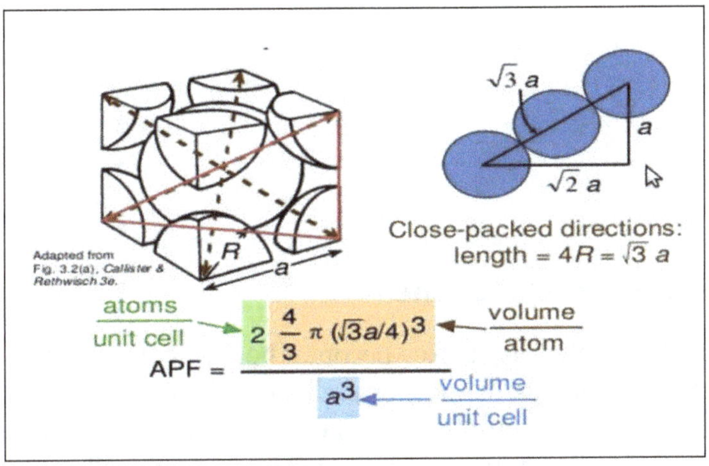

- Let a be the side of unit cell Coordinate on number: 8.
- Nearest neighbor distance: √3 a /2 units.
- Relationship between atomic radius (R) and side of unit cell: $4R = \sqrt{3}a$.
- Number of atoms per unit cell, n = 2.

Atomic packing fraction: $APF = \frac{n.(4/3)\pi R^3}{a^3} = 0.68.$

Structure of FCC

- Let a be the side of unit cell.
- Coordination number: 8.
- Nearest neighbor distance: $a/\sqrt{2}$ units.

- Relationship between atomic radius (R) and side of unit cell: $4R = \sqrt{2}\,a$.
- Number of atoms per unit cell, n = 4.

Atomic packing fraction: $APF = \dfrac{n.(4/3)\pi R^3}{a^3} = 0.74.$

Therefore, based on the above packing fractions information, FCC is the most closely packed structure among all the three cubic crystal lattices SC, BCC and FCC.

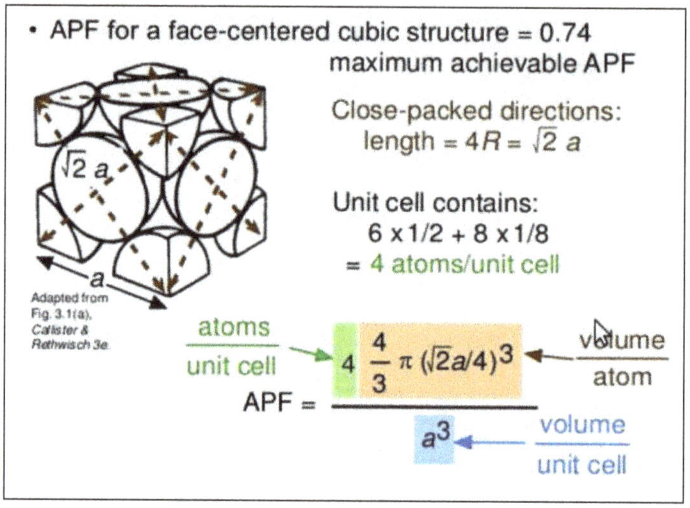

Miller Indices

Crystal Directions and Crystal Planes

When dealing with crystalline materials, it often becomes necessary to specify a particular point within a unit cell, a crystallographic direction, or some crystallographic plane of atoms. Labeling conventions have been established in which three numbers (integers) are used to designate point locations, directions, and planes. The basis for determining index values is the unit cell, with a right-handed coordinate system consisting of three (x, y, and z) axes situated at one of the corners and coinciding with the unit cell edges. For some crystal systems - namely, hexagonal, rhombohedral, monoclinic, and triclinic - the three axes are not mutually perpendicular, as in the familiar Cartesian coordinate scheme.

A crystallographic direction is defined as a line between two points, or a vector. The following steps are utilized in the determination of the three directional indices:

- A vector of convenient length is positioned such that it passes through the origin of the coordinate system. Any vector may be translated throughout the crystal lattice without alteration, if parallelism is maintained.
- The length of the vector projection on each of the three axes is determined; these are measured in terms of the unit cell dimensions a, b, and c.
- These three numbers are multiplied or divided by a common factor to reduce them to the smallest integer values.

- The three indices, not separated by commas, are enclosed in square brackets, thus: [uvw]. The u, v, and w integers correspond to the reduced projections along the x, y, and z axes, respectively.

Crystal planes are defined as some imaginary planes inside a crystal in which large concentration of atoms are present. Inside the crystal, there exists certain direction along which large concentration of atoms exists. These directions are called crystal directions. Figure below shows a two-dimensional lattice with different orientations of crystal planes.

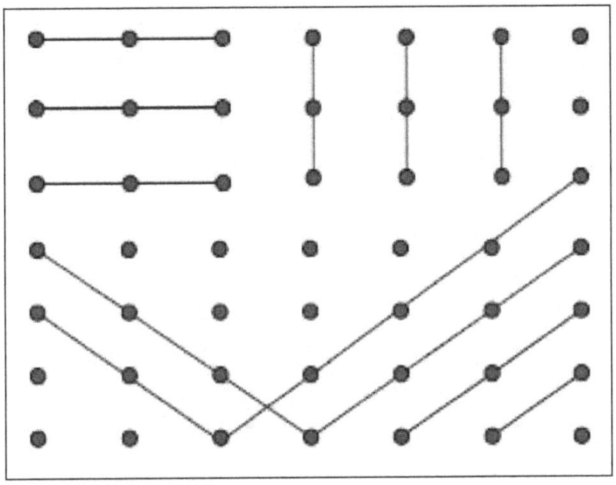

A two-dimensional lattice with crystal planes.

Crystal planes and directions can be represented by a set of three small integers called Miller indices. These integers are represented in general as h, k and l. If these integers are enclosed in round brackets as (hkl), then it represents a plane. On the other hand, if they are enclosed in square brackets as [hkl], then it represents crystal direction perpendicular to the above-said plane.

Procedure for drawing the given plane having Miller indices (hkl):

- A unit cell is drawn with the given lattice parameters. After taking any convenient point as the origin O, the OX, OY and OZ crystallographic axes are to be marked. If lattice parameters are not given, then unit cubic cell of arbitrary lattice constant will be taken.
- The reciprocals of the Miller indices 1/h, 1/k, 1/l are to be taken. These values provide the intercepts of the given plane on OX, OY and OZ axes, respectively.
- The intercepts are marked in the unit cell and the plane is drawn.
- If the plane is parallel to any of the axis, then intercept is ∞ on that axis.

Reciprocal Lattice and Problems

Reciprocal Lattice

It is easy to think a set of parallel crystal planes in terms of their normals, since as the planes are two dimensional, their normal will be one dimensional in nature. The normal represent orientation of crystal plane and its length denotes inter-planar spacing. The length of normal is proportional to reciprocal of the respective inter-planar spacing. The end points of these normals form a lattice array. Since the distances of end points in this array is reciprocal to distances of planes in the crystal, the array of points formed by the end points of the normals is called reciprocal lattice. It is also called as k-space. Geometrical construction of reciprocal lattice:

- Fix up some point in the direct lattice as a common origin.
- From this common origin draw normals to each and every set of parallel planes in the direct lattice.
- Fix the length of each normal equal to the reciprocal of the inter-planar spacing (1/dhkl) of the set of parallel planes (h k l) it represents.
- Put a point at the end of each normal.

The collection of all these points in space is reciprocal lattice space.

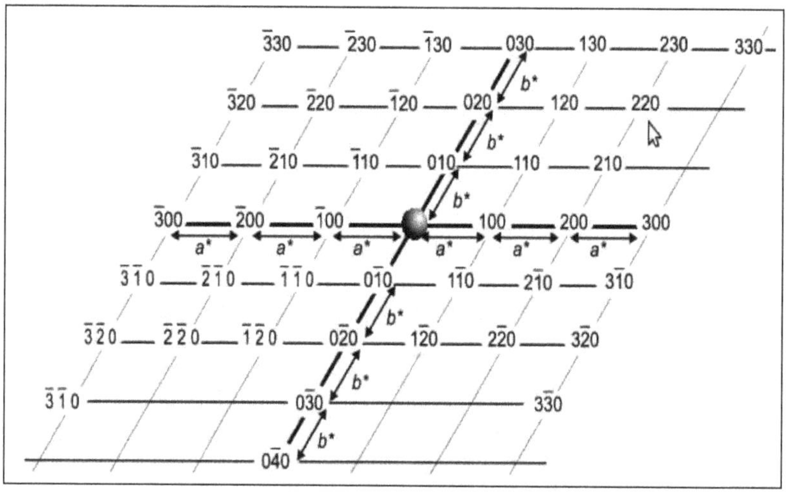

Reciprocal lattice.

Properties of Reciprocal Lattice

- Any diffraction pattern of a crystal is a map of the reciprocal lattice of the crystal whereas the microscopic image is a map of direct lattice.
- The concept of reciprocal lattice is useful to understand the X-ray diffraction phenomena in a simple way and wave mechanical behavior of electrons in a periodic crystal lattice.
- When crystal is made to rotate both the direct and reciprocal lattice rotate.
- While primitive vectors of a direct lattice have the dimensions of length, those of the reciprocal lattice have the dimensions of $(\text{length})^{-1}$.

Problems:

Within a cubic unit cell sketch the following directions:

$[101]$, $[211]$, $[\overline{313}]$, $[\overline{111}]$, $[212]$, $[\overline{31}\,2]$, $[3\ 0\ 1]$.

- Sketch the following planes in cubic unit cell: (101).
- Determine the Miller indices of the planes shown in the following figure:

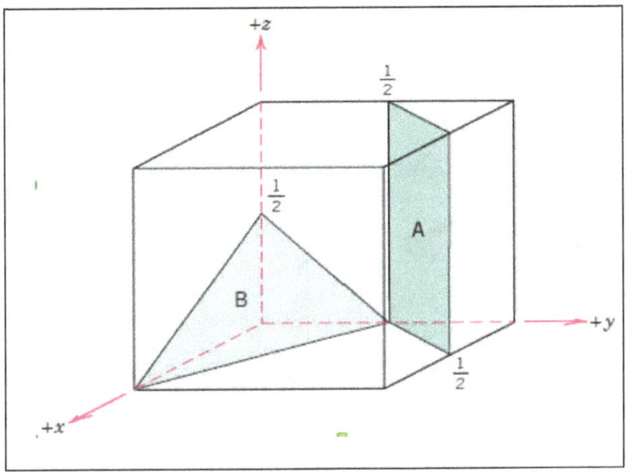

- Obtain Miller indices of a plane which has intercepts at a, b/2 and 3c of cubic unit cell.

Ionic Solids

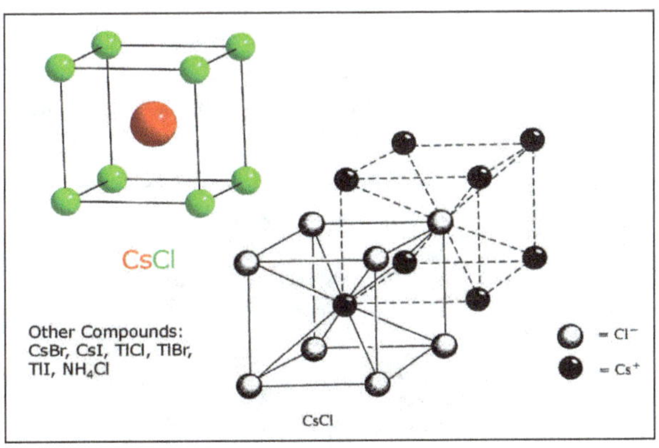

Other Compounds: CsBr, CsI, TlCl, TlBr, TlI, NH_4Cl.

Rock Salt – Nacl

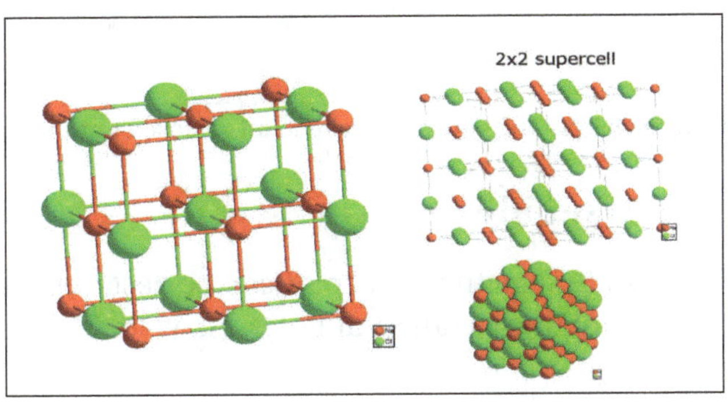

Other Compounds (200+): Most alkali halides, all the alkali hydrides, AgF, AgCl, AgBr, MO (M = Mg, Ca, Sr, Ba), MS (M = Mg, Ca, Sr, Ba).

Linked Octahedra

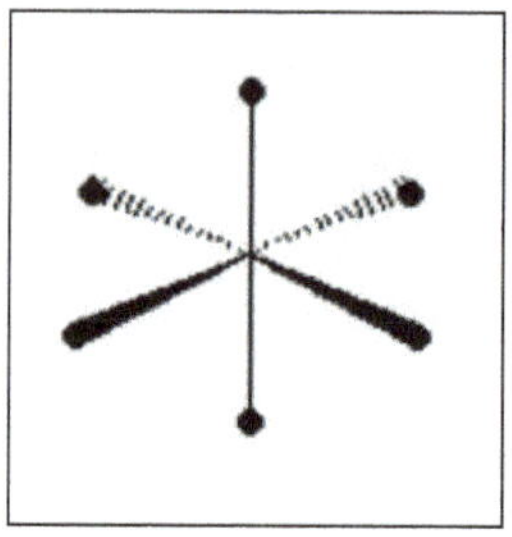

$[MX_6]$ Octahedron.

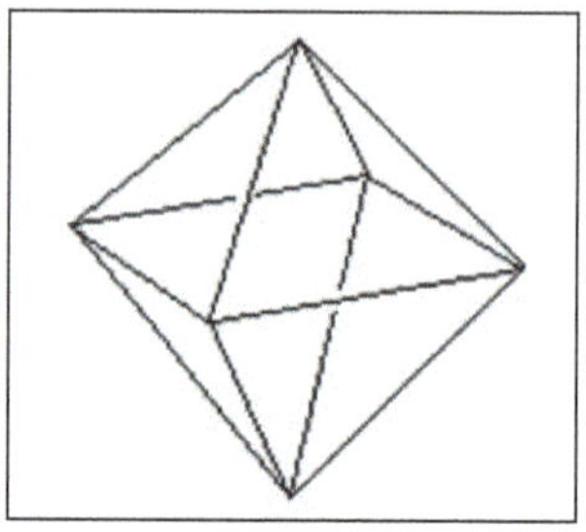

Solid Octahedron (polyhedral).

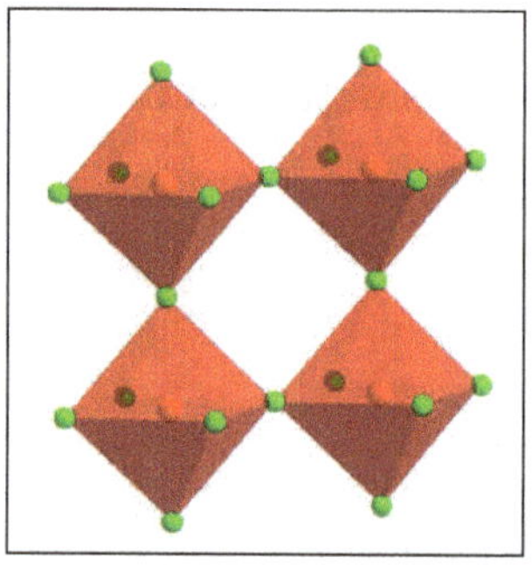

Corner-sharing: Two octahedra = M_2O_{11}.

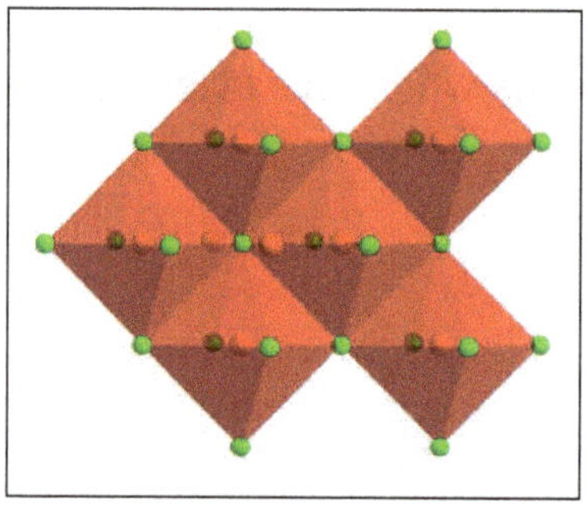

Edge-sharing: Two octahedra = M_2O_{10}.

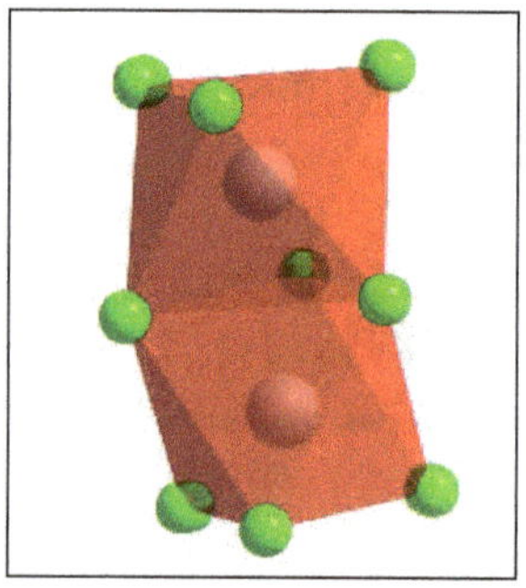

Face-sharing: Two octahedral = M_2O_9.

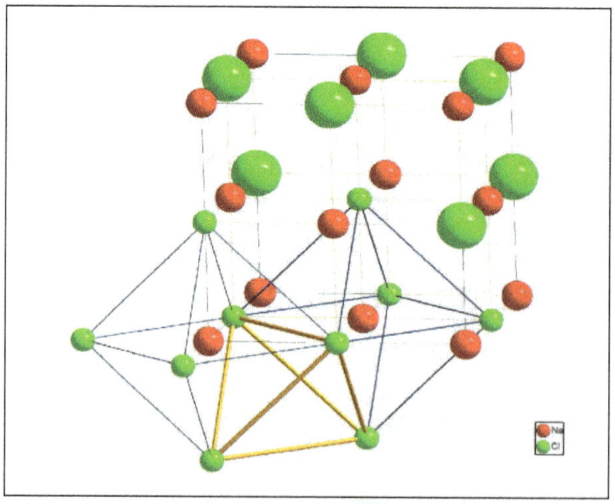

NaCl showing edge-sharing of octahedral and an enclosed tetrahedral space.

Structure Type - Nickel Arsenide (NiAs)

Equivalence to the sodium chloride structure in hcp. It can be described as an hcp array of arsenic atoms with nickel atoms occupying the octahedral holes. As is located in the center of a trigonal prism of six nickel atoms.

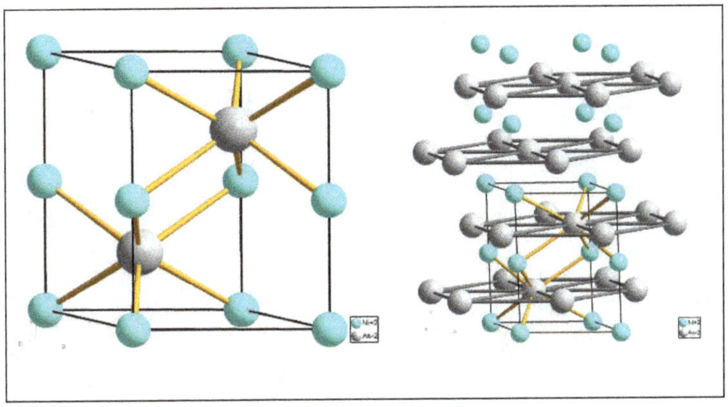

Structure Type – Fluorite (CaF_2)

The structure can be described as a ccp array of calcium ions, with fluoride ions occupying all the tetrahedral holes.

- Ca^{2+} ions are smaller than F^-, so it is not fully described as a ccp array, but this description gives an exact description of the relative positions of the ions.
- The larger octahedral holes are unoccupied in this structure, which is important when considering the movement of ions in defect structures.
- By drawing cubes of fluoride ions, the structure is now divided into octants with alternate octants occupied by calcium.
- In antifluorite, the positions of cation and anions are reversed.

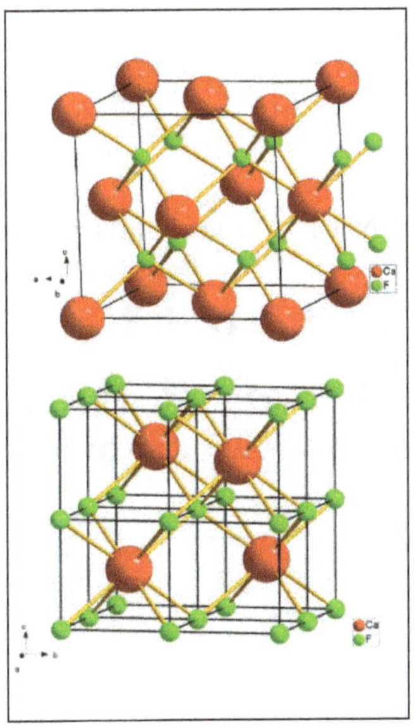

Structure Type – Zinc Blende/Sphalerite (ZnS)

A ccp array of sulfide ions with zinc ions occupying every other tetrahedral hole in an ordered manner.

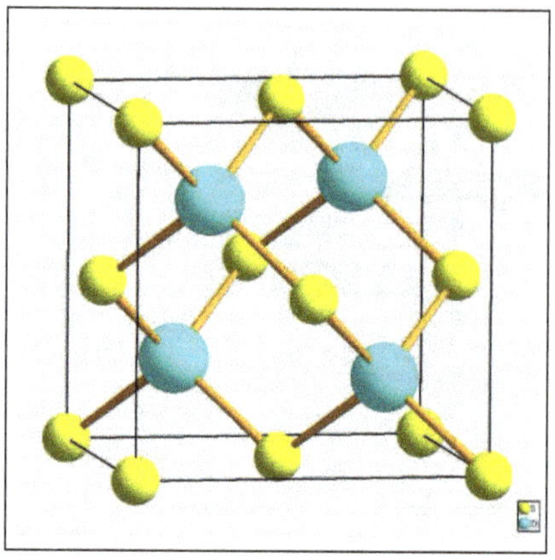

Compounds adopting this structure include copper halides, Zn, Cd, and Hg sulfides.

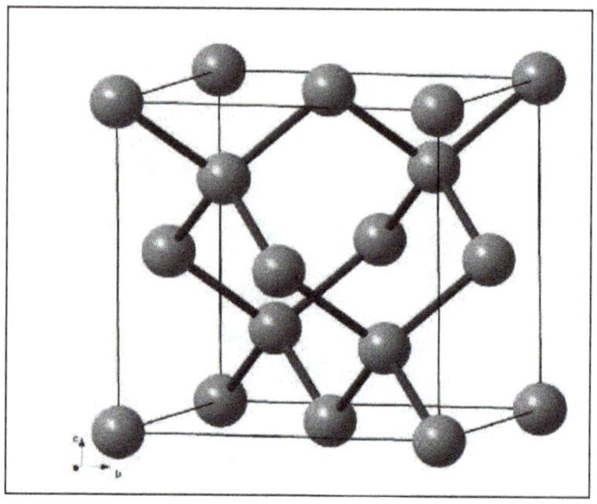

If all atoms are the same, the structure would be identical to C (diamond).

Structure Type – Wurtzite (ZnS)

The Wurtzite structure type is composed of an hcp array of sulfide ions with alternate tetrahedral holes occupied by zinc ions.

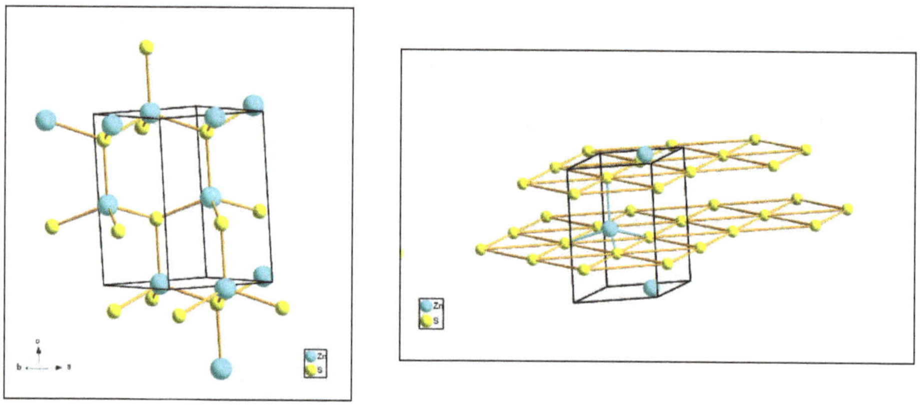

BeO, ZnO, NH_4F also adopt this structure type.

Structure Type – Cadmium Chloride ($CdCl_2$)

Close packing of anions (ccp) with half of the octahedral holes occupied by cations. The cations occupy all the octahedral holes in every other layer, giving an overall layer structure with 6:3 coordination.

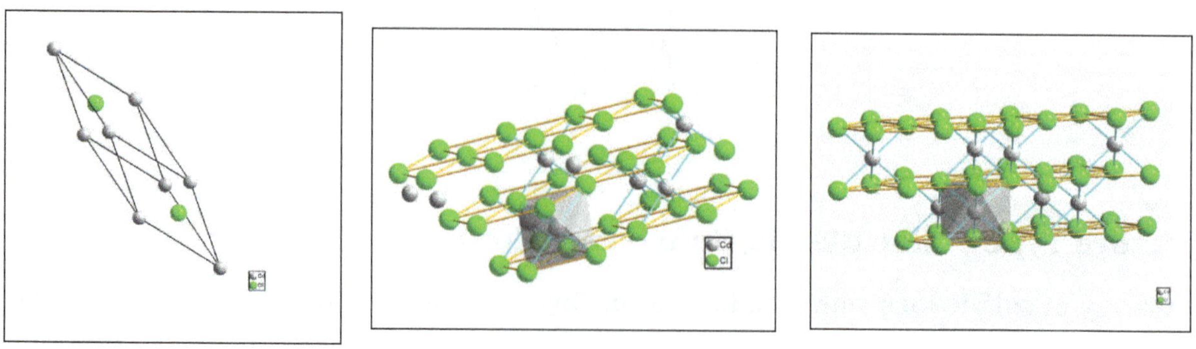

Structure Type – Cadmium Iodide (CdI_2)

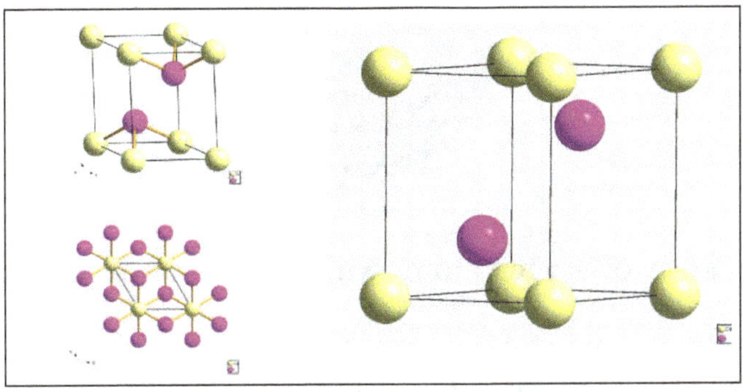

The CdI_2 structure can be viewed as a hexagonal close packing (hcp) of iodine, with half of the octahedral holes occupied by cadmium.

- There is a weak van der Waals attraction between neighboring iodine layers.
- This structure is unstable if the cation-anion bonds have significant ionic character, because of significant anion-anion repulsion.
- This structure is seen primarily when the anion is one of the larger halides or chalcogenides.

Structure Type – Rutile (TiO_2)

High refractive index and widely used as white pigments in paints and plastics.

- Tetragonal unit cell, but not close packed. Each Ti^{4+} is coordinated by six O^{2-}.
- Each oxygen is surrounded by three planar titanium, which lie at the corners of an almost equilateral triangle.

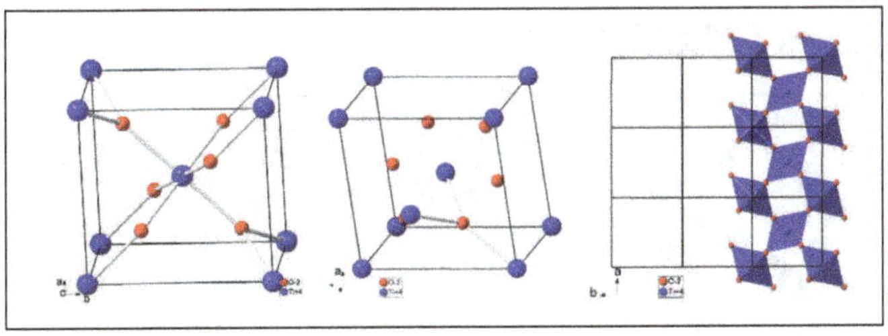

Structure Type – Corundum (Al2O_3)

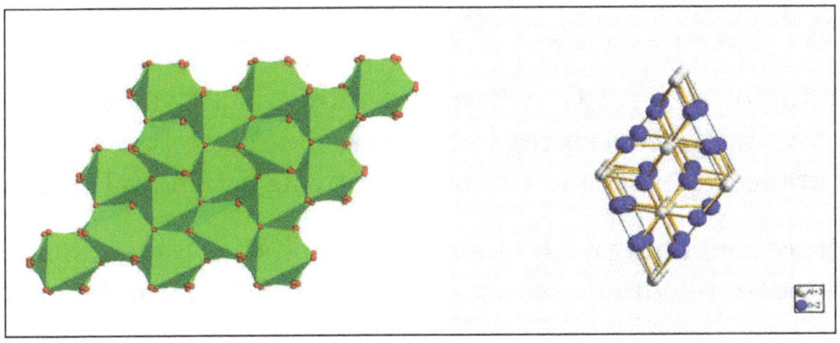

Mineral form is the basis for ruby (Cr) and sapphire (Ti).

It is composed of an hcp array of oxygen atoms with two-thirds of the octahedral holes occupied by Al^{3+}.

Structure Type – Rhenium Trioxide (ReO_3)

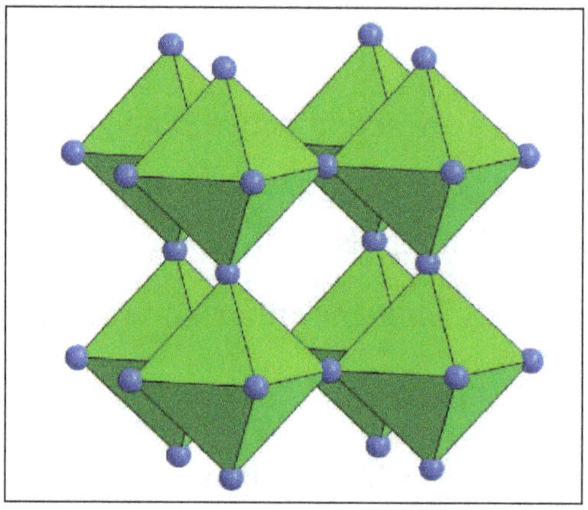

Also called the aluminum fluoride structure, which is adopted by the fluoride of Al, Sc, Fe, Co, Rh, and Pd, the high temperature form of WO_3, and ReO_3. Corner-sharing network of $[ReO_6]$ that give a highly symmetrical structure with cubic symmetry.

Structure Type – Spinel ($MgAl_2O_4$)

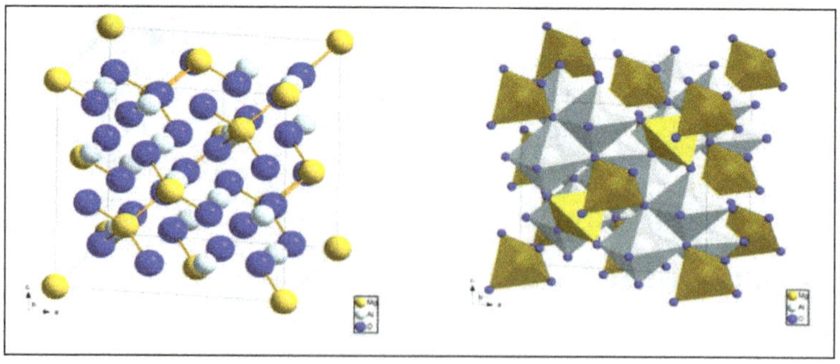

Spinels have the general formula AB_2O_4, where generally A is a divalent ion A^{2+}, B is trivalent ion B^{3+}. The structure can be described as being based on a cubic close-packed array of oxide ions, with A^{2+} ions occupying tetrahedral holes and B^{3+} ions occupying octahedral holes.

A spinel crystal structure containing n AB_2O_4 structural units has 8n tetrahedral holes and 4n octahedral holes. 1/8th of the tetrahedral holes are occupied by A^{2+} ions and ½ of the octahedral holes by the B^{3+} ions.

Structure Type – Perovskite ($CaTiO_3$)

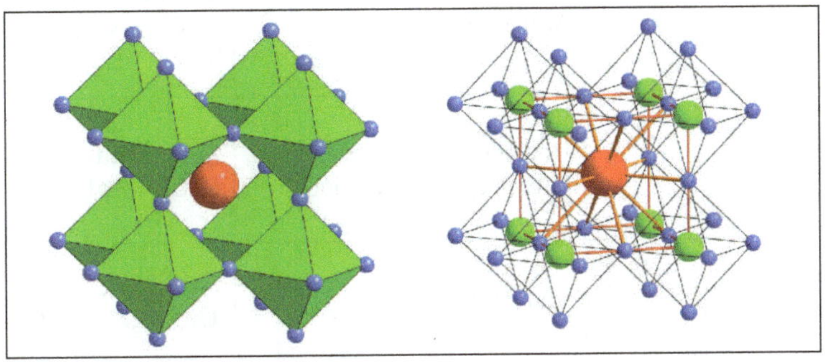

The structure is named after the mineral $CaTiO_3$:

- It has the same octahedral network as the ReO_3 structure, with the A ion added into the cavity.
- The structure may also be visualized as a ccp array of A and X ions with the B ions occupying the octahedral holes. (Compare with NaCl).
- Structures of high-temperature superconductors are based on perovskite.

Structure Type – Ilmenite (ABO_3)

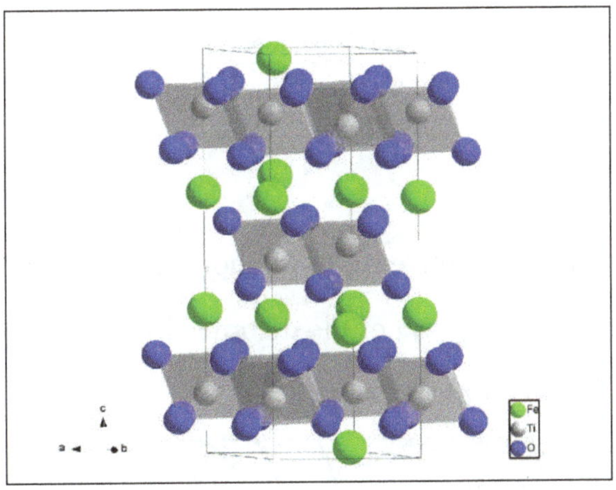

Typically adopted for ABO_3 structures where A and B are similar in size and their total charge adds up to +6.

- Structure is named after the mineral $Fe^{II}Ti^{IV}O_3$.
- Similar to corundum structure type, where there is a hcp array of oxygen ions, but in ilmenite two different cations are present occupying two-thirds of the octahedral holes.

Table: Structures related to close-packed arrangements of anions.

Formula	Cation:anion Coordination	Type and number of holes occupied	Examples: Cubic close packing	Examples: Hexagonal close packing

MX	6:6	All octahedral	Sodium chloride: NaCl, FeO, MnS, TiC	Nickel arsenide: NiAs, FeS, NiS
	4:4	Half tetrahedral; every alternate site occupied	Zinc blende: ZnS, CuCl, γ-AgI	Wurtzite: ZnS, β-AgI
MX_2	8:4	All tetrahedral	Fluorite: CaF_2, ThO_2, ZrO_2, CeO_2	None
	6:3	Half octahedral; alternate layers have fully occupied sites	Cadmium chloride: $CdCl_2$	Cadmium iodide: CdI_2
MX_3	6:2	One-third octahedral; alternate pairs of layers have two-thirds of the octahedral sites occupied.		Bismuth iodide: BiI_3, $FeCl_3$, $TiCl_3$, VCl_3.
M_2X_3	6:4	Two-thirds octahedral		Corundum: α-Al_2O_3, α-Fe_2O_3, V_2O_3, Ti_2O_3, α-Cr_2O_3
ABO_3		Two-thirds octahedral		Ilmenite: Fe_2O_3
AB_2O_4		One-eighth tetrahedral and one-half octahedral		Olivine: Mg_2SiO_4

Ionic Radii

How does one determine when one atom 'ends' and another 'begins'?

- Easy for the same atom (e.g. F_2), ½ of the distance between the atoms.
- What about Na^+-Cl^- (for example)?

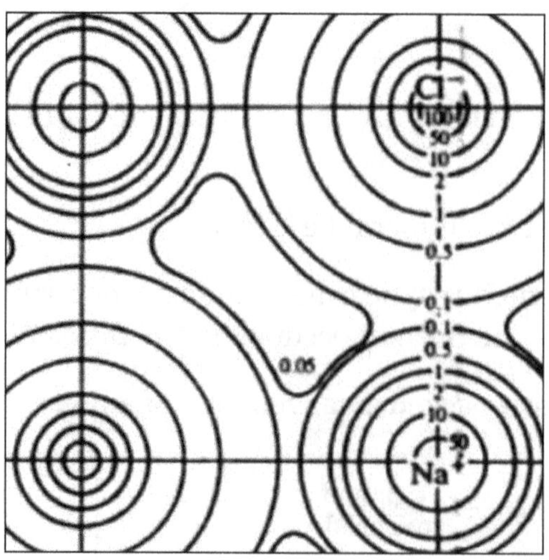

High Resolution X-ray diffraction contour map of electron density. The minimum of electron density along the interionic distances enables accurate determination of the values of the radius for the cation and anion.

Table: Shannon-Prewitt Ionic Radii in Angstrom Units (Å) of Cations.

Shannon–Prewitt Ionic Radii in Angstrom Units (Å) of Cations

C.N.	1A	2A	3B	4B	5B	6B	7B		1B	2B	3A	4A	5A	6A	7A
	Li	Be									B	C	N		
3	—	0.30									0.15	0.06	0.044		
4	0.73	0.41									0.25	0.29	—		
6	0.90	0.59									0.41	0.30	0.27		
	Na	Mg									Al	Si	P	S	Cl
4	1.13	0.71									0.53	0.40	0.31	0.26	0.22
6	1.16	0.86									0.675	0.540	0.52	0.43	0.41
8	1.32	1.03									—	—	—	—	—
	K	Ca	Sc	Ti	V	Cr	Mn		Cu	Zn	Ga	Ge	As	Se	Br
4	1.51	—	—	0.56	0.495	0.40	0.39		0.74	0.74	0.61	0.530	0.475	0.42	0.39
6	1.52	1.14	0.885	0.745	0.68	0.58	0.60		0.91	0.880	0.760	0.670	0.60	0.56	0.53
8	1.65	1.26	1.010	0.88	—	—	—		—	1.040	—	—	—	—	—
	Rb	Sr	Y	Zr	Nb	Mo	Tc		Ag	Cd	In	Sn	Sb	Te	I
2	—	—	—	—	—	—	—		0.81	—	—	—	—	—	—
4	—	—	—	0.73	0.62	0.55	0.51		1.14	0.92	0.76	0.69	—	0.57	0.56
									1.16						
6	1.66	1.32	1.040	0.86	0.78	0.73	0.70		1.29	1.09	0.940	0.830	0.74	0.70	0.67
8	1.75	1.40	1.159	0.98	0.88	—	—		1.42	1.24	1.06	0.95	—	—	—
	Cs	Ba	La	Hf	Ta	W	Re		Au	Hg	Tl	Pb	Bi	Po	At
4	—	—	—	0.72	—	0.56	0.52		—	1.10	0.89	—	—	—	—
6	1.81	1.49	1.172	0.85	0.78	0.74	0.67		1.51	1.16	1.025	0.79	0.90	0.81	0.76
8	1.88	1.56	1.300	0.97	0.88	—	—		—	1.28	1.12	0.915	—	—	—
	La	Ce	Pr	Nd	Pm	Sm	Eu	Gd	Tb	Dy	Ho	Er	Tm	Yb	Lu
6	1.172	1.15	1.13	1.123	1.11	1.098	1.087	1.078	1.063	1.052	1.041	1.030	1.020	1.008	1.001
7	1.24	1.21	—	—	—	1.16	1.15	1.14	1.12	1.11	—	1.085	—	1.065	—
8	1.300	1.283	1.266	1.249	1.233	1.219	1.209	1.193	1.180	1.167	1.155	1.144	1.134	1.125	1.117
	Ac^{III}	Th^{IV}	Pa^{V}	U^{VI}	Np^{IV}	Pu^{IV}	Am^{IV}	Cm^{IV}	Bk^{IV}	Cf^{IV}				No^{II}	
6	1.26	1.08	0.92	0.87	1.01	1.00	0.99	0.99	0.97	0.961				1.24	
8	—	1.19	1.05	1.00	1.12	1.10	1.09	1.09	1.07	1.06					

Note: Charge on cation is the same as the group number except as noted in actinides

Table: Shannon-Prewitt Ionic Radii in Angstrom Units (Å) of Cations with Variable Oxidation States.

Shannon–Prewitt Ionic Radii in Angstrom Units (Å) of Cations with Variable Oxidation States

Oxidation number													
	Ti	V	Cr	Mn	Fe	Co	Ni	Cu	Zn	Ga	Ge	As	Se
2	1.00	0.93	0.94 0.87^{LS}	0.970 0.81^{LS}	0.77(4) 0.92 0.75^{LS}	0.72(4) 0.88.5 0.79^{LS}	0.69(4) $0.63(4)^{Sq}$ 0.83	0.71(4) $0.71(4)^{Sq}$ 0.87	0.74(4) 0.88				
3	0.810	0.780	0.755	0.785 0.72^{LS}	0.785 0.69^{LS} 0.63(4)	0.75 $0.68.5^{LS}$	0.74 0.70^{LS}	0.68^{LS}		0.76		0.72	
4	0.745	0.72	0.69	0.67	0.725	0.67	0.62^{LS}				0.67		0.64
	Zr	Nb	Mo	Tc	Ru	Rh	Pd	Ag	Cd	In	Sn	Sb	Te
2	—	—	—	—	—	—	$0.78(4)^{Sq}$	$0.93(4)^{Sq}$	1.09	—	1.22*(8)	—	—
3	—	0.86	0.83	—	0.82	0.805	0.90	$0.81(4)^{Sq}$	—	0.94	—	0.94(5)	—
4	0.86	0.82	0.79	0.785	0.76	0.74	0.755	—	—	—	0.83	—	1.11
	Hf	Ta	W	Re	Os	Ir	Pt	Au	Hg	Tl	Pb	Bi	Po
1	—	—	—	—	—	—	—	1.51	1.33 1.11(3)	1.64	—	—	—
2	—	—	—	—	—	—	$0.74(4)^{Sq}$	—	1.16	—	1.33	—	—
3	—	0.86	—	—	—	0.82	—	$0.82(4)^{Sq}$	—	1.025	—	1.17	—
4	0.85	0.82	0.80	0.77	0.770	0.765	0.765	—	—	—	0.915	—	1.08
	Ce^{IV} 1.01	Pr^{IV} 0.99	Sm^{II} 1.41(8)	Eu^{II} 1.31	Tb^{IV} 0.90	Tm^{II} 1.17	Yb^{II} 1.16						
	Pa^{IV} 1.04	U^{IV} 1.03	Np^{VI} 0.86	Pu^{VI} 0.85	Am^{III} 1.115	Cm^{III} 1.11	Bk^{III} 1.10	Cf^{III} 1.09					

Size of ions depends on oxidation state.

Table: Shannon-Prewitt Ionic Radii in Angstrom Units (Å) for Common Anions.

	OH	H^-
	1.23	1.53
N^{3-}	O^{2-}	F^-
1.32	1.26	1.19
(C.N.4)	S^{2-}	CI^-
	1.70	1.67
	Se^{2-}	Br^-
	1.84	1.82
	Te^{2-}	I^-
	2.07	2.06

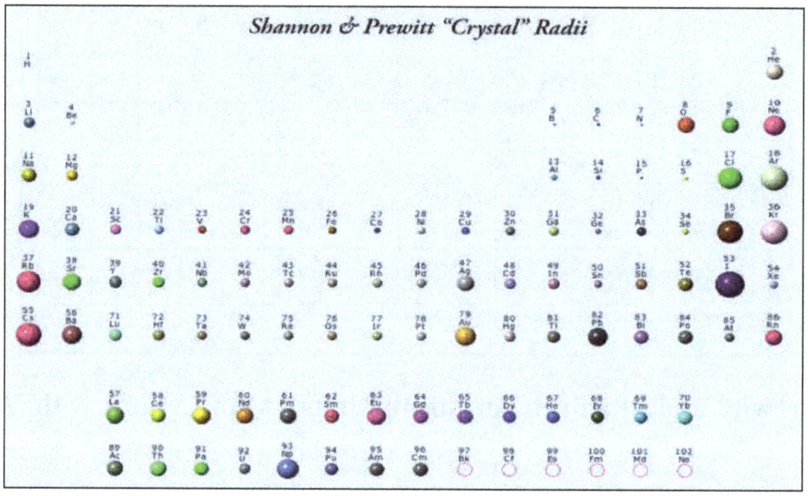

Shannon and Prewitt "Crystal" Radii.

Ionic Radii

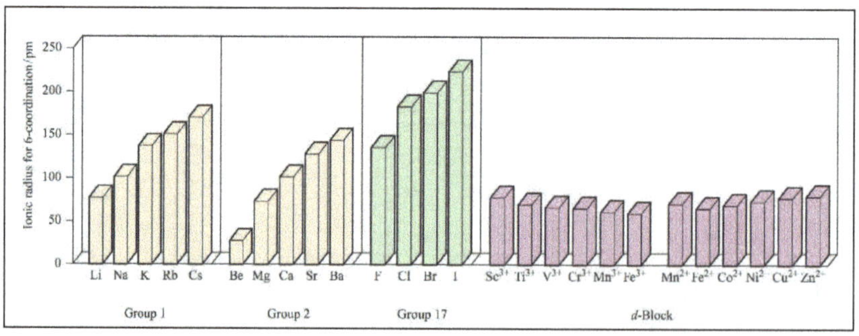

Trends in iconic radii, r_{ion}, within the metal ions of groups 1 and 2, the anions of group 17 and metal ions from the first row of the d-block.

Trends in Sizes of Ions

- The radii of ions within a group increase with atomic number Z. More electrons are present and the outer electrons are further away from the nucleus.

- In a series of isoelectronic cations (e.g. Na^+, Mg^{2+}, Al^{3+}) the radius decreases with increasing positive charge.
- In a series of isoelectronic anions (e.g. F^-, O^{2-}) the radius increases with increasing negative charge.
- For elements with more than one oxidation state (e.g. Ru^{4+}, Ru^{5+}), the radii decease as the oxidation state increases.
- As you move from left to right across a row of the periodic table, there is an overall decrease in radius for similar ions with the same charge.
- The spin state (high or low spin) affects the ionic radius of transition metals.
- The radii increase with an increase in coordination number.

These trends in the sizes of ions may be explained by consideration of the shielding and effective nuclear charge. $Z_{eff} = Z - S$.

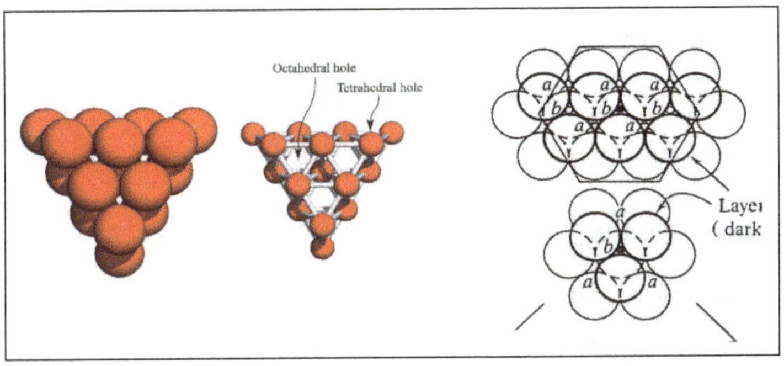

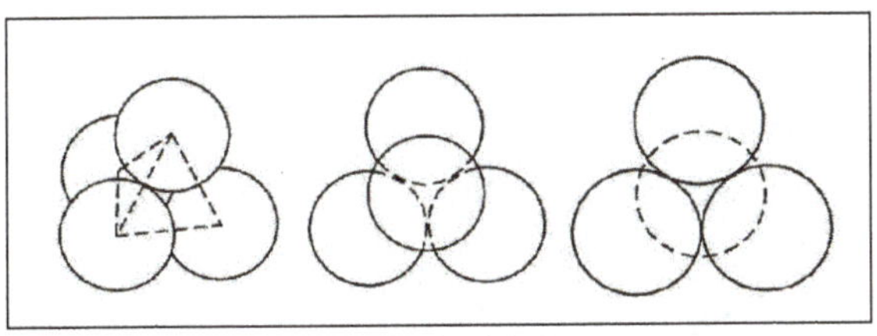

Tetrahedral hole.

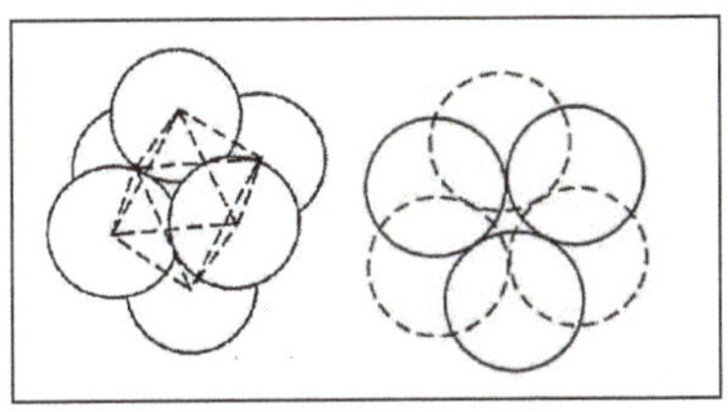

Octahedral hole.

Radius Ratios

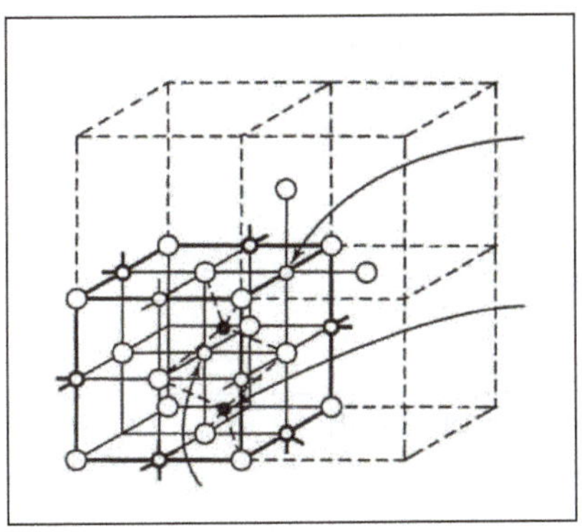

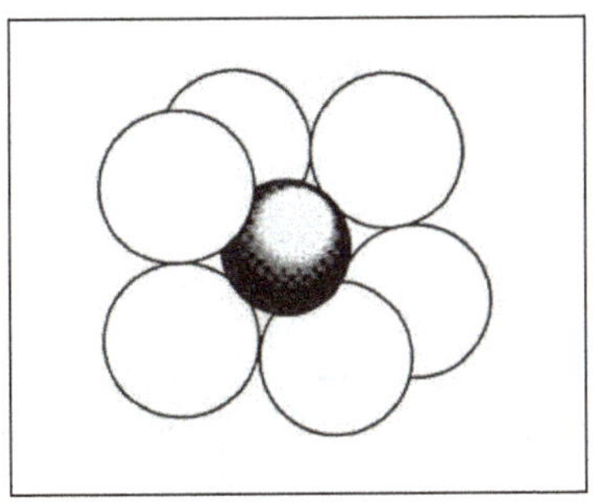

What are the relative sizes and coordination numbers of the cubic, octahedral, and tetrahedral holes?

Size of hole	Cubic >	Octahedral >	Tetrahedral
Coord. no.	8	6	4

If a cation were to occupy one of these holes, it would have the coordination number indicated. The coordination number is proportional to the size of the hole. How does one calculate the size of a hole? Enables one to predict which size cation would fit into that hole.

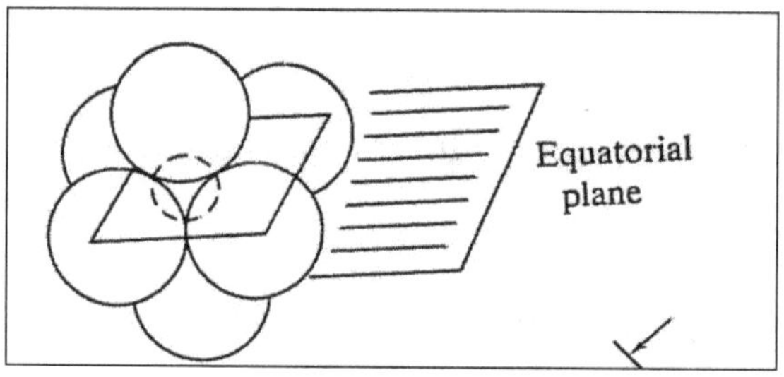

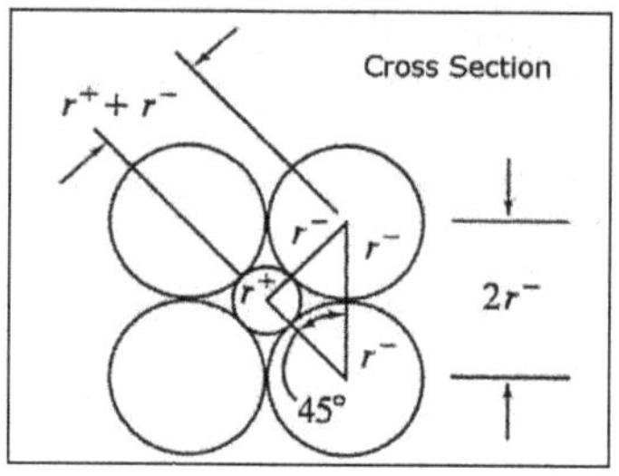

$$\sin(45°) = \frac{r^+ + r^-}{2r^-} = 0.707$$

$$\frac{r^+}{r^-} = 0.414$$

Radius ratio for an octahedral hole.

Calculate radius ratio for a trigonal hole:

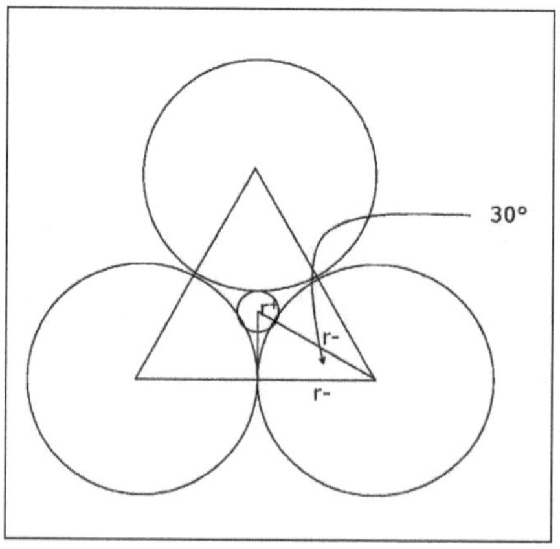

Relatively uncommon coordination environment.

$$\cos(30°) = \frac{r^-}{r^+ + r^-} = \frac{\sqrt{3}}{2}$$

$$\frac{r^+}{r^-} = 0.155$$

Value of $\frac{r_+}{r_-}$	Predicted coordination number of cation	Predicted coordination geometry of cation	Cation r_+/pm	Li^+ 76	Na^+ 102	k^+ 135	Rb^+ 149	Cs^+ 170

<0.15	2	Linear						
0.15-0.22	3	Trigonal planar	Anion r_+/pm	F^- 133	Cl^+ 181	Br^- 196	I^- 220	
0.22-0.41	4	Tetrahedral						
0.41-0.73	6	Octahedral						
>0.73	8	Cubic						

Use this information to predict which type of hole will be occupied with given ions:

Electron-electron repulsion also plays a role in determining the hole the atom occupies. For example, consider the ionic compound NaCl. The 6 coordinate ionic radius (Shannon) of Na^+ is 1.16Å and Cl^- is 1.67Å. The radius ratio is 1.16/1.67Å = 0.695, therefore falls in the range 0.41 to 0.73 so the Na^+ is expected to occupy an octahedral hole.

The Cl^- is expected to form an A-type lattice – it has a ccp type. The sodium cations occupy the octahedral holes in that lattice.

AB Structures

Radius ratios, coordination numbers, and coordination environment for both the cations and anions for four representative AB-type ionic lattices.

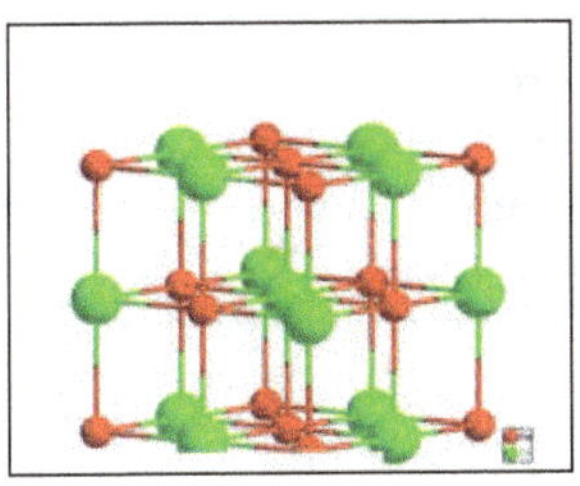

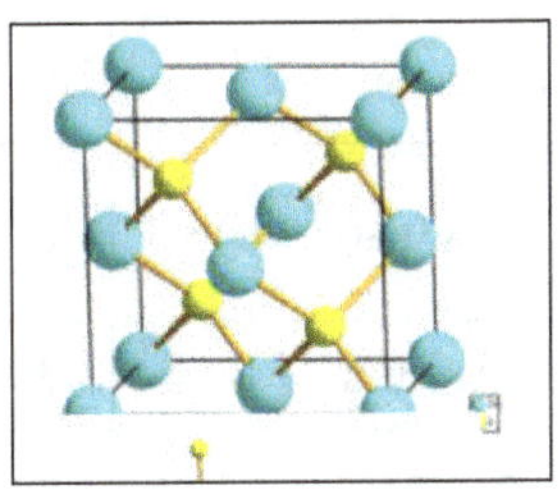

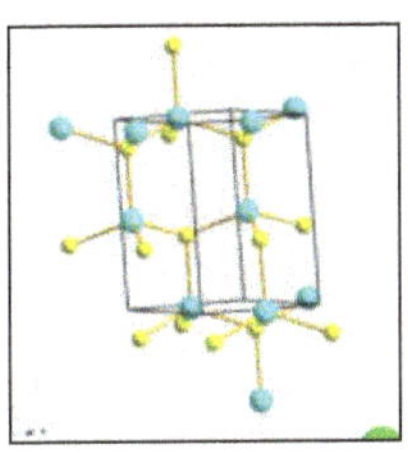

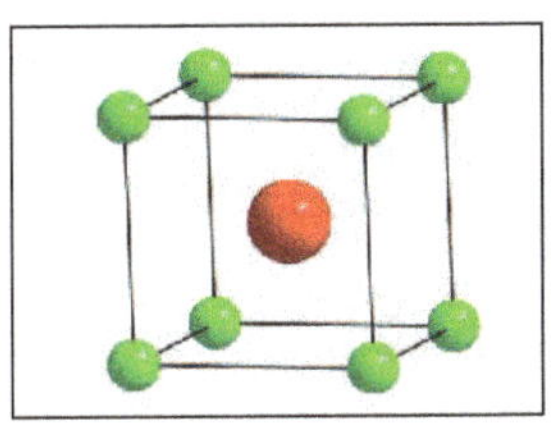

		Anions		Cations	
Structure	r^+/r^-	Structure	C.N.	Structure	C.N.

Sodium chloride (NaCl)	$\frac{1.16}{1.67}$	ccp (fcc)	6	All octahedral holes	6
Zinc blend (ZnS)	$\frac{0.88}{1.70}$	ccp (fcc)	4	Half tetrahedral holes	4
Wurtzite (ZnS)	$\frac{0.88}{1.70}$	hcp	4	Half tetrahedral holes	4
Cesium chloride	$\frac{1.81}{1.67}$	All cubic holes	8	Simple cubic	8

Table: Compounds with common AB structures.

Structure	Compounds	Correlation %
Rock Salt	Alkali halides and hydrides, AgF, AgCl, AgBr, MO (M = Mg, Ca, Sr, Ba, Mn, Fe,Co, Ni, Cd), MS (M = Mg, Ca, Sr, Ba, Mn, Pb), TiC, VC, InP, InAs, SnP, SnAs, NH_4I, KOH, KSH, KCN	58
Zinc Blende	BeX, ZnX, CdX, HgX (X = S, Se, Te), diamond, silicon, germanium, gray tin, SiO_2 (cristabolite), BN, BP, SiC, CuX (X = F, Cl, Br, I), XY (X = Al, Ga, In; Y = P, As, Sb)	33
Wurtzite	ZnX (X = O, S, Te), CdX (X = S, Se), BeO, MgTe, NH_4F, MN (M = Al, Ga, In), AgI	33
Cesium Chloride	CsX (X = Cl, Br, I, SH, CN), TlX (X = Cl, Br, I, CN), NH_4X (X = Cl, Br,CN)	100

Table: (C.N. of A)(no. of A in formula)=(C.N. B)(no. of B in formula).

AB2 –type structures		Anions		Cations	
Structure	r^+/r^-	Structure	C.N.	Structure	C.N.
Fluorite (CaF_2)	1.14/1.19 = 0.96	Simple cubic. All tetra.	4	1. ½ cubic holes 2. fcc	8
Cadmium Iodide (CdI_2)	1.09/2.06 = 0.53	hcp	3	½ of octahedral holes	6
Rutile (TiO_2)	0.745/1.26 = 0.59	Simple tetra.	3	Octahedral holes	6

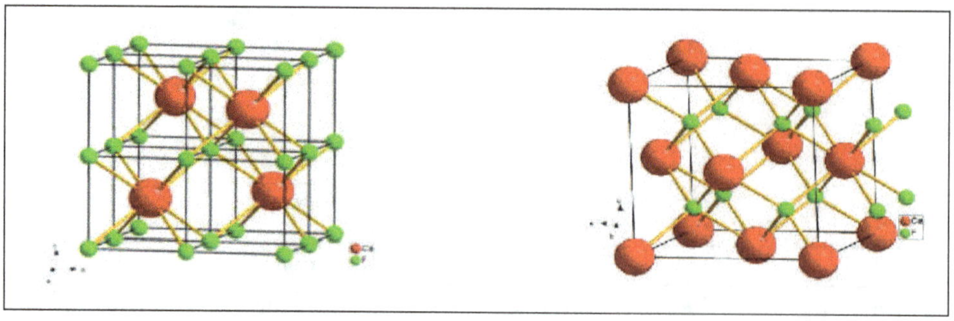

Extended Covalent Arrays

Several elements for covalently bonded solids at room temperature, including B, C, Si, Ge, P, As, Se, Te.

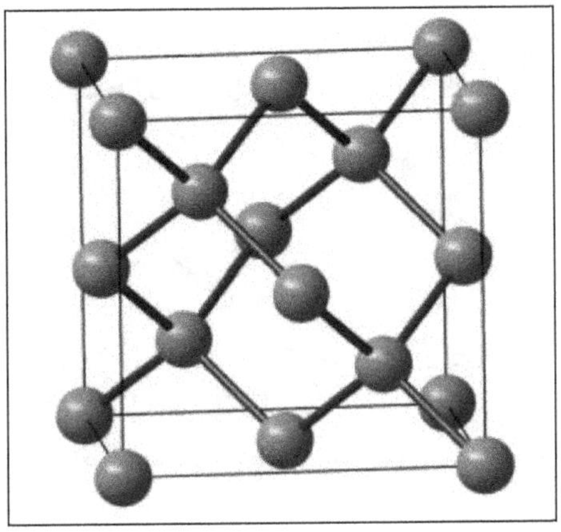

- Unit cell of C (diamond).
- All C-C bond length is identical (1.54 Å).
- Covalent C-C bonds are strong and the rigid 3d network of atoms give diamond the hardest substance known and it also has a high melting point.

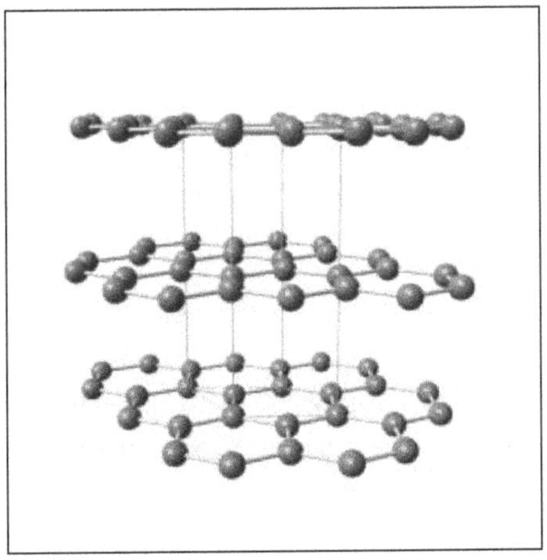

- Unit cell of C (graphite).
- All C-C bond length is identical (1.42 Å).
- Two dimensional layers of carbon atoms in a hexagonal layer.
- The distance between layers is 3.40 Å.

Covalent Network Crystals

A covalent network crystal is composed of atoms or groups of atoms arranged into a crystal lattice, which is held together by an interlocking pair of covalent bonds.

Covalent bonds, the sharing of one or more pairs of electrons in a region of overlap between two or more atoms, are directional interactions as opposed to ionic and metallic bonds that are non-directional.

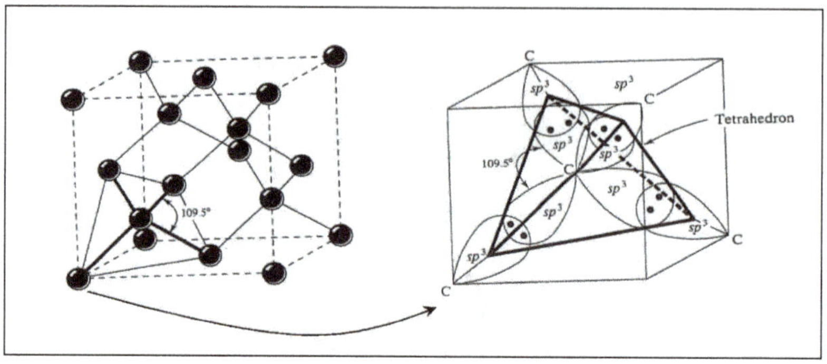

These interactions are directional in nature. Compounds that form covalent network crystals include SiO_2, C, Si, BN.

Metallic Crystals

The structure of metals may be visualized as a lattice of cations held together by a Fermi sea ("of electrons").

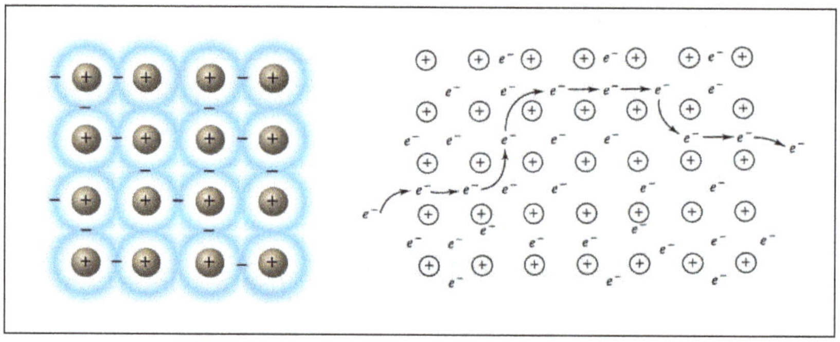

The electrons are no longer associated with any particular cation. Valence electrons are delocalized about the lattice of cations. This type of electron movement enables high conductivity typically associated with metals.

Ionic Crystals

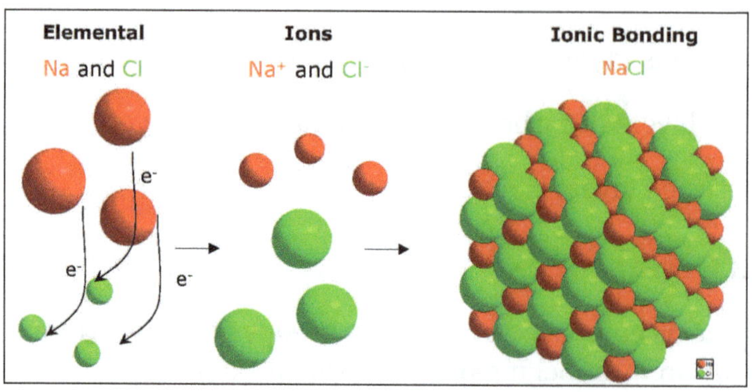

Picking any Na^+, the chloride ion could approach from any direction. Ionic bonds are nondirectional – the electrostatic forces dominate. Many salts form ionic crystals CsCl, CaF_2, KNO_3, and NH_4Cl.

Charge-Dipole Interactions

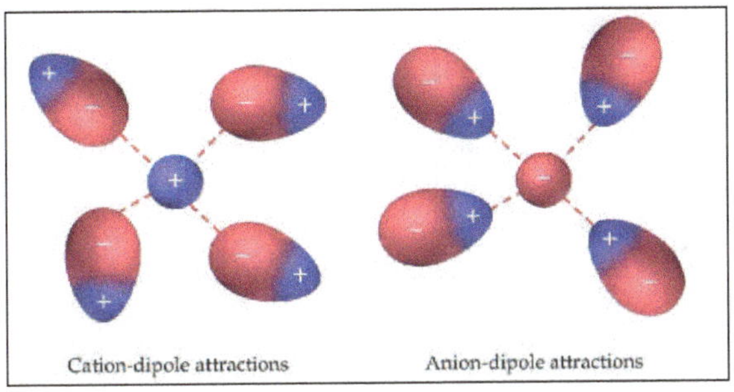

Dipole-Dipole Interactions

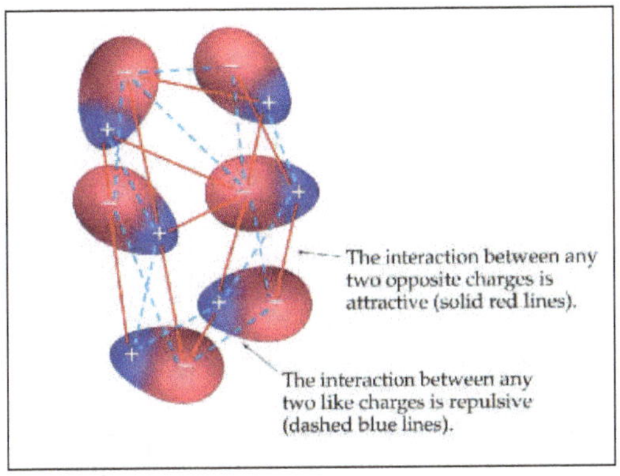

London Dispersion Interactions

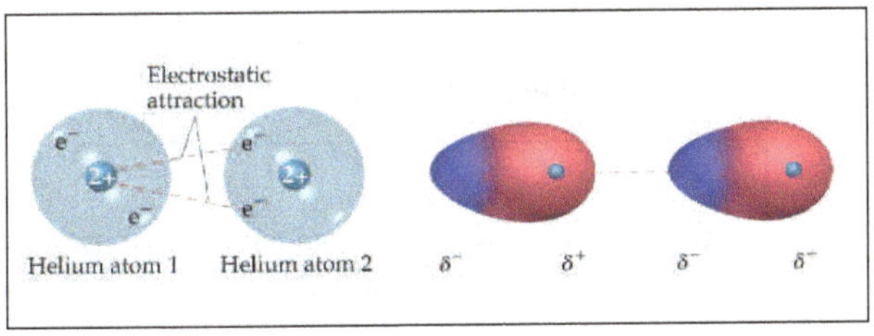

Hydrogen Bonding Interactions

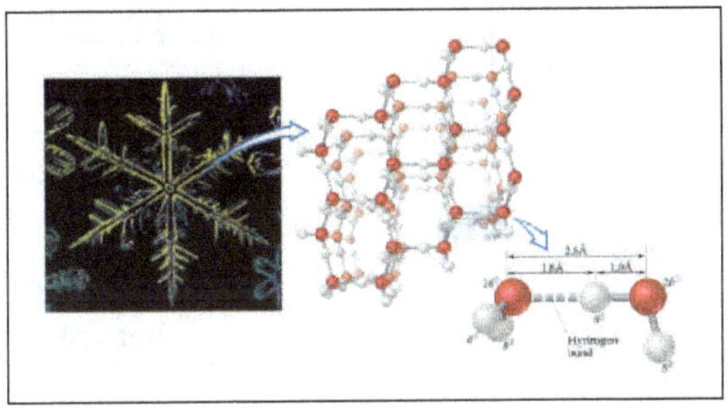

Molecular Structures

Solid CO_2 (dry ice) is a face-centered cubic array of CO_2 molecules.

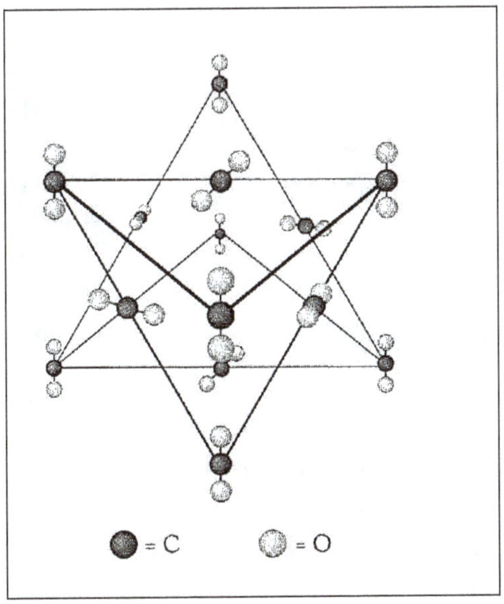

Solid H_2O (ice) has several forms depending on P and T.

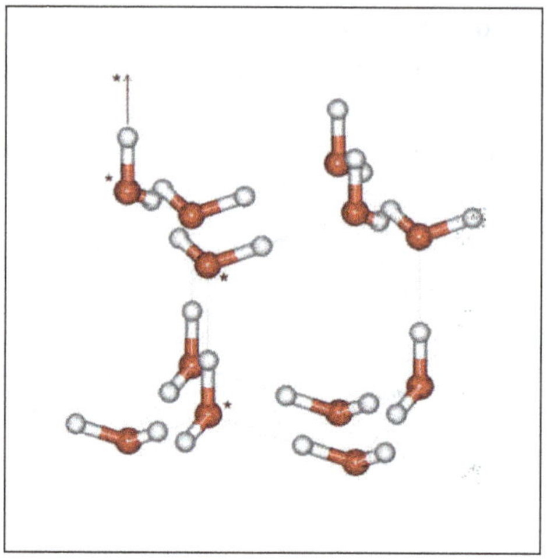

Type	Structural Unit	Bonding	Characteristics	Examples
Ionic	Cations and anions	Electrostatic, nondirectional	Hard, brittle, high m.p.; moderate insulators, melts are conducting	Alkali metal halides
Extended covalent array	Atoms	Mainly covalent	Strong, hard crystals high m.p.	Diamond, silica
Molecular	Molecules	Mainly covalent between atoms in a molecule, van der Waals of H- bonding between molecules	Soft crystals of low m.p; large coefficient of expansion; insulators	Ice, organic compounds
Metallic	Metal atoms	Band model	m.p. varies, good conductors	Fe, Al

Silicates – SiO_2

SiO_2 is polymorphic:

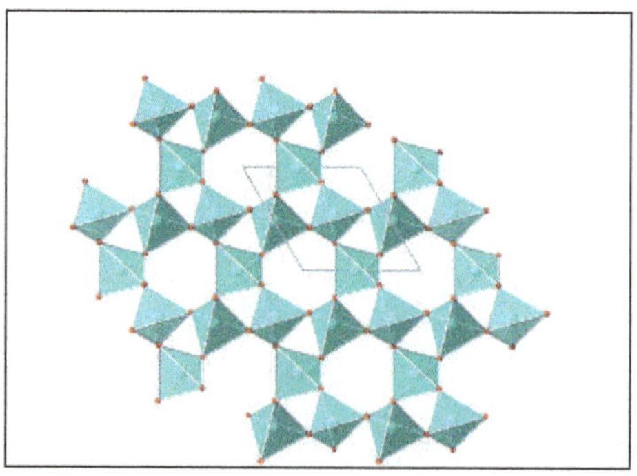

Quartz

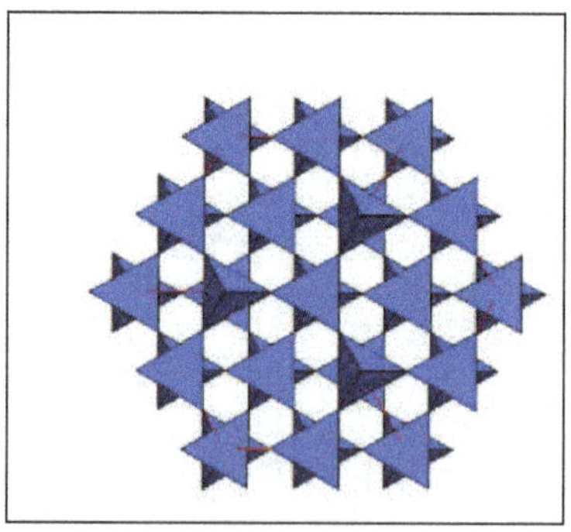

Cristobalite

Table: Crystallographic Data for different forms of silca.

Name Formula	Crystal System	Space Group	Lattice Parameters (nm)
Low-Quartz	Trigonal	$P3_221$	$a = 0.49138 \pm 0.00002$ $c = 0.54052 \pm 0.00002$

High-Quartz	Hexagonal	$P6_222$	a = 0.5038 c = 0.5460
Low-Tridymite	Monoclinic	C2/c	a = 1.854 b = 0.501 c = 2.579 β = 117° 40'
Middle-Tridy-mite	Orthorhombic	$C222_1$	a = 0.994 b = 1.721 c = 4.092
High-Tridymite	Hexagonal	$P6_3$/nunc	a = 0.503 c = 0.822
Low-Cristobalite	Tetragonal	$P4_12_11$	a = 0.4971 ± 0.0003 c = 0.6918 ± 0.0003
High-Cristob-alite	Cubic	Fd3m	a = 0.71297 ± 0.00008
Coesite	Monoclinic	C2/c	a = 0.7173 ± 0.0004 b = 1.2328 ± 0.0006 c = 0.7175 ± 0.0004 β = 120.00° ± 0.04°
Keatite	Tetragonal	$P4_12_12$	a = 0.746 c = 0.861
Stishovite	Tetragonal	$P4_2$/mnm	a = 0.41790 ± 0.00004 c = 0.26649 ± 0.00004

Silicates – Structural Classification,

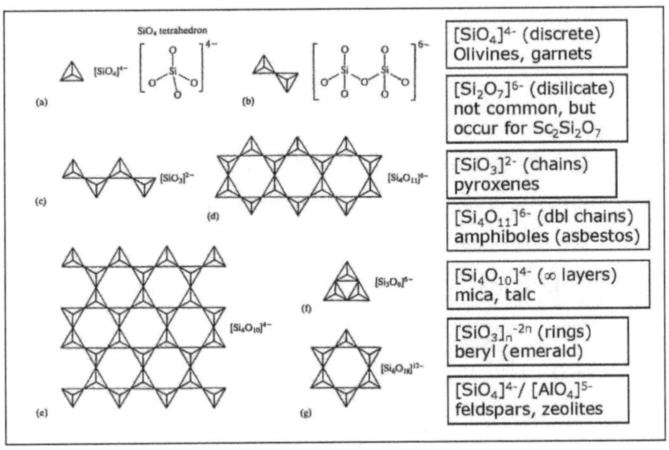

Quasicrystals - Ordered, but aperiodic,

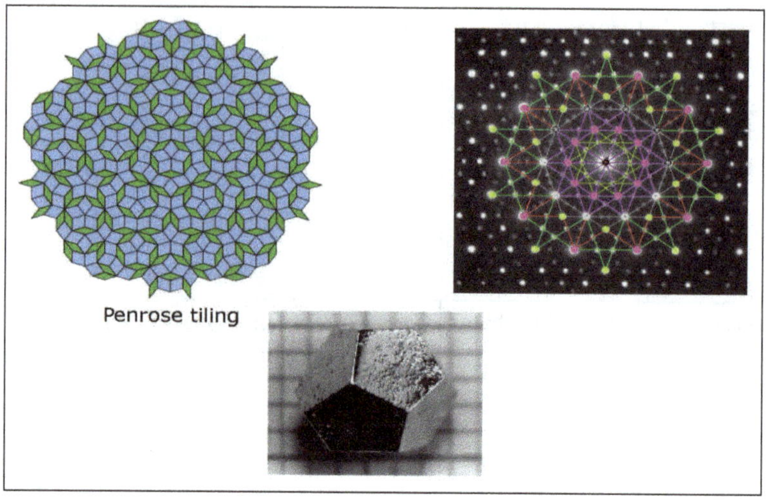

Born-Haber Cycle

By accounting for three energies (ionization energy, electron affinity, and lattice energy), we can get a good idea of the energetics of the enthalpy of formation of an ionic solid.

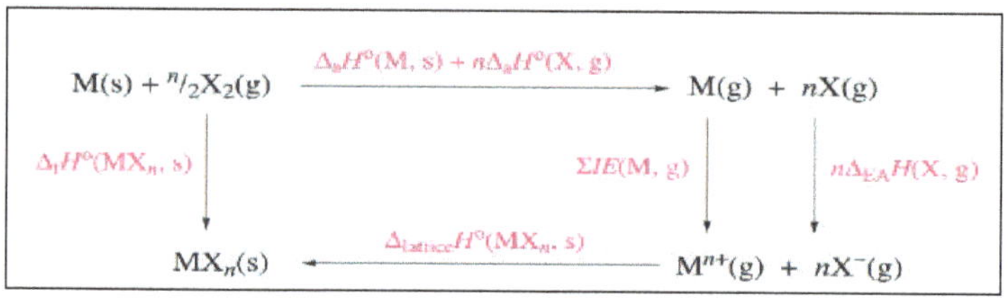

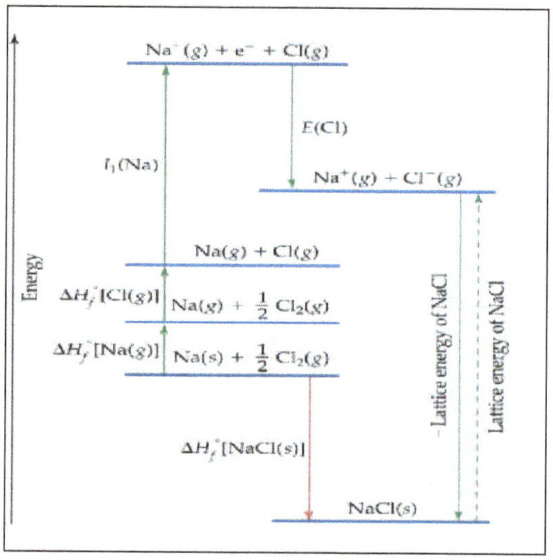

- $\Delta_a H°(M, s)$ = Enthalpy of atomization of metal M.
- $\Delta_a H°(X, g)$ = Enthalpy of atomization of X.
- ΣIE (M, g) = Sum of the ionization energies for the processes $M(g) \rightarrow M^+(g) \rightarrow M^{2+}(g) ... \rightarrow M^{n+}(g)$.
- $\Delta_{EA}H$ (X, g) = Enthalpy change associated with the attachment of an electron.
- $\Delta_f H°(MX_n, s)$ = Standard enthalpy of formation.
- $\Delta_{lattice} H°(MX_n, s)$ = Lattice enthalpy change.

Lattice Energy

The energy required to completely separating a mole of a solid ionic compound into its gaseous ions; or the energy released when ions (gaseous) form a crystal.

Coulomb's Law (ion pair M_+, X_-): $E = -\frac{e^2}{4\pi\varepsilon_0 r}$.

The magnitudes of the charges on the ions are Z_+ and Z_-.

$$E = -\frac{Z_+ Z_- e^2}{4\pi\varepsilon_0 r}$$

e is the electronic charge, 1.6×10^{-19} C.

ε_o is the permittivity of vacuum, 8.854×10^{-12} F m^{-1}.

The energy due to the Coulomb interactions in a crystal is calculated for a particular structure by summing all the ion-pair interactions, thus producing an infinite series.

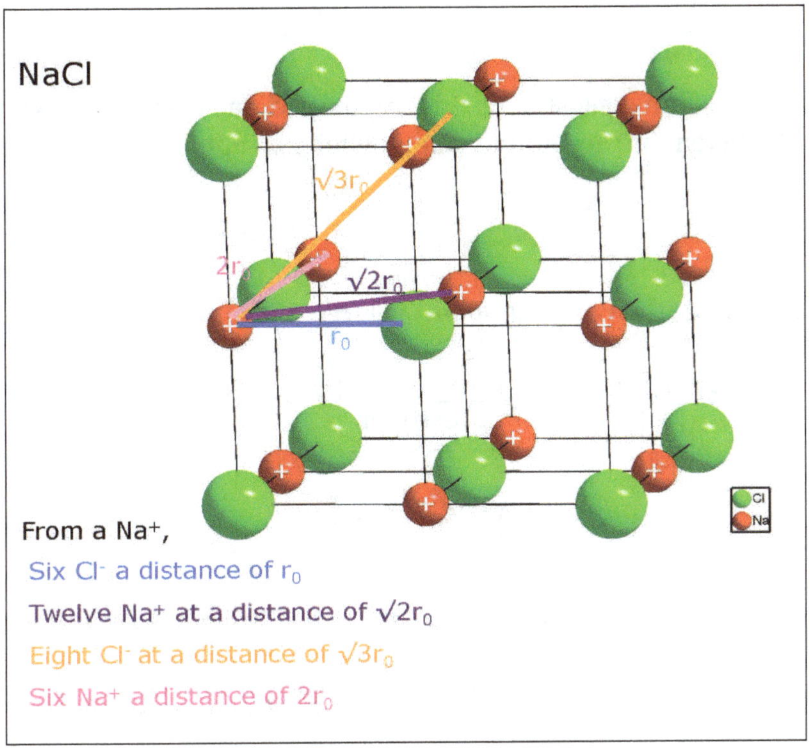

$$E_C = -\frac{e^2}{4\pi\varepsilon_o r}\left(6 - \frac{12}{\sqrt{2}} + \frac{8}{\sqrt{3}} - \frac{6}{2} + \frac{24}{\sqrt{5}} \ldots\right)$$

$$E_C = -\frac{e^2}{4\pi\varepsilon_o r}\left(\frac{6}{\sqrt{1}} - \frac{12}{\sqrt{2}} + \frac{8}{\sqrt{3}} - \frac{6}{\sqrt{4}} + \frac{24}{\sqrt{5}} \ldots\right)$$

The term (series) inside the brackets is known as the Madelung constant, A. For one mole of NaCl, the energy due to the coulombic interactions is:

$$\boxed{E_C = -\frac{N_A A e^2}{4\pi\varepsilon_o r}} \quad N_A = \text{Avogadro's number}$$

Madelung constants have been computed for many of the simple ionic structures.

Structure	Madelung Constant, A	Number of ions in formula unit, v	A/v	Coordination
Cesium chloride, CsCl	1.763	2	0.882	8:8
Sodium chloride, NaCl	1.748	2	0.874	6:6

Fluorite, CaF_2	2.519	3	0.840	8:4
Zinc Blende, ZnS	1.638	2	0.819	4:4
Wurtzite, ZnS	1.641	2	0.821	4:4
Corundum, Al_2O_3	4.172	5	0.835	6:4
Rutile, TiO_2	2.408	3	0.803	6:3

Born Exponent

Ions are not point charges, but consist of positively charged nuclei surrounded by electron clouds. Repulsion needs to be taken into account at small distances repulsion can be expressed by:

$$E_R = \frac{B}{r^n}$$

where B is a constant and n (the Born exponent) is large and also a constant.

$$L = E_C + E_R = -\frac{N_A A Z_+ Z_- e^2}{4\pi\varepsilon_o r} + \frac{B}{r^n} \quad \boxed{\text{Derivation in textbook}} \quad L = -\frac{N_A A Z_+ Z_- e^2}{4\pi\varepsilon_o r}\left(1 - \frac{1}{n}\right)$$

Born-Landé Equation: $L = -\frac{1.389 \times 10^5\ A Z_+ Z_-}{r_o}\left(1 - \frac{1}{n}\right)$

Calculate average (n), r_o in pm, units of L are kJ/mol, e.g. RbCl is 9.5 (Average of 9 and 10).

Ion Type	Constant
[He]	5
[Ne]	7
[Ar]	9
[Kr]	10
[Xe]	12

The large dependence on the charge of the ions: $L = -\frac{1.389 \times 10^5\ A Z_+ Z_-}{r_o}\left(1 - \frac{1}{n}\right)$.

Multiply charged ions tend to have larger lattice energies.

It was noted by A.F. Kapustinskii that the Madelung constant divided by the number of ions in one formula unit (v) of the structure (A/v) was almost constant (0.88 to 0.80). A general equation can be developed to set up a general lattice energy equation and use the resulting equation to calculate the lattice energy of an unknown structure.

$$L(\text{kJ/mol}) = -\frac{1.214 \times 10^5\ vZ_+Z_-}{r_+ + r_-(\text{in\_pm})}\left(1 - \frac{1}{n}\right) \quad \text{Kapustinskii equations}$$

Considering all of the approximations, there is good agreement with values obtained using a Born-Haber cycle, except with large polarizable ions.

It is not possible to measure lattice energy directly. Typically a thermochemical cycle is involved, but not all of the data (electron affinities, etc.) is necessarily available. Estimations are useful. Neil Bartlett used a similar approach to prepare $XePtF_6$.

Table: Lattices energies of some alkali and alkaline earth metal halides at 0 KL/kJ mol^{-1}.

Compound	Structure	Born-Haber cycle	Born-Landé Equation	Extended calculation	Kapunstinskii Equation
LiF	NaCI	-1025	-	-1033	-1033
LiL	NaCI	-756	-	-738	-729
NaF	NaCl	-910	-904	-906	-918
NaCI	NaCI	-772	-757	-770	-763
NaBr	NaCI	-736	-720	-735	-724
Nal	NaCI	-701	-674	-687	-670
KCl	NaCI	-704	-690	-702	-677
Kl	NaCI	-646	-623	-636	-603
CsF	NaCI	-741	-724	-734	-719
CsCl	CsCl	-652	-623	-636	-620
Csl	CsCl	-611	-569	-592	-558
MgF_2	Rutile	-2922	-2883	-2914	-3158
CaF_2	Fluorite	-2597	-2594	-2610	-2779
$CaCl_2$	Deformed rutile	-2226	-	-2223	-2304

Inelastic Interaction of X-Rays with Matter

Wilhelm Conrad Roentgen discovered x-rays at the University of Wurzburg in 1895. In the twentieth century, x-rays have played an important role in gaining an understanding of matter on the atomic scale and quantum mechanics in general. The discovery of x-rays unfolded a new era of science. One of the major accomplishments was the exploration of the microscopic structural details of molecules, liquids and solids at length scales of inter-atomic distances. X-rays can penetrate thick objects that are opaque to visible light. The following diagram illustrates the range of frequencies, energy and wavelength for x-rays.

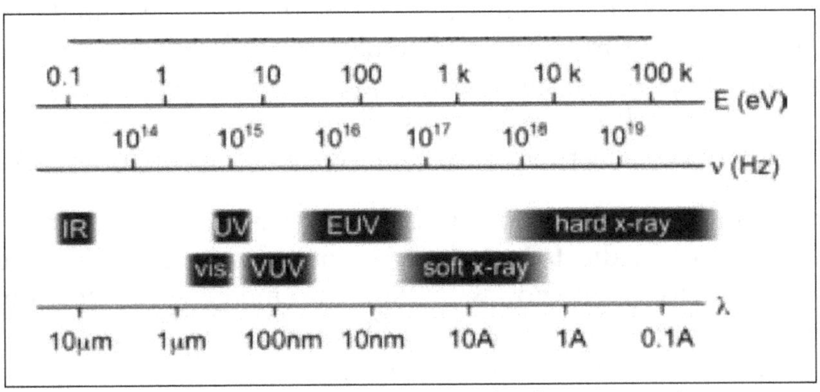

Diagram illustrating soft and hard x-rays in the electromagnetic spectrum.

On the basic of penetrating power and applications the x-rays are usually categorized as soft x-rays and hard x-rays. On the basis of this property, atomic-resolution three-dimensional real-space imaging methods are used to obtain the invisible interior structure of objects. Several x-ray imaging techniques have been developed for this purpose. For example, real-space x-ray microscopy is similar to visible light microscopy, except that traditional refractive lenses are replaced by x-ray lenses such as zone plates. Lens less imaging gets rid of lenses altogether and instead reciprocal-space coherent diffraction patterns are recorded. They are subsequently inverted using computational algorithms to obtain a real-space image. These techniques allow the recording of high-resolution images that can be sensitive to elemental composition, chemical-state, and state of magnetism.

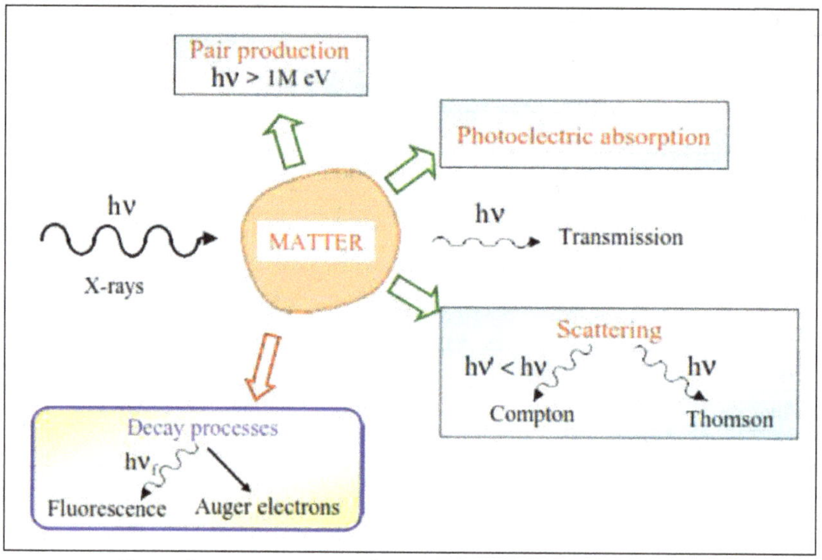

Several consequences of x-ray interaction with matter.

The interaction of x-rays with matter results in several processes which are illustrated in the above diagram. Two high energy x-rays can produce matter particles (a pair of positron and electron). A photon can be completely absorbed to produce a photoelectron. Inelastic scattering results in a partial loss of photon energy via Compton and Raman scattering. In an elastic scattering the photon is scattered without losing its energy. The excitation of atoms and molecules results in decay process through fluorescence and Auger electrons. With the advent of high-intensity short-pulse x-ray sources, ultrafast processes have become accessible for investigations. These sources will allow the study of materials on time scales comparable to the motion of electrons circling around atoms, on spatial scales of inter-atomic bonds, and on energy scales that hold electrons in correlated motion with their neighbors. Ultra-short electromagnetic radiation sources are a critical tool for studying material properties, with the ultimate outlook of recording femtosecond movies of atomic and chemical processes. The earliest x-ray sources were x-ray tubes. Even though their light output is of relatively low brightness, x-ray tubes have enabled numerous important discoveries.

The laser was invented more than half century ago and has led to steep progress in the optical sciences. Similarly, the advent of dedicated synchrotron sources around 1970 has enabled enormous advances in the x-ray sciences. We are now witnessing the emergence of short-pulse high-intensity x-ray sources such as x-ray free-electron lasers (XFELs), high-harmonic generators, x-ray lasers, and laser-plasma sources. With these new sources we are at the dawn of a very exciting time in x-ray science. One can expect progress of similar grandeur as resulted from the introduction of lasers and synchrotrons. Reversible interaction mechanisms of x-rays with matter, such as elastic x-ray scattering, are often used to probe materials. Since these interactions are usually relatively weak, a large photon flux is required to obtain sufficiently intense probe signals.

Since the absorption of x-rays is typically much stronger than the elastic scattering strength, high-intensity x-ray radiation also modifies the structure of materials. How x-rays can be used to probe and modify matter. We will now provide examples for the application of x-ray matter interaction and discuss methods to produce x-ray radiation. X-rays and their application in diverse disciplines including life sciences, crystallography, atomic physics, plasma physics, materials science, chemistry, and astronomy. X-ray radiation can be prepared to have a high spectral purity, which is useful for x-ray spectroscopy, a very short pulse duration, which enables time-resolved studies, and high intensity, which can lead to nonlinear x-ray–matter interaction phenomena. X-ray astronomy has enabled detailed studies of supernovae, pulsars, and black holes. Since the Earth's atmosphere is opaque to x-ray radiation, x-rays can only be observed from outer space.

Absorption of x-Rays

When the x-rays hit a sample, the oscillating electric field of the electromagnetic radiation interacts with the electrons bound in an atom. Either the radiation will be scattered by these electrons or absorbed and excite the electrons.

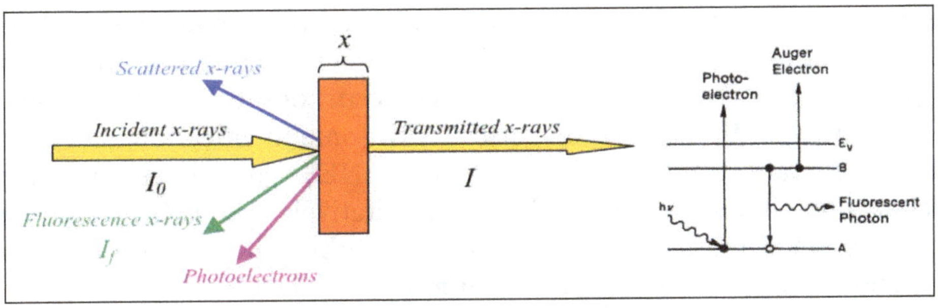

Illustration of x-ray absorption.

A narrow parallel monochromatic x-ray beam of intensity I_o passing through a sample of thickness x will get a reduced intensity I according to the expression:

$$\ln(I_o / I) = \mu x$$

where μ is the linear absorption coefficient, which depends on the types of atoms and the density of the material. At certain energies where the absorption increases drastically and gives rise to an absorption edge. Each such edge occurs when the energy of the incident photons is just sufficient to cause excitation of a core electron of the absorbing atom to a continuum state, i.e. to produce a photoelectron. Thus, the energies of the absorbed radiation at these edges correspond to the binding energies of electrons in the K, L, M, etc. shells of the absorbing elements. The absorption edges are labeled in the order of increasing energy, K, L_I, L_{II}, L_{III}, M etc. corresponding to the excitation of an electron from the 1s ($^2S_{1/2}$), 2s ($^2S_{1/2}$), 2p ($^2P_{1/2}$), 2p ($^2P_{3/2}$), 3s ($^2S_{1/2}$) orbital (states), respectively.

X-Ray Absorption Edge

An absorption edge by itself is of little value beyond elemental identification. However, if one examines any of the edges, they are found to contain a wealth of information. This is illustrated by the schematic absorption edge shown in figure. The absorption edge is not simply a discontinuous increase in absorption, but in fact shows significant structure both in the immediate vicinity of the edge jump and well above the edge. The structure in the vicinity of the edge is sometimes referred to as X-ray absorption near-edge structure (XANES). The oscillations above the edge, which can extend for 1,000 eV or more, are often referred to as extended X-ray absorption fine structure (EXAFS).

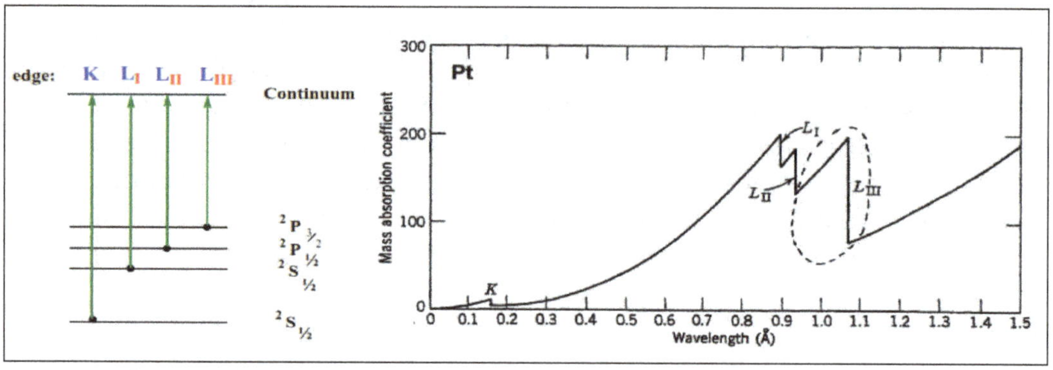

Absorption lines from different orbitals and absorption edge in the spectrum.

The distinction between XANES and EXAFS is arbitrary, since the same fundamental physical principles govern photo-absorption over the entire XAS region and there is ambiguous definition that distinguishes between "near-edge" and "extended" structure. In an attempt to emphasize the essential similarity of these regions, the term XAFS (X-ray absorption fine structure) has gained some currency as a reference to the entire structured absorption region. Nevertheless, the terms EXAFS and XANES remain the most widely used, with some justification, since the XANES and EXAFS regions are generally analyzed differently. The XANES region is sensitive to oxidation state and geometry, but is not, in most cases, analyzed quantitatively. The EXAFS region is sensitive to the radial distribution of electron density around the absorbing atom and is used for quantitative determination of bond length and coordination number.

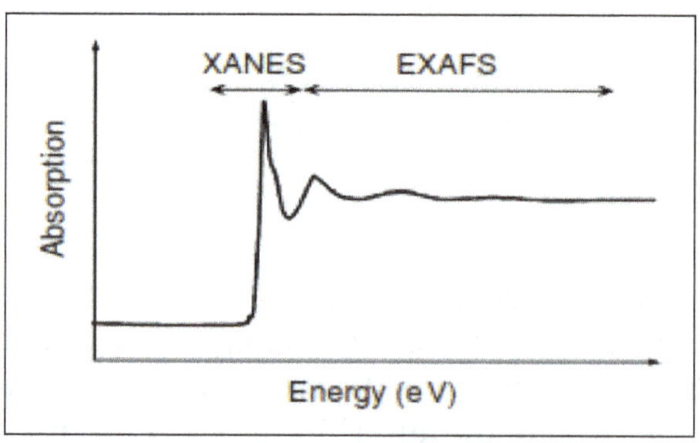

Illustration of x-ray absorption edge.

X-Ray Fluorescence

Absorption of an ionizing X-ray results in photoelectron ejection, leaving behind a highly excited core-hole state. This can relax by a variety of mechanisms, with the two most important being emission of an Auger electron and X-ray fluorescence. For lower-energy excitation, Auger emission can be the dominant relaxation process. For higher-energy excitation (e.g., for the K edges of elements with atomic numbers greater than 40), X-ray fluorescence is the primary relaxation process, with X-ray fluorescence yields approaching 1. For light elements, the X-ray fluorescence spectrum is quite simple. However, for heavy elements, a large number of X-ray emission lines are observed.

The nomenclature associated with X-ray fluorescence lines predates a modern, quantum understanding of the origins of X-ray fluorescence and consequently there is not a simple relationship between the names of different emission lines and the origin of the line. Some of the possible emission lines are shown in figure. Like all emission spectroscopy, X-ray fluorescence is governed by a series of selection rules. Consequently, only certain transitions, referred to as "diagram lines," are allowed. As with other spectroscopies, a variety of forbidden transitions (non-diagram lines) is also observed, and can yield important information. Each element has unique "characteristic" X-ray emission energies which are, in most cases, well resolved from neighboring emission lines.

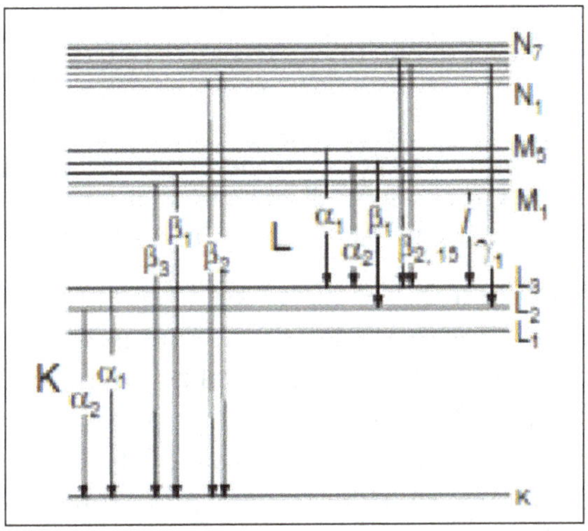

Nomenclature for selected X-ray emission lines.

Measurement of X-Ray Absorption

In the simplest case, measurement of an X-ray absorption spectrum involves only measurement of the incident and the transmitted X-ray flux. This can be accomplished, for example, with an ionization chamber in front of and behind the sample, using equation $\ln(I_o / I) = \mu x$ to convert to absorption coefficient. This approach is limited to moderately concentrated samples (greater than 500 ppm) and, depending on the energy of the absorption edge, even these concentrations may not be accessible. For example, sulfur or chlorine containing solvents are nearly opaque to lower energy X-rays and thus interfere with XAS measurements. To avoid the limitations of absorption, XAS spectra are frequently measured as fluorescence excitation spectra. This is particularly important for dilute samples such as catalysts, biological samples, or environmental samples. Providing the sample is dilute (absorbance due to the element of interest is much smaller than the background absorbance) or thin (total absorbance), the intensity of the fluorescence X-rays is proportional to the X-ray absorption cross-section. In most cases, the sample will emit a variety of X-rays, both the fluorescence X-rays of interest and a background of scattered X-rays. In order to have good sensitivity, the fluorescence detector needs some kind of energy resolution to distinguish between the signals and background X-rays.

In some cases, energy resolution can be provided by a simple low-pass filter although for the ultimate sensitivity it is necessary to use higher resolution in order to more effectively exclude background radiation. This is typically an energy resolving solid-state fluorescence detector, although recent advances with wavelength resolving detectors (i.e., multilayer diffraction gratings) may be important in special cases. In principle, any physical property that changes in proportion to X-ray absorption could be used to measure XAS spectra. In addition to X-ray fluorescence, properties that have been used include photoconductivity, optical luminescence, and electron yield, although only the latter is widely used. Electron yield detection of XAS is particularly important for studies of surfaces. Since the penetration depth of an electron through matter is quite small, electron yield can be used to make XAS measurements surface sensitive. Although XAS can be studied for virtually any X-ray absorption edge, experiments are simplest.

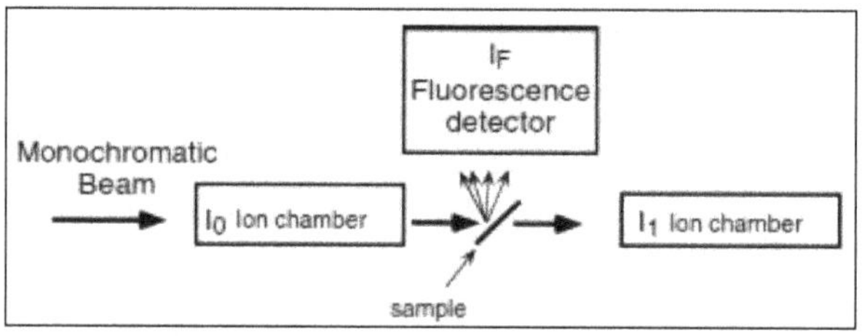

Typical experimental apparatus for XAS measurements: Incident and transmitted intensities are typically measured using an ion chamber; a variety of detectors can be used to measure X-ray fluorescence intensity for dilute samples.

To avoid the limitations of absorption, XAS spectra are frequently measured as fluorescence excitation spectra. This is particularly important for dilute samples such as catalysts, biological samples, or environmental samples. The basic experimental geometry provided the sample is dilute or thin the intensity of the fluorescence X-rays is proportional to the X-ray absorption cross-section. In most cases, the sample will emit a variety of X-rays, both the fluorescence X-rays of interest and a background of scattered X-rays. In order to have good sensitivity, the fluorescence detector needs some kind of energy resolution to distinguish between the signals and background X-rays. In some cases, energy resolution can be provided by a simple low-pass filter although for the ultimate sensitivity it is necessary to use higher resolution in order to more effectively exclude background radiation. This is typically an energy-resolving solid-state fluorescence detector, although recent advances with wavelength resolving detectors may be important in special cases. In principle, any physical property that changes in proportion to X-ray absorption could be used to measure XAS spectra. In addition to X-ray fluorescence, properties that have been used include photoconductivity, optical luminescence, and electron yield, although only the latter is widely used. Electron yield detection of XAS is particularly important for studies of surfaces. Since the penetration depth of an electron through matter is quite small, electron yield can be used to make XAS measurements surface sensitive.

Although XAS can be studied for virtually any X-ray absorption edge, experiments are simplest when they can be performed at atmospheric pressure. This limits the accessible X-ray energies to those greater than approximately 5 keV (for air) or 2 keV (for a He atmosphere). Lower energy measurements (i.e., measurements of the K edges for elements lighter than phosphorus) require that the sample be in vacuum in order to avoid excessive attenuation of the incident X-ray beam. Similarly, it is difficult, although not impossible, to make XAS measurements at energies above approximately 30 ke.

However, this does not limit XAS significantly, since elements that are heavy enough to have K edge energies >30 keV have readily accessible L edge energies (the L edge for Sn is at 4 keV). This means that XAS spectra can be measured for virtually every element, although measurements for elements lighter than phosphorus generally require that the sample be made vacuum compatible. X-ray absorption spectra can be measured for solids, liquids, or gases and do not require that samples have long-range order (i.e., be crystalline) or that samples possess particular magnetic properties (e.g., nonzero electron spin or specific isotopes). Measurements can be made at low temperature for studies of unstable samples, or at high temperature and/or pressure, for example for studies of catalysts under reaction conditions or of geochemical samples under conditions that approximate the inner mantle. This flexibility, combined with near universality, has made XAS a widely utilized technique in all areas of coordination chemistry.

The critical experimental detail that limits the utility of XAS, and that accounts for XAS having been an obscure technique prior to about 1975, is the need for an intense, tunable X-ray source. Conventional X-ray sources work much the same as the X-ray tube that was invented by Rontgen: an electron beam strikes a target which emits both "characteristic" radiation (X-ray fluorescence lines) and a broad continuous background of bremsstrahlung radiation. Only the latter is useful for XAS, since XAS measurements require a broad band of X-ray energies. The intensity of monochromatic radiation that can be obtained from the bremsstrahlung radiation is too low for most XAS measurements.

Elastic Interaction of X-Rays with Matter

X-rays are scattered elastically by electrons, which is named Thomson scattering. In this process the electron oscillates like a Hertz dipole at the frequency of the incoming beam and becomes a source of dipole radiation. The wavelength λ of x-rays is conserved for Thomson scattering in contrast to the two inelastic scattering processes mentioned above. It is the Thomson component in the scattering of x-rays that is made use of in structural investigations by x-ray diffraction.

Materials are made of atoms. Knowledge of how atoms are arranged into crystal structures and microstructures is the foundation on which we build our understanding of the synthesis, structure and properties of materials. In a day to day work we talk about x-ray reflections from a series of parallel planes inside the crystal. The orientation and inter-planner spacing of these planes are defined by the three integers h, k, l called Miller indices. A given set of planes with indices h, k and l cut the a-axis of the unit cell in h sections, the b-axis in k sections and the c-axis in l sections. A zero indicates that the planes are parallel to the corresponding axis. For example the (220) planes cut the a-axis and the b-axis in half, but are parallel to c-axis. The procedure employed in determination of the h, k, and l index numbers is as follows:

- If the plane passes through the selected origin, either another parallel plane must be constructed within the unit cell by an appropriate translation or a new origin must be established at the corner of another unit cell.
- At this point the crystallographic plane either intersects or parallels each of the three axes; the length of the planar intercept for each axis is determined in terms of the lattice parameters a, b, and c.

- The reciprocals of these numbers are taken. A plane that parallels an axis may be considered to have an infinite intercept, and, therefore, a zero index.
- If necessary, these three numbers are changed to the set of smallest integers by multiplication or division by a common factor.
- Finally, the integer indices, not separated by commas, are enclosed within parentheses, thus: (hkl).

An intercept on the negative side of the origin is indicated by a bar or minus sign positioned over the appropriate index. Furthermore, reversing the directions of all indices specifies another plane parallel to, on the opposite side of and equidistant from, the origin. Mirror indices of different planes are shown in the figure below:

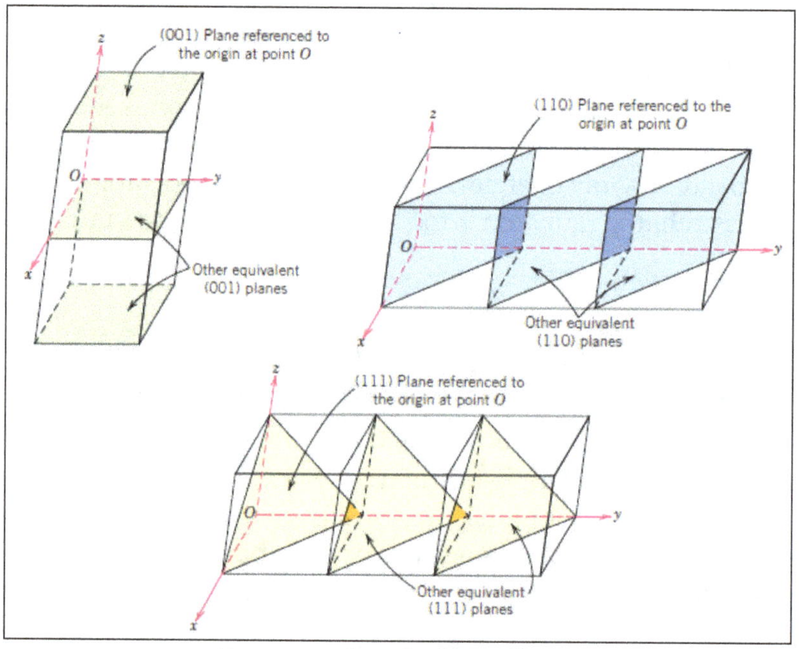

Representation of Miller Indices.

Construct a $(0\,\bar{1}\,1)$ plane within a cubic unit cell. To begin, the indices are removed from the parentheses, and reciprocals are taken, which yields ∞, -1, and 1. This means that the particular plane parallels the x axis while intersecting the y and z axes at –b and c, respectively. This plane has been drawn in another sketch in figure below. A plane is indicated by lines representing its intersections with the planes that constitute the faces of the unit cell or their extensions. For example, in this figure, line ef is the intersection between the $(0\,\bar{1}\,1)$ plane and the top face of the unit cell; also, line gh represents the intersection between this same $(0\,\bar{1}\,1)$ plane and the plane of the bottom unit cell face extended. Similarly, lines eg and fh are the intersections between $(0\,\bar{1}\,1)$ and back and front cell faces, respectively.

There are many different methods for measuring structure across this wide range of distances, but the more powerful experimental techniques involve diffraction. To date, most of our knowledge about the spatial arrangements of atoms in materials has been gained from diffraction experiments. In a diffraction experiment, an incident wave is directed into a material and a detector is typically moved about to record the directions and intensities of the outgoing diffracted waves.

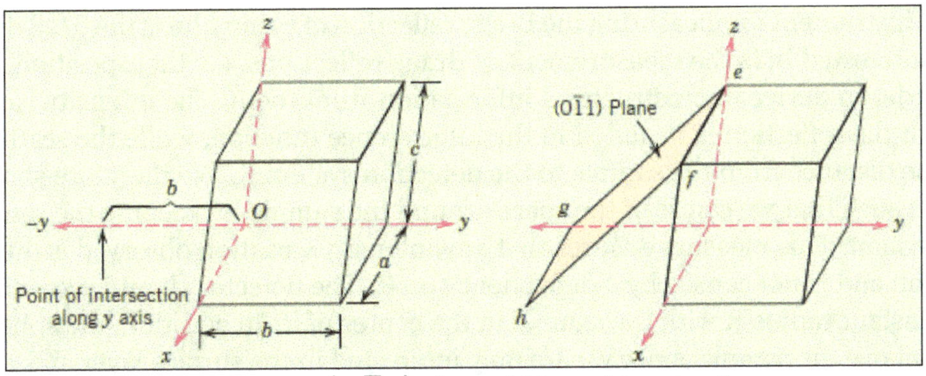

Construct a $(0\ \bar{1}\ 1)$ plane within a cubic unit cell.

Diffraction effects are observed when electromagnetic radiation impinges on periodic structures with geometrical variations on the length scale of the wavelength of the radiation. The inter-atomic distances in crystals and molecules amount to 0.15–0.4 nm which correspond in the electromagnetic spectrum with the wavelength of x-rays having photon energies between 3 and 8 keV. Accordingly, phenomena like constructive and destructive interference should become observable when crystalline and molecular structures are exposed to x-rays.

Coherent scattering preserves the precision of wave periodicity. Constructive or destructive interference then occurs along different directions as scattered waves are emitted by atoms of different types and positions. There is a profound geometrical relationship between the directions of waves that interfere constructively, which comprise the "diffraction pattern," and the crystal structure of the material. The diffraction pattern is a spectrum of real space periodicities in a material. Atomic periodicities with long repeat distances cause diffraction at small angles, while short repeat distances cause diffraction at high angles. It is not hard to appreciate that diffraction experiments are useful for determining the crystal structures of materials. Much more information about a material is contained in its diffraction pattern, however. Crystals with precise periodicities over long distances have sharp and clear diffraction peaks. Crystals with defects such as impurities, dislocations, planar faults, internal strains, or small precipitates are less precisely periodic in their atomic arrangements, but they still have distinct diffraction peaks. Their diffraction peaks are broadened, distorted, and weakened, however, and "diffraction line shape analysis" is an important method for studying crystal defects. Diffraction experiments are also used to study the structure of amorphous materials, even though their diffraction patterns lack sharp diffraction peaks.

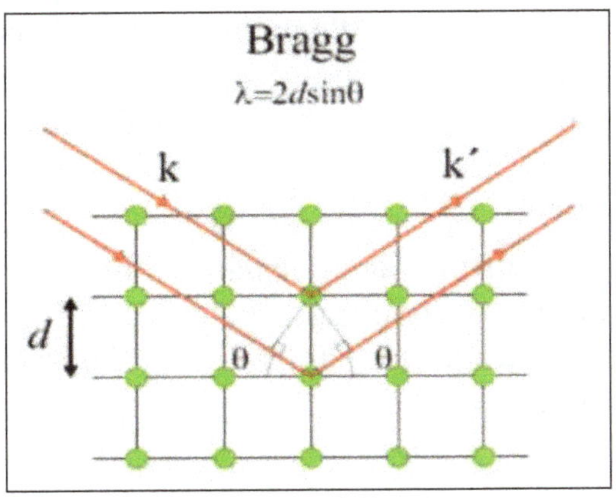

Bragg's X-ray diffraction.

An often-used instrument for measuring the Bragg reflection of a thin film is the θ/2θ diffractometer. What we are interested in is the measurement of Bragg reflections, i.e. their position, shape, intensity, etc., in order to derive microstructural information from them. The intensity variation that is associated with the reflection is included in the interference function, while the scattered intensity depends on the distance from the sample to the detection system R. We therefore should configure the instrument such that we can scan the space around the sample by keeping the sample–detector distance R constant. This measure ensures that any intensity variation observed is due to the interference function and is not caused by a dependency on R. The detector should accordingly move on a sphere of constant radius R with the sample in the center of it. In addition, the sphere reduces to a hemisphere above the sample, since we are only interested in the surface layer and data collection will be performed in reflection mode. The geometry is shown in above figure. Because the scattering of x-rays depends sensitively on the orientation of the crystal with respect to the scattering vector, we carefully have to define the various coordinate systems with which we are dealing.

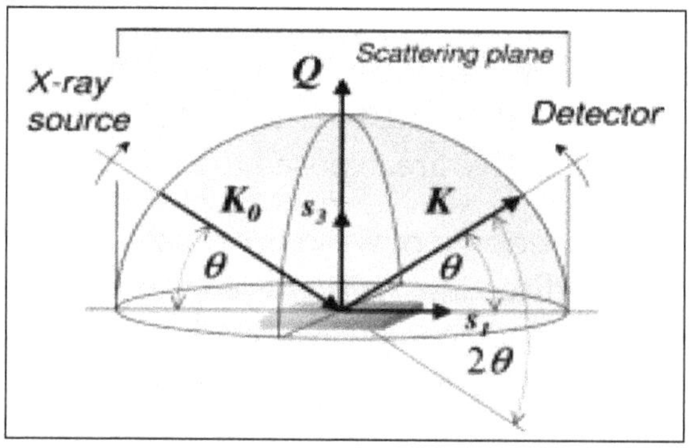

The above given figure is schematic representation of a θ/2θ scan from the viewpoint of the sample reference frame. K_o is the incident wave vector and K is the reflected wave vector.

The inter-planar spacing, d, sets the difference in path length for the ray scattered from the top plane and the ray scattered from the bottom plane. Figure shows that this difference in path lengths is 2d sin θ. Constructive wave interference occurs when the difference in path length for the top and bottom rays is equal to one wavelength, λ:

$$2d\sin\theta = n\lambda$$

The right hand side is multiplied by an integer, n, since this condition also provides constructive interference. For first order diffraction, n = 1. When there is a path length difference of nλ between adjacent planes, we changed even though this new d may not correspond to a real interatomic distance. For example, when our diffracting planes are (100) cube faces, and,

$$2d_{100}\sin\theta = 2\lambda$$

then we speak of a (200) diffraction from planes separated by d_{200} = (d)/2. A diffraction pattern from a material typically contains many distinct peaks, each corresponding to a different inter-planar spacing, d. For cubic crystals with lattice parameter a_o, the inter-planar spacing, d, of planes labeled by Miller indices (hkl) are,

$$d_{hkl} = \frac{a_o}{\sqrt{h^2 + k^2 + l^2}}$$

which can be proved by the definition of Miller indices and the 3D Pythagorean theorem. From Bragg's law we find that the (hkl) diffraction peak occurs at the measured angle $2\theta_{hkl}$:

$$2\theta_{hkl} = 2\arcsin\left(\frac{\lambda\sqrt{h^2+k^2+l^2}}{2a_o}\right)$$

There are often many individual crystals of random orientation in the sample, so all possible Bragg diffractions can be observed in the "powder pattern." There is a convention for labeling, or "indexing," the different Bragg peaks in a powder diffraction pattern using the numbers (hkl). An example of an indexed diffraction pattern is shown in figure below. The intensities of the different diffraction peaks vary widely, and are zero for some combinations of h, k, and l. For this example of polycrystalline silicon, notice the absence of all combinations of h, k, and l that are mixtures of even and odd integers, and the absence of all even integer combinations whose sum is not divisible by 4.

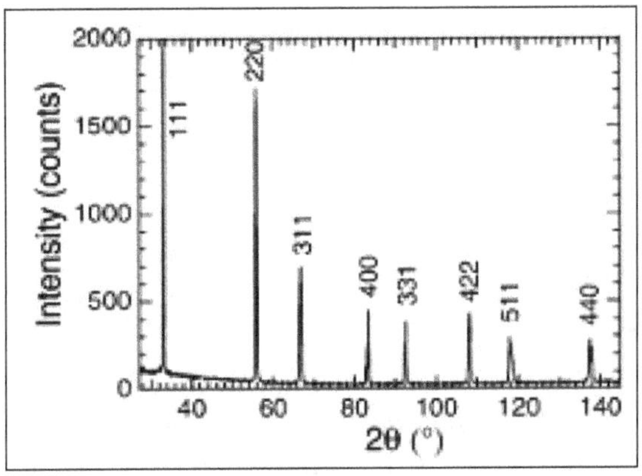

Indexed powder x-ray diffraction pattern of polycrystalline silicon.

One important use of x-ray powder diffractometry is for identifying unknown crystals in a sample. The idea is to match the positions and the intensities of the peaks in the observed diffraction pattern to a known pattern of peaks from a standard sample or from a calculation. There should be a one-to-one correspondence between the observed peaks and the indexed peaks in the candidate diffraction pattern. For a simple diffraction pattern, this tentative indexing still needs to be checked. To do so, the θ-angles of the diffraction peaks are obtained, and used with equation $2d_{100}\sin\theta = 2\lambda$ to obtain the interplanar spacing for each diffraction peak. For cubic crystals it is then possible to use equation $d_{hkl} = \frac{a_o}{\sqrt{h^2+k^2+l^2}}$ to convert each interplanar spacing into a lattice parameter. The indexing is consistent if all peaks provide the same lattice parameters. Non-cubic crystals usually require an iterative refinement of lattice parameters and angles.

Table: The relation between lattice constants and miller indices for non-cubic symmetry.

Crystal System	Constraints	$\frac{1}{d_{hkl}^2} =$
Cubic	$a = b = c$ $\alpha = \beta = \gamma = 90°$	$\frac{h^2+k^2+l^2}{a^2}$

Tetragonal	$a = b$ $\alpha = \beta = \gamma = 90°$	$\frac{h^2 + k^2}{a^2} + \frac{l^2}{c^2}$
Orthorhombic	$\alpha = \beta = \gamma = 90°$	$\frac{h^2}{a^2} + \frac{k^2}{b^2} + \frac{l^2}{c^2}$
Hexagonal	$a = b$ $\alpha = \beta = 90°$ $\gamma = 120°$	$\frac{4}{3}\frac{h^2 + hk + k^2}{a^2} + \frac{l^2}{c^2}$
Trigonal/Rhombo-hedral	$a = b = c$ $\alpha = \beta = \gamma$	$\frac{(h^2 + k^2 + l^2)\sin^2\alpha + 2(hk + hl + kl)(\cos^2\alpha - \cos\alpha)}{a^2(1 - 3\cos^2\alpha + 2\cos^3\alpha)}$
Monoclinic	$\alpha = \gamma = 90°$	$\frac{h^2}{a^2\sin^2\beta} + \frac{k^2}{b^2} + \frac{l^2}{c^2\sin^2\beta} - \frac{2hl\cos\beta}{ac\sin^2\beta}$
Triclinic	None	

Crystal Defects

Atoms are orderly arranged without any disruptions in ideal solids. Under mechanical loads, all solids tend to reshape themselves. This occurs usually by process called slip, the translation of one plane of atoms over another under implied shear stresses. Consider two planes of atoms subjected a shear stress under applied load. Let's assume, as shown in figure, distance between atoms in slip direction is b, and spacing between the two planes is a.

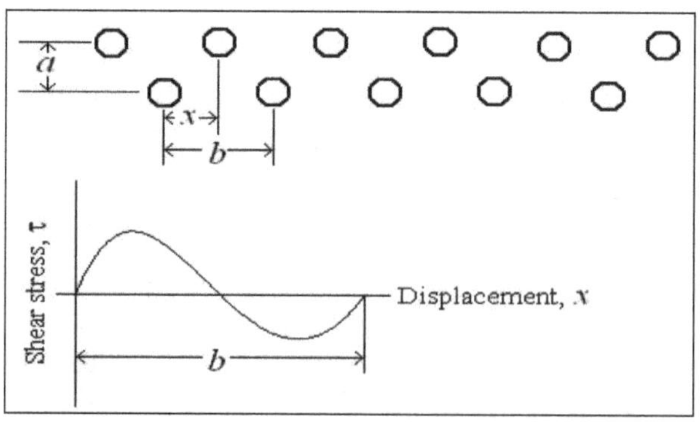

Shear displacement of atom planes and variation of shear stress.

The shear stress is zero initially and also once planes moved one identity distance, b. In between, it can be said that shear stress is a periodic function of displacement as each atom is attracted toward the nearest atom of the other row. Thus the relation of shear stress with displacement can be expressed as follows:

$$\tau = \tau_m \sin\frac{2\Pi x}{b}$$

where x is the amplitude of sine wave, and b is the period. For small values of x/b, after simplification,

$$\tau \approx \tau_m \frac{2\Pi x}{b}$$

It is safe to approximate at small values of displacement Hooke's law is applicable. Thus,

$$\tau = G\gamma = \frac{Gx}{a}$$

Combing the above two equations, maximum shear stress at which slip should occur is,

$$\tau_m = \frac{G}{2\Pi}\frac{b}{a}$$

If it is assumed as b ≈ a, approximate value of shear stress needed for initiation of slip is,

$$\tau_m = \frac{G}{2\Pi}$$

Shear modulus, G, of solids is in the range of 20-150 GPa, and thus theoretical shear strength could be in range 3-30 GPa. However, actual values of the shear stress required for plastic deformation is only about 0.5-10 MPa i.e. about 100 times less than the theoretical value. Because of this discrepancy, it must be said that a mechanism other than shearing of planes must be operative under application of loads thus causing plastic deformation.

Point Defects

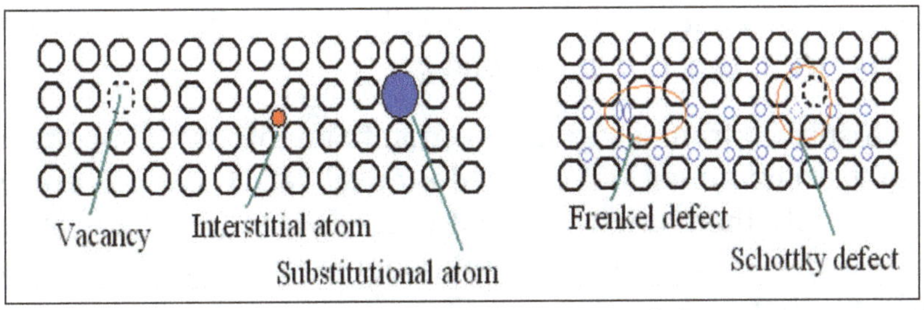

Schematic depiction of various point defects.

Defects exists any all solid materials. For ease of their characterization, defects are classified on the basis of their geometry, which is realistic as defects are disrupted region in a volume of a solid. Defects are:

- Point defects (zero-dimensional).
- Line defects (single dimensional).
- Surface defects (two dimensional).
- Volume defects (three dimensional).

Point defects, as the name implies, are imperfect point-like regions in the crystal. Typical size of a point defect is about 1-2 atomic diameters. Different point defects are explained in the following paragraphs. Figure depicts various point defects.

A vacancy is a vacant lattice position from where the atom is missing. It is usually created when the solid is formed by cooling the liquid. There are other ways of making a vacancy, but they also occur naturally as a result of thermal excitation, and these are thermodynamically stable at temperatures greater than zero. At equilibrium, the fractions of lattice sites that are vacant at a given temperature (T) are:

$$— \quad /_{kT}$$

where n is the number of vacant sites in N lattice positions, k is gas or Boltzmann's constant, T is absolute temperature in kelvins, and Q is the energy required to move an atom from the interior of a crystal to its surface. It is clear from the equation that there is an exponential increase in number of vacancies with temperature. When the density of vacancies becomes relatively large, there is a possibility for them to cluster together and form voids.

An interstitial atom or interstitialcy is an atom that occupies a place outside the normal lattice position. It may be the same type of atom as the rest surrounding it (self-interstitial) or a foreign impurity atom. Interstitialcy is most probable if the atomic packing factor is low. Another way an impurity atom can be fitted into a crystal lattice is by substitution. A substitutional atom is a foreign atom occupying original lattice position by displacing the parent atom. In the case of vacancies and foreign atoms (both interstitial and substitutional), there is a change in the coordination of atoms around the defect. This means that the forces are not balanced in the same way as for other atoms in the solid, which results in lattice distortion around the defect.

In ionic crystals, existence of point defects is subjected to the condition of charge neutrality. There are two possibilities for point defects in ionic solids:

- When an ion displaced from a regular position to an interstitial position creating a vacancy, the pair of vacancy-interstitial is called Frenkel defect. Cations are usually smaller and thus displaced easily than anions. Closed packed structures have fewer interstitials and displaced ions than vacancies because additional energy is required to force the atoms into the interstitial positions.
- A pair of one cation and one anion can be missing from an ionic crystal, without violating the condition of charge neutrality when the valency of ions is equal. The pair of vacant sites, thus formed, is called Schottky defect. This type of point defect is dominant in alkali halides. These ion-pair vacancies, like single vacancies, facilitate atomic diffusion.

Line Defects or Dislocations

Line imperfections (one-dimensional defects) are also called Dislocations. They are abrupt changes in the regular ordering of atoms along a line (dislocation line) in the solid. They occur in high densities and strongly influence the mechanical properties of material. They are characterized by the Burgers vector (b), whose direction and magnitude can be determined by constructing a loop around the disrupted region and noticing the extra inter-atomic spacing needed to close the loop. The Burgers vector in metals points in a close packed lattice direction. It is unique to a dislocation.

Dislocations occur when an extra incomplete plane is inserted. The dislocation line is at the end of the plane. Dislocations can be best understood by referring to two limiting cases - Edge dislocation and Screw dislocation.

Edge dislocation or Taylor-Orowan dislocation is characterized by a Burger's vector that is perpendicular to the dislocation line. It may be described as an edge of an extra plane of atoms within a crystal structure. Thus regions of compression and tension are associated with an edge dislocation. Because of extra incomplete plane of atoms, the atoms above the dislocation line are squeezed together and are in state of compression whereas atoms below are pulled apart and experience tensile stresses. Edge dislocation is considered positive when compressive stresses present above the dislocation line, and is represented by ⊥. If the stress state is opposite i.e. compressive stresses exist below the dislocation line, it is considered as negative edge dislocation, and represented by ⊤. A schematic view of edge dislocations are shown in figure below.

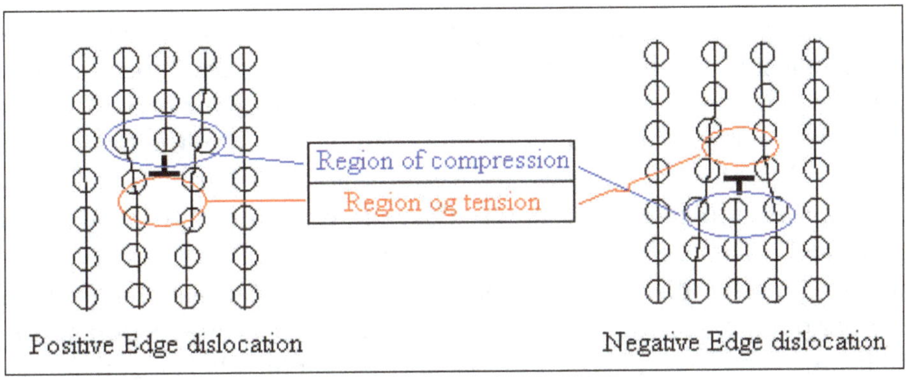

Edge Dislocations.

A pure edge dislocation can glide or slip in a direction perpendicular to its length i.e. along its Burger's vector in the slip plane (made of b and t vectors), on which dislocation moves by slip while conserving number of atoms in the incomplete plane. It may move vertically by a process known as climb, if diffusion of atoms or vacancies can take place at appropriate rate. Atoms are added to the incomplete plane for negative climb i.e. the incomplete plane increases in extent downwards, and vice versa. Thus climb motion is considered as non-conservative, the movement by climb is controlled by diffusion process.

Screw dislocation or Burgers dislocation has its dislocation line parallel to the Burger's vector. A screw dislocation is like a spiral ramp with an imperfection line down its axis. Screw dislocations result when displacing planes relative to each other through shear. Shear stresses are associated with the atoms adjacent to the screw dislocation; therefore extra energy is involved as it is in the case of edge dislocations. Screw dislocation is considered positive if Burger's vector and t-vector or parallel, and vice versa. (t-vector – an unit vector representing the direction of the dislocation line). A positive screw dislocation is represented by '⊙' a dot surrounded by circular direction in clock-wise direction", whereas the negative screw dislocation is represented by '⊙' a dot surrounded by a circular direction in anti-clock-wise direction". A schematic view of a negative screw dislocation is shown in figure.

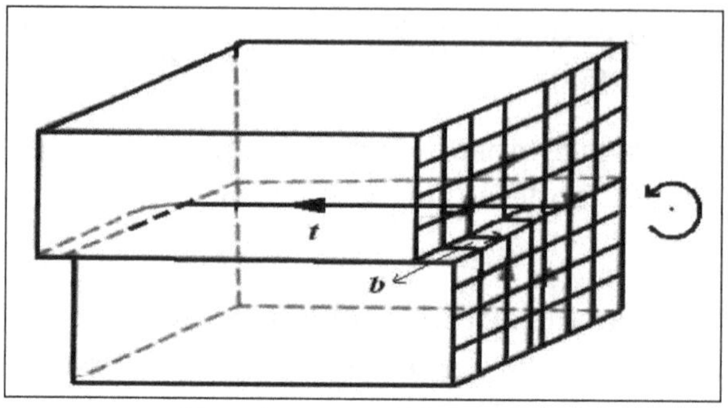

Negative screw dislocation.

A screw dislocation does not have a preferred slip plane, as an edge dislocation has, and thus the motion of a screw dislocation is less restricted than the motion of an Edge dislocation. As there is no preferred slip plane, screw dislocation can cross-slip on to another plane, and can continue its glide under favorable stress conditions. However, screw dislocation cannot move by climb process, whereas edge dislocations cannot cross-slip.

Dislocations more commonly originate during plastic deformation, during solidification, and as a consequence of thermal stresses that result from rapid cooling. Edge dislocation arises when there is a slight mismatch in the orientation of adjacent parts of the growing crystal. A screw dislocation allows easy crystal growth because additional atoms can be added to the 'step' of the screw. Thus the term screw is apt, because the step swings around the axis as growth proceeds. Unlike point defects, these are not thermodynamically stable. They can be removed by heating to high temperatures where they cancel each other or move out through the crystal to its surface. Virtually all crystalline materials contain some dislocations. The density of dislocations in a crystal is measures by counting the number of points at which they intersect a random cross-section of the crystal. These points, called etch-pits, can be seen under microscope. In an annealed crystal, the dislocation density is the range of 10^8-10^{10} m^{-2}.

Any dislocation in a crystal is a combination of edge and screw types, having varying degrees of edge and screw character. Figure below depicts a schematic picture of a common dislocation.

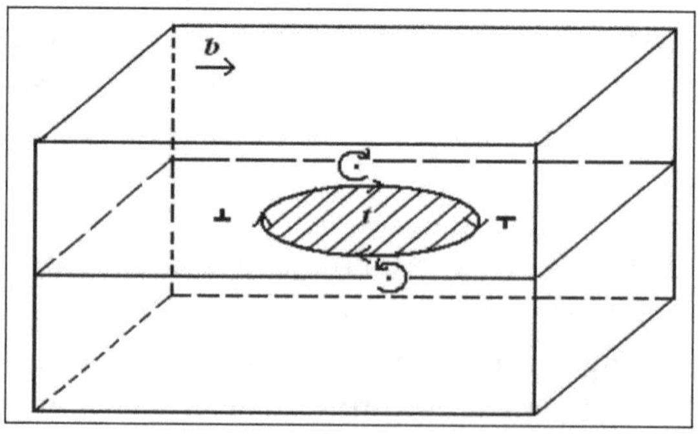

Typical dislocation in a crystal.

The Burger's vector will have the same magnitude and direction over the whole length of dislocation irrespective of the character of the dislocation (edge, screw, or mixed) i.e. Burger's vector is invariant. Other geometrical characteristics of dislocations are:

- Vectorial sum of Burger's vectors of dislocations meeting at a point, called node, must be zero.

- T-vectors of all the dislocations meeting at a node must either point towards it or away from it.
- A dislocation line cannot end abruptly within the crystal. It can close on itself as a loop, or ends either at a node or at the surface.

Dislocations have distortional energy associated with them as is evident from the presence of tensile/compressive/shear stresses around a dislocation line. Strains can be expected to be in the elastic range, and thus stored elastic energy per unit length of the dislocation can be obtained from the following equation:

$$E \cong Gb^2/2$$

where G is shear modulus and b is Burger's vector.

Dislocations in the real crystal can be classified into two groups based on their geometry – full dislocations and partial dislocations. Partial dislocation's Burger's vector will be a fraction of a lattice translation, whereas Burger's vector is an integral multiple of a lattice translation for full dislocation. Elastic energy associated with a dislocation is proportional to square of its Burger's vector; dislocation will tend to have as small a Burger's vector as possible. This explains the reason for separation of dislocations that tend to stay away from each other.

Interfacial Defects, Bulk or Volume Defects and Atomic Vibrations

Interfacial Defects

Interfacial defects can be defined as boundaries that have two dimensional imperfections in crystalline solids, and have different crystal structures and/or crystallographic orientations on either side of them. They refer to the regions of distortions that lie about a surface having thickness of a few atomic diameters. For example: external surfaces, grain boundaries, twin boundaries, stacking faults, and phase boundaries. These imperfections are not thermodynamically stable, rather they are meta-stable imperfections. They arise from the clustering of line defects into a plane.

External Surface

The environment of an atom at a surface differs from that of an atom in the bulk; especially the number of neighbors (coordination) at surface is less. Thus the unsaturated bonds of surface atoms give rise to a surface energy. This result in relaxation (the lattice spacing is decreased) or reconstruction (the crystal structure changes). To reduce the energy, materials tend to minimize, if possible, the total surface area.

Grain Boundaries

Crystalline solids are, usually, made of number of grains separated by grain boundaries. Grain boundaries are several atoms distances wide, and there is mismatch of orientation of grains on either side of the boundary as shown in figure. When this misalignment is slight, on the order of few degrees ($< 10°$), it is called low angle grain boundary. These boundaries can be described in terms of aligned dislocation arrays. If the low grain boundary is formed by edge dislocations, it is called tilt boundary, and twist boundary if formed of screw dislocations. Both tilt and twist boundaries are planar surface imperfections in contrast to high angle grain boundaries. For high angle grain boundaries, degree of disorientation is of large range (> 15°). Grain boundaries are chemically more reactive because of grain boundary energy. In spite of disordered orientation of atoms at grain boundaries, polycrystalline solids are still very strong as cohesive forces present within and across the boundary.

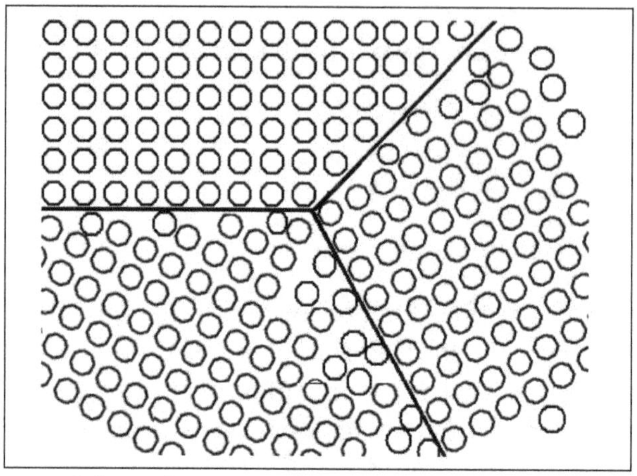

Schematic presentation of grain boundaries.

Twin Boundaries

It is a special type of grain boundary across which there is specific mirror lattice symmetry. Twin boundaries occur in pairs such that the orientation change introduced by one boundary is restored by the other. The region between the pair of boundaries is called the twinned region. Twins which forms during the process of recrystallization are called annealing twins, whereas deformation twins form during plastic deformation. Twinning occurs on a definite crystallographic plane and in a specific direction, both of which depend on the crystal structure. Annealing twins are typically found in metals that have FCC crystal structure (and low stacking fault energy), while mechanical/deformation twins are observed in BCC and HCP metals. Annealing twins are usually broader and with straighter sides than mechanical twins. Twins do not extend beyond a grain boundary.

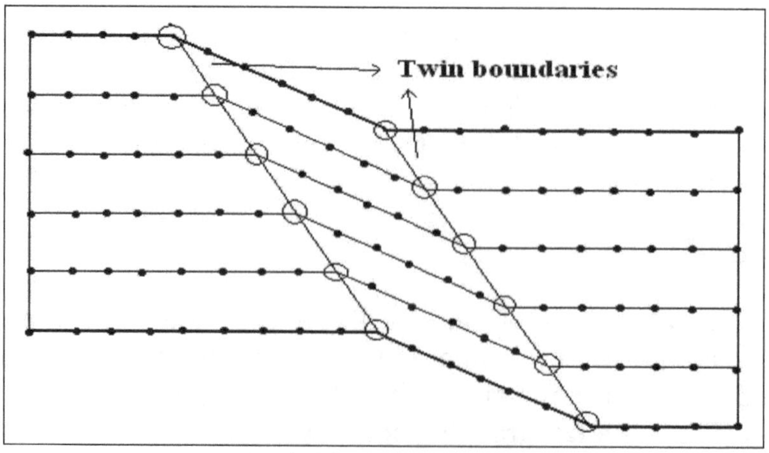

A pair of twin boundaries.

Stacking Faults

They are faults in stacking sequence of atom planes. Stacking sequence in an FCC crystal is ABC ABC ABC ..., and the sequence for HCP crystals is AB AB AB.... When there is disturbance in the stacking sequence, formation of stacking faults takes place. Two kinds of stacking faults in FCC crystals are: (a) ABC AC ABC...where CA CA represent thin HCP region which is nothing but stacking fault in FCC, (b) ABC ACB CABC is called extrinsic or twin stacking fault. Three layers ACB constitute the twin. Thus stacking faults in FCC crystal can also be considered as submicroscopic twins. This is why no microscopic twins appear in FCC crystals as formation of stacking faults is energetically favorable. Stacking fault energy varies in range 0.01-0.1 J/m^2. Lower the stacking fault energy, wider the stacking fault, metal strain hardens rapidly and twin easily. Otherwise, metals of high stacking fault energy i.e. narrower stacking faults show a deformation structure of banded, linear arrays of dislocations. Phase boundaries exist in multiphase materials across which there is sudden change in physical/chemical characteristics.

Bulk or Volume Defects

Volume defects as name suggests are defects in 3-dimensions. These include pores, cracks, foreign inclusions and other phases. These defects are normally introduced during processing and fabrication steps. All these defects are capable of acting as stress raisers, and thus deleterious to parent metal's mechanical behavior. However, in some cases foreign particles are added purposefully to strengthen the parent material. The procedure is called dispersion hardening where foreign particles act as obstacles to movement of dislocations, which facilitates plastic deformation. The second-phase particles act in two distinct ways – particles are either may be cut by the dislocations or the particles resist cutting and dislocations are forced to bypass them. Strengthening due to ordered particles is responsible for the good high-temperature strength on many super-alloys. However, pores are detrimental because they reduce effective load bearing area and act as stress concentration sites.

Atomic Vibrations

Atomic vibrations occur, even at zero temperature (a quantum mechanical effect) and increase in amplitude with temperature. In fact, the temperature of a solid is really just a measure of average vibrational activity of atoms and molecules. Vibrations displace transiently atoms from their regular lattice site, which destroys the perfect periodicity. In a sense, these atomic vibrations may be thought of as imperfections or defects. At room temperature, a typical vibrational frequency of atoms is of the order of 10^{13} vibrations per second, whereas the amplitude is a few thousandths of a nanometer. Many properties and processes in solids are manifestations of this vibrational atomic motion. For example: melting occurs once the atomic bonds are overcome by vigorous vibrations.

Movement of Dislocations

The most useful definition of a dislocation is given in terms of the Burgers circuit. A burgers circuit is any atom-to-atom path taken in a crystal containing dislocations which forms a closed loop. Such a path, MNOPQ, is illustrated in the following figure,

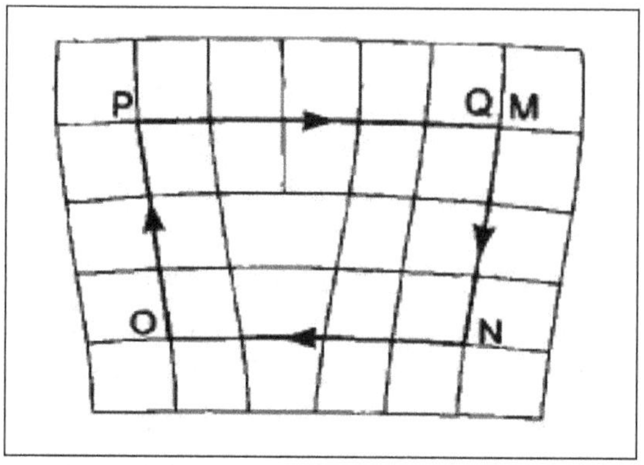

Burger circuit round an edge dislocation.

If the same atom-to-atom sequence is made in a dislocation-free crystal, and the circuit does not close, then the first circuit, the one showed in above figure, must enclose one or more dislocations. The vector required to complete the circuit is called the Burgers vector.

- The Burgers vector of an edge dislocation is normal to the line of the dislocation.
- The Burgers vector of a screw dislocation is parallel to the line of the dislocation.

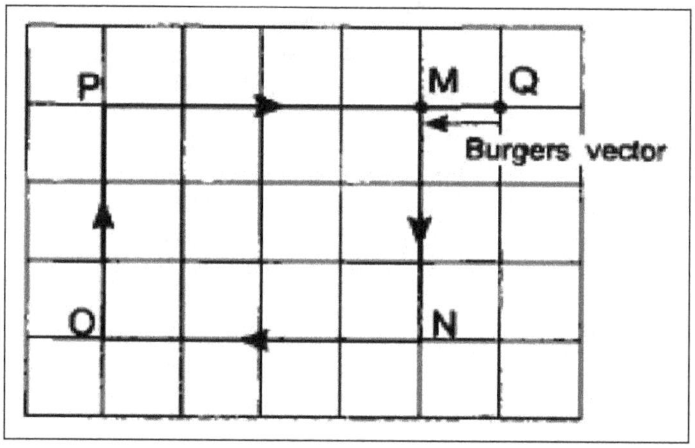

In the most general case the dislocation line lies at an arbitrary angle to its Burgers vector and the dislocation line has a mixed edge and screw character. However, the Burgers vector of a single dislocation has fixed length and direction, and is independent of the position and orientation of the dislocation line.

Burgers circuit taken around other defects, such as vacancies and interstitials, do not lead to closure failures. Two roles are used defining Burgers vectors and Burger circuits. First, when looking along the dislocation line, which defines the positive line sense or direction of the dislocation, the circuit is taken in a clockwise fashion. Second, the Burgers vector is taken to run from the finish to the star point of the reference circuit in the perfect crystal. This defines the right-hand/finish-start (RH/FS) convention. Reversing the line sense reverses the direction of the Burgers vector for a given dislocation.

Furthermore, dislocations with the same line sense but opposite Burgers vectors (or alternatively with opposite line senses and the same Burgers vector) are physical opposites. Dislocations which are physical opposites of each other annihilate and restore perfect crystal if brought together. The Burgers vector b is conveniently described using Miller's indices. Dislocations lines can end at the surface of a crystal and at grain boundaries, but never inside a crystal. Thus, dislocations must either form closed loops or branch into other dislocations. When three or more dislocations meet at a point, or node, it is a necessary condition that the Burgers vector is conserved. A schematic example of this condition is shown in figure below:

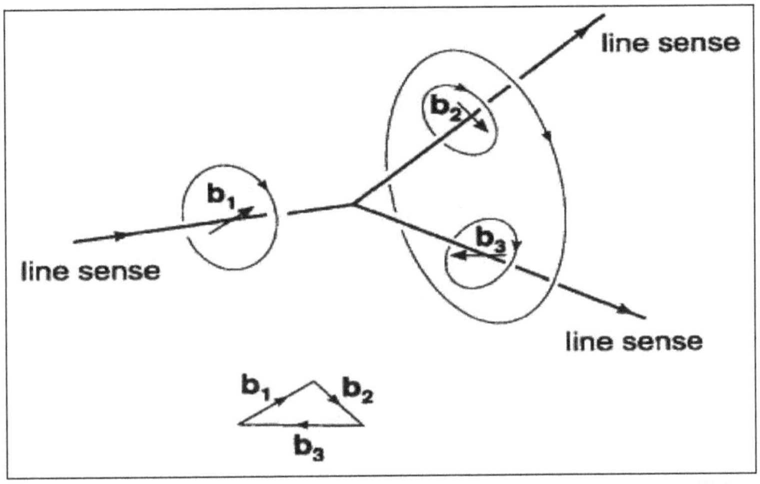

Three dislocations forming a node and Burgers' conservation condition.

The mathematical expression of this condition can be expressed:

$$b_1 = b_2 + b_3$$

The dislocation density ρ is defined as the total length of dislocation line per unit volume of crystal. An alternative definition, which is sometimes more convenient to use, is the number of dislocations intersecting a unit area, again measured in units of m^{-2}. If all the dislocations are parallel, the two density values are the same, but for a completely random arrangement the volume density is twice the surface density. The average distance between dislocations in a network of density ρ is of the order of $1/\sqrt{\rho}$.

There are two basic types of dislocation movement. Glide or conservative motion occurs when the dislocation moves in the surface which contains both its line and Burgers vector. Climb or non-conservative motion occurs when the dislocation moves out of the glide surface, and thus normal to the Burgers vector.

Concept of Slip

Glide of many dislocations results in slip, which is the most common manifestation of plastic deformation in crystalline solids. It can be envisaged as sliding or successive displacement of one plane of atoms over another on so-called slip planes. This is uniquely defined as the plane which contains both the line and the Burgers vector of the dislocation. Further deformation occurs either by more movement on existing slip planes or by the formation of new slip planes.

The slip planes are normally the planes with the highest density of atoms, and the direction of slip is the shortest lattice translation vectors. Often, this direction is one in which the atoms are most closely spaced. A slip plane and a slip direction in the plane constitute a slip system. Slip results in the formation of steps on the surface of the crystal. The slip direction is necessarily always parallel to the Burgers vector of the dislocation responsible for slip. The movement of one dislocation across the slip plane to the surface of the crystal produces a surface slip step equal to the Burgers vector. A characteristic shear stress is required for slip. It can be schematically shown in figure.

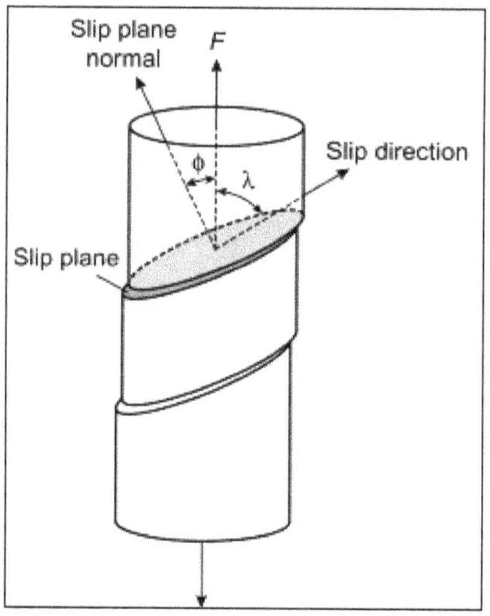

Illustration of the geometry of slip in crystalline materials.

The illustrated crystal is being deformed in tension by an applied force F along the axis of the cylindrical crystal. The cross sectional Area is A and so the tensile stress parallel to F is:

$$\sigma = \frac{F}{A}$$

The force has a component F cosλ in the slip direction, where λ is the angle between F and the slip direction. This force acts over the slip surface which has an area A/cos φ, where φ is the angle between F and the normal to the slip plane. Thus the shear stress τ, resolved on the slip plane in the slip direction, is:

$$\tau = \frac{F}{A}\cos\phi\cos\lambda$$

If F_c is the tensile force required to start slip, the corresponding value of the shear stress τ_c is called the critical resolved shear stress for slip. It has been found in some crystals which deform on a single slip system that τ_c is independent of the orientation of the crystal.

Dislocations and Slip

With the concept of slip just introduced, dislocation can be defined as the boundary between the slipped and unslipped parts of the crystal. The theoretical shear stress for slip was many times greater than the experimentally observed stress. The low value can be accounted for by the movement of dislocations. Only a relatively small applied stress is required to move the dislocation along the slip plane, as it is demonstrated in the following scheme for an edge dislocation figure below where the arrow indicate the applied shear stress.

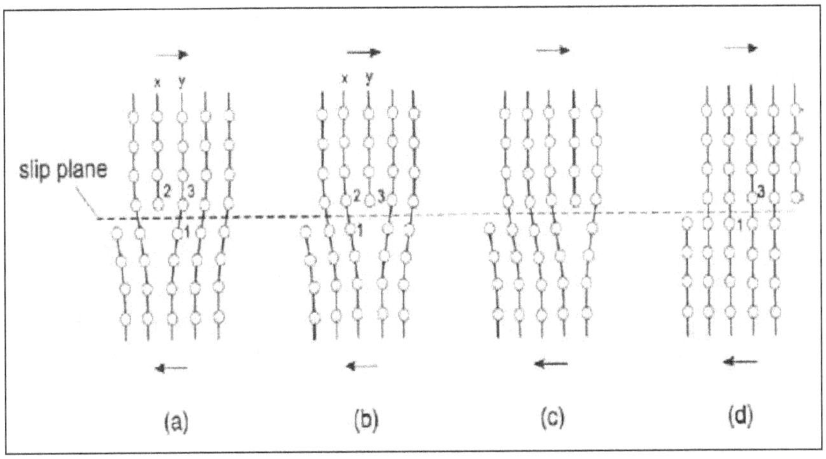

Movement of an edge dislocation.

Well away from the dislocation, the atom spacing are close to the perfect crystal values, and a shear stress as high as the theoretical value would be required to slide them all past each other. Near the dislocation line itself, some atom spacing is far away from the ideal values, and small relative changes in position of only a few of atoms are required for the dislocation to move. Thus the applied stress required to overcome the lattice resistance to the movement of the dislocation is much smaller than the theoretical shear stress of a perfect lattice, due to the successive glide of the extra half-plane.

It is also possible to see that two neighboring atoms on sites adjacent across the slip plane, like the atoms "1" and "3" in above figure are displaced relative to each other by the Burgers vector when the dislocation glides past. The global effect of many thousands of such dislocations is a surface step on the material.

Single crystal of cadmium deforming by dislocation slip, forming steps on the surface.

It is important to remark two particular rules for dislocation glide:

- Dislocations of opposite sign glide in opposite directions under the same stress.
- For dislocation glide a shear stress must act on the slip plane in the direction of the Burgers vector, irrespective of the direction of the dislocation line.

Those properties are schematically shown for an edge and screw dislocation:

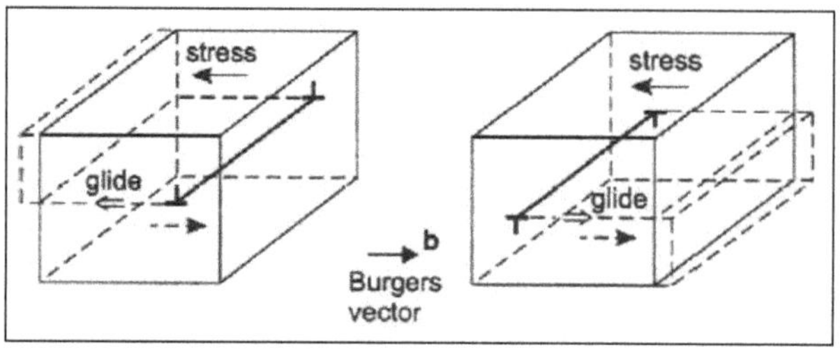

Plastic deformation by glide of edge dislocation under the applied shear stress.

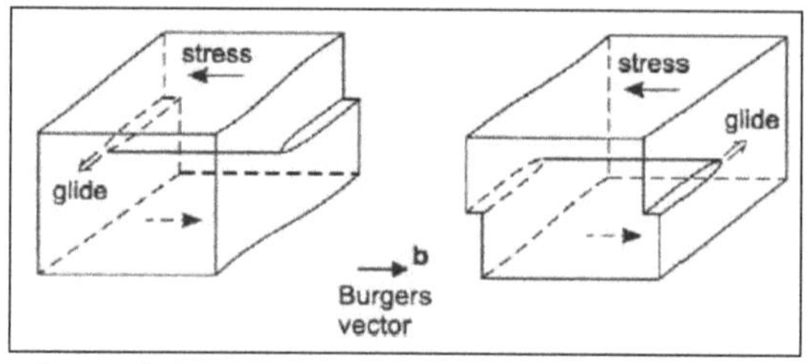

Plastic deformation by glide of screw dislocation under the applied shear stress.

Climb

At low temperatures where diffusion is difficult, and in the absence of a non-equilibrium concentration of point defects, the movement of dislocations is restricted almost entirely to glide. However, at higher temperatures and edge dislocation can move out of its slip plane by a process called climb.

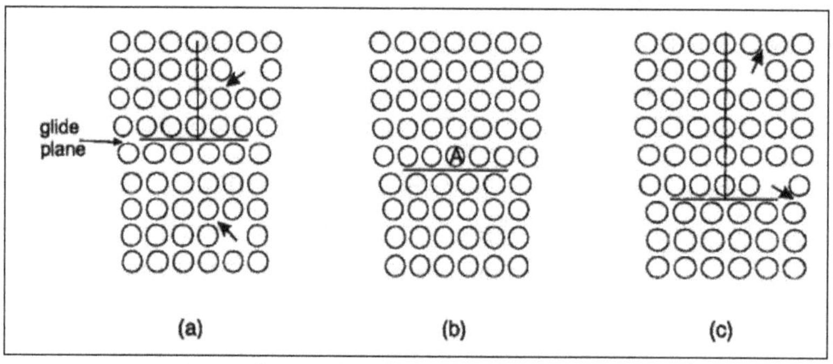

Positive and negative climb of an edge dislocation.

The row of atoms A normal to the plane of the diagram is removed, the dislocation line moves up one atom spacing out of its original slip plane: this is called positive climb. Similarly, if a row of atoms is introduced below the extra half-plane the dislocation line moves down one atom spacing, negative climb. For this kind of movements, volumes are not conserved. The mass transport involved occurs by diffusion and therefore climb requires thermal activation.

It has been implied above that a complete row of atoms is removed simultaneously, whereas in practice individual vacancies or small clusters of vacancies diffuse to the dislocation. The effect of this is that the climb of a short section of a dislocation line results in the formation of two steps called jogs.

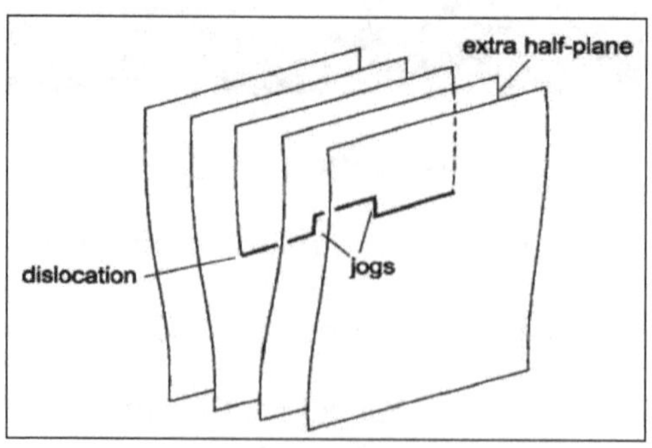

Single jogs on an edge dislocation.

Both positive and negative climb proceeds by the nucleation and motion of jogs. Conversely, jogs are sources and sinks for vacancies. Jogs are steps on the dislocation which move it from one atomic slip plane to another. Steps which displace it on the same slip plane are called kinks. The two are distinguished below for and edge and screw dislocation:

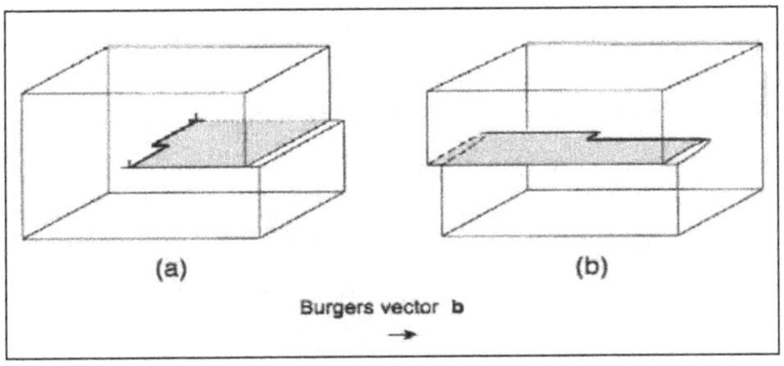

Kinks in edge (a) and screw (b) dislocation.

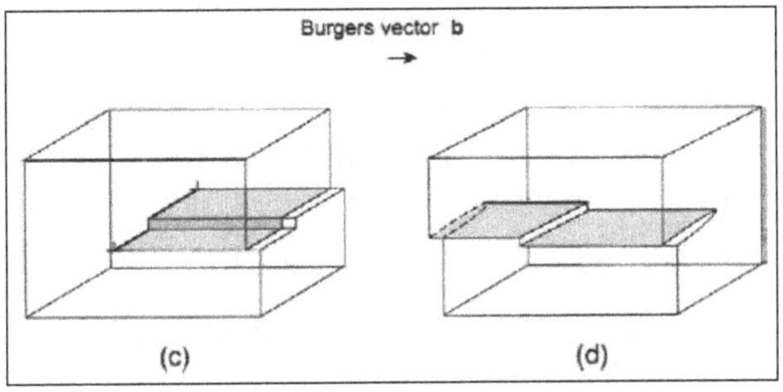

Jogs in edge (c) and screw (d) dislocation.

Jogs and kinks are short elements of dislocation with the same Burgers vector as the line on which they lie, and the usual rules apply for their conservative and non-conservative movements.

Coherence and Finite Size Effects

Consider an inhomogeneous alloy, in which the composition and the lattice constants vary on the scale of millimetres. Clearly, the different regions of the crystal will scatter independently, each region being in essence a "perfect crystal". The result, therefore, with be the incoherent superposition of different patterns. Each hkl will produce a set of independent Bragg peaks (corresponding to the different lattice constants), which will typically appear as a broadening of the scattered reflections. Let us now imagine to "shrink" the length scale of the composition and lattice parameters fluctuations down to nanometre sizes. Here, the different compositions will scatter coherently, and the appropriate picture is that of an average lattice. In the different regions, the atoms will be displaced away from the average positions, and this, as we know by now, will produce a reduction of the Bragg intensity at high q, in complete analogy to the Debye-Waller factors. Furthermore, the displacements will be correlated (i.e., nearby atoms will tend to be displaced in the same direction). At different lengthscales, he diffraction patterns will be therefore qualitatively different. At what point, between millimetres and nanometres, does this qualitative transition occur? We can intuit that the lengthscale where the transition occurs might be set by the probe, i.e., that there may be intrinsic coherence lengths for X-rays, neutrons, electrons etc.

A very similar argument may be construed for timescales. If the positions of the atoms in the crystal "fluctuate", e.g., due to phonons, do the different configurations occurring at different times contribute coherently or incoherently to the diffraction pattern? If the timescale of the fluctuations if of the order of seconds, the latter will be true, but, as the timescale of the fluctuations is reduced, this will no longer be the case. What is the typical coherence time of the different probes?

A complete treatment of the correlation lengths and times is too complex to be presented here, but the following points should provide a good idea of the issues involved:

1. One distinguishes between transverse and longitudinal coherence lengths, perpendicular and parallel, respectively, to the direction of the beam.

2. For a quasi-parallel beam geometry (as it is typical of a diffraction instrument), the transverse coherence length ξ_t is proportional to the wavelength and inversely proportional to the beam divergence α ($\xi_t = \lambda/\alpha$). Here, the beam divergence is defined as the angle subtended by the source as seen from the sample. The derivation is analogous to that of the double-slit experiment, and corresponds to the distance between slits where the interference pattern is lost. For typical diffraction instruments, the beam divergence varies between a few mrad (lab diffractometers) down to a tenth of mrad (synchrotrons), so for λ = 1Å, ξ_t varies between a few tens of nm up to about 1 μm. Highly coherent X-ray beams (several tens of μm) are employed for special studies (lensless imaging, "speckle" patterns), revealing the shape and internal structure of large crystal domains. Neutron beams have comparatively relaxed divergences, and typical coherence lengths are 1000- 2000 Å.

3. For geometries employing focusing, (e.g., electron diffraction) it is possible to manipulate the coherence domain by varying the focal plane where the diffraction pattern is formed (see below). If the diffraction pattern is in focus, the coherence length is essentially limited by the aberration of the lenses, and can be as large as for X-rays, in spite of the fact that the wavelength employed are 1-2 orders of magnitude smaller. If the sample is in focus, the transverse coherence length is much smaller (down to atomic sizes), so that each detector pixel receives a coherent contribution of a column of atoms in the direction of the beam. In this case, the coherence domain coincides with the resolving power of the instrument.

4. The longitudinal coherence length is inversely proportional to the relative wavelength spread of the beam: $\xi_l = \frac{1}{2}\lambda/(\Delta\lambda/\lambda)$. High-resolution monochromator at synchrotron sources typically yield $\Delta\lambda/\lambda \sim 0.5-1\times10^{-4}$, so at 1 Å ξ_l is about 1 μm. The wavelength spread of neutron and electron beams is typically 10 times and 100 times larger, respectively.

5. The coherence time can be calculated using the uncertainty principle as $\tau = \hbar/\Delta E$ or from the relation $\tau = \xi_l/v$, yielding the same result apart for a factor of the order 1. Adopting the first approach, we obtain,

$$\tau = \frac{1}{\pi^2}\frac{m\lambda^2}{\hbar}\left(\frac{\Delta\lambda}{\lambda}\right)^{-1} \text{ for particles with mass}$$

$$\tau = \frac{1}{2\pi}\frac{\lambda}{c}\left(\frac{\Delta\lambda}{\lambda}\right)^{-1} \text{ for photons}$$

The prefactor is 1.6 × 10^{-14} [secÅ^{-2}] for neutrons, 0.9 × 10^{-17} [secÅ^{-2}] for electrons and 5 × 10^{-20} [secÅ^{-1}] for photons. Using the wavelengths and monochromaticity values mentioned above, we find that coherence times for neutrons, photons and 100 Kev electrons are in the picosecond, femtosecond and tens of attosecond ranges, respectively.

Finite Size Effects

The cross section becomes then a series of delta functions, centered on the RL nodes. We need not make any particular approximation other than the fact that the crystal should be composed of N_1, N_2 and N_3 unit cell in the a_1, a_2 and a_3 directions, respectively.

$$\sum_{n=-N}^{N} e^{-inx} = \frac{\sin\left((2N+1)\frac{x}{2}\right)}{\sin\frac{x}{2}}$$

after some straightforward math (in each direction, $x = q \cdot a_i$) we obtain,

$$\frac{d\sigma}{d\Omega} = \left[\prod_i \frac{\sin^2\left(N_i \frac{1}{2} q \cdot a_i\right)}{\sin^2\left(\frac{1}{2} q \cdot a_i\right)}\right] |F(q)|^2 [\varepsilon \cdot \varepsilon']^2$$

The diffraction pattern from a small single crystal taken with coherent radiation, i.e., radiation with coherence length (=wavepacket length/width) larger than the crystal, will display 3-dimensional fringes, as typical of the function $\sin^2(Nx)/\sin^2 x$. However, for typical experiments with incoherent radiation, the number of unit cells is set by the coherence length of each photon rather than by the crystal size, and the oscillations will be smeared out. In this case, we can replace the oscillatory functions with a Gaussian function with the same maximum and same area:

$$\frac{\sin^2 Nx}{\sin^2 x} \to N^2 e^{-(Nx)^2/\pi}$$

Observing that $\prod_i N_i = N_c$, the total number of unit cell in the crystal, we obtain,

$$\frac{d\sigma}{d\Omega} = N_c^2 \left[\prod_i e^{-\left(N_i \frac{1}{2} q . a_i\right)^2/\pi}\right] |F(q)|^2 [\varepsilon \cdot \varepsilon']^2$$

where the Gaussian functions have variance and FWHM,

$$\sigma_i^2 = \frac{2\pi}{N_i^2 a_i^2}$$

$$FW\ HM = \frac{4\sqrt{\pi \ln 2}}{N_i a_i}$$

We can therefore conclude that:

- The cross section at a given q is proportional to N_c^2.
- The width in q is inversely proportional to the number of unit cells along that direction.

- The integrated cross section in three dimensions (remember the Gaussian integral $\sqrt{2\pi\sigma^2}$) is therefore proportional to N_c, which reproduces the result we obtained for the infinite crystal.

Beyond the Perfect-Crystal Approximation: Diffuse Scattering

Finite-size effects only generate scattering in the vicinity of the "ideal" Bragg positions, leaving the integrated intensity of the reflection unaltered. Additional scattering is generated whenever either the atomic scattering factors or the positions of individual atoms deviate from the average values and/or the "ideal" lattice sites. In general, this additional scattering is present throughout reciprocal space, and is therefore known as diffuse scattering. We will imagine that the scattering process occurs coherently from a frozen snapshot of the crystal. To estimate the full scattered intensity, at the end, we will undertake a time/thermal averaging process on the scattered intensities. This is a very good approximation in the case of X-rays (and electrons), since the coherence times are much shorter than phonon frequencies. For neutrons, where this is clearly not the case, a more complex treatment is required, but qualitatively similar considerations (with some caveats) can be applied.

Classification of Disorder in Crystals

We will here consider two types of disorder:

- Substitutional disorder is typical of alloys and, more generally, of solid solutions. It occurs when more than one atonic species can reside on the same crystallographic site, and gives rise to a fluctuation of the atomic scattering factor around an "average" value, the latter reflecting the average composition of the alloy. For example, in a 67%Cu-33%Au "random" alloy, Cu and Au occupy the same crystallographic sites of the FCC structure. To calculate the Bragg scattering from such an alloy, one should use the average atomic scattering factor ($f_{ave}(0) = 45.7$). However, locally, the scattering factor varies between that of Cu ($f_{Cu}(0) = 29$) and Au ($f_{Au}(0) = 79$). This local fluctuation, as gives rise to diffuse scattering. Substitutional disorder is almost always static, since atoms in alloys do not usually move from one site to the next, but counter-examples do exist (e.g., in the case of fast ion conductors).
- Dynamic displacive disorder is present in all materials even at zero temperatures, due to phonons and zero-point motion. In addition to scattering contrast fluctuations, substitutional disorder can itself produce static displacive disorder, since different species have in general different atomic radii.

Simple Diffuse Scattering Calculations

We shall now calculate the full scattering cross section for an (infinite) crystal in the presence of both substitutional and displacive disorder. To simplify the formalism, we will perform this calculation in the case of a crystal with one atom per unit cell, but the extension to more complex structures is completely straightforward.

$$\frac{d\sigma}{d\Omega} = r_0^2 \mathcal{P}(\gamma)\left(\sum_j \sum_i e^{-iq\cdot(r_i - r_j)}\right) f_i(q) f_j^*(q)$$

In above equation, r_i represent the actual positions of the atoms, while we indicate with R_i the position of the lattice nodes. Moreover, we will write,

$$r_i = R_i + u_i$$

where the u_i are the displacements away from the lattice nodes and are "small". With the additional notation $R_{ij} = R_i - R_j$ and $u_{ij} = u_i - u_j$ we can write:

$$\frac{d\sigma}{d\Omega} = r_o^2 \mathcal{P}(\gamma) \sum_{R_{ij}} e^{-iq \cdot R_{ij}} \sum_{\text{Pairs with } R_{ij}} \left[f_i(q) f_j^*(q) e^{-iq.u_{ij}} \right]$$

In above equation, we have rearranged the double summation, so that the rightmost sum runs over all pairs of unit cells joined by a given vector R_{ij}, whereas the leftmost sum runs over these vectors. Using, for example, periodic boundary conditions, we find that for each R_{ij}, there are exactly N pairs of sites joined by it (N being the total number of atoms in the crystal. We can therefore rewrite above equation as:

$$\frac{d\sigma}{d\Omega} = N r_o^2 \mathcal{P}(\gamma) \sum_{R_{ij}} e^{-iq \cdot R_{ij}} \left\langle f_i(q) f_j^*(q) e^{-iq \cdot u_{ij}} \right\rangle R_{ij}$$

We can subtract from above equation the Bragg scattering cross section to obtain the expression for the non-Bragg or diffuse scattering (we stress, once again, that the average atomic scattering factor is used to calculate the Bragg scattering):

$$\left(\frac{d\sigma}{d\Omega}\right)_{diff} = N r_o^2 \mathcal{P}(\gamma) \sum_{R_{ij}} e^{-iq \cdot R_{ij}} \left\langle f_i(q) f_j^*(q) e^{-iq \cdot u_{ij}} - \left| f_{ave}(q) \right|^2 e^{-2W} \right\rangle_{R_{ij}}$$

where the sign $\langle \ \rangle_{R_{ij}}$ indicates the average of the quantity within it over all the pairs in the coherence domain joint by the vector R_{ij}.

Diffuse Scattering from Uncorrelated Substitutional Disorder

Here, we will make the slightly unrealistic assumption that the atoms are not displaced at all from their ideal lattice sites, and consider the diffuse scattering arising from purely substitutional disorder ($u_{ij} = 0$). We can write,

$$\left(\frac{d\sigma}{d\Omega}\right)_{diff} = N r_o^2 \mathcal{P}(\gamma) \sum_{R_{ij}} e^{-iq \cdot R_{ij}} \left\langle f_i(q) f_j^*(q) - \left| f_{ave}(q) \right|^2 \right\rangle_{R_{ij}}$$

If the sites have random occupancy, for i ≠ j the average is identically zero for all R_{ij}, so only the autocorrelation term i = j survives, yielding,

$$\left(\frac{d\sigma}{d\Omega}\right)_{diff} = N r_o^2 \mathcal{P}(\gamma) \left[\left(\left| F(q) \right|^2 \right)_{ave} - \left| f_{ave}(q) \right|^2 \right]$$

For example, for a binary alloy with species A and B and concentrations C_A and $1 - C_A$, we obtain,

$$\left(\frac{d\sigma}{d\Omega}\right)_{diff} = N r_o^2 \mathcal{P}(\gamma) C_A (1 - C_A) \left| f_A(q) - f_B(q) \right|^2$$

These cross sections are slowly varying in q, since they are only modulated by the atomic scattering factors, and have their maximum near q = 0.

Diffuse Scattering from Uncorrelated Displacements

Let us now assume that we have a single atomic species ($f_i = f_j = f_{ave}$) and that the displacements of atoms i and j are not correlated with each other. If i ≠ j,

$$\left\langle e^{-iq\cdot u_{ij}}\right\rangle = \left\langle e^{-iq\cdot u_i}\right\rangle\left\langle e^{-iq\cdot u_j}\right\rangle = e^{-2W}$$

so R_{ij} vectors joining different atoms do not give any contributions to the diffuse scattering. The only surviving term is the autocorrelation term, for which u_{ij} = 0. Therefore,

$$\left(\frac{d\sigma}{d\Omega}\right)_{diff} = Nr_o^2\mathcal{P}(\gamma)\left|F(q)\right|^2\left(1-e^{-2W}\right)$$

The diffuse scattering from uncorrelated displacements is a slowly varying function of q; it is zero near the origin of the reciprocal space and increases with q (it is in the first approximation proportional to q^2), only to decay again with the decay of the atomic scattering factors.

General Treatment of Displacive Disorder and Thermal Diffuse Scattering

In the two simple examples we proposed above, we have treated the deviations from the average structure as completely uncorrelated from site to site. The result, in both cases, is a form of diffuse scattering that has very little structure in reciprocal space. In the extreme case of neutron studies of substitutional disorder, the atomic form factor is a constant, and the resulting diffuse scattering is completely flat.

In realistic cases, correlations are almost always present: even in disordered alloys, a given species may have a "preference" to be surrounded by atoms of the same kind or, quite frequently, of a different kind, so near-neighbour probabilities are generally different from the average. These correlations are lost after a few unit cells, and they are referred to as short-range order. In this case, beside the autocorrelation term, additional terms in the summation will be non-zero, giving rise to a more structured diffuse scattering pattern. Each R_{ij} pair that corresponds to a definite correlation will contribute with a Fourier component $\exp(R \cdot R_{ij})$ (or $\cos R \cdot R_{ij}$ in the centrosymmetric case) to the diffuse scattering.

The correlated displacements due to thermal vibrations are of even longer range: an ideal plane-wave phonon involves all sites of the crystal, and therefore involves all terms in the summation (equation $\left(\frac{d\sigma}{d\Omega}\right)_{diff} = Nr_o^2\mathcal{P}(\gamma)\sum_{R_{ij}} e^{-iq\cdot R_{ij}}\left\langle f_i(q)f_j^*(q)e^{-iq\cdot u_{ij}} - \left|f_{ave}(q)\right|^2 e^{-2W}\right\rangle_{R_{ij}}$). The correlation terms u_{ij} will typically involve sums and differences of sines and cosines, and since they appear in the exponential, the summation over an infinite number of pairs cannot be performed analytically. The typical approach is to expand the exponential in Taylor series. This is done as follows: for simplicity, let us rewrite equation $\left(\frac{d\sigma}{d\Omega}\right)_{diff} = Nr_o^2\mathcal{P}(\gamma)\sum_{R_{ij}} e^{-iq\cdot R_{ij}}\left\langle f_i(q)f_j^*(q)e^{-iq\cdot u_{ij}} - \left|f_{ave}(q)\right|^2 e^{-2W}\right\rangle_{R_{ij}}$ in the case of pure displacive disorder (no substitutional disorder):

$$\left(\frac{d\sigma}{d\Omega}\right)_{diff} = Nr_o^2\mathcal{P}(\gamma)\sum_{R_{ij}} e^{-iq\cdot R_{ij}}\left|f(q)\right|^2\left[\left\langle e^{-iq\cdot u_{ij}}\right\rangle_{R_{ij}} - e^{-2W}\right]$$

which is valid for small displacements or harmonic displacements:

$$\left\langle e^{-iq\cdot u_{ij}} \right\rangle = e^{-\frac{1}{2}\left\langle [q\cdot u_{ij}]^2 \right\rangle}$$

Finally, we expand the exponential in above equation in Taylor series:

$$\left(\frac{d\sigma}{d\Omega}\right)_{diff} = Nr_o^2 \mathcal{P}(\gamma) \sum_{R_{ij}} e^{-iq\cdot R_{ij}} |f(q)|^2$$

$$\left\{ \left[-\frac{1}{2}\left\langle [q\cdot u_{ij}]^2 \right\rangle + 2W \right] + \left[\frac{1}{8}\left\langle [q\cdot u_{ij}]^2 \right\rangle^2 - 2W^2 \right] + \ldots \right\}$$

$$= I_1 + I_2 + \ldots$$

The different terms in above equation are known as first-order, second-order etc., diffuse scattering.

First-Order Thermal Diffuse Scattering

In well-ordered crystals, the most important contribution to the diffuse scattering is that from the displacement patterns due to phonons. This type of diffuse scattering is known as Thermal Diffuse Scattering or TDS. The first-order term in above equation is often the dominant contribution, particularly at moderate temperatures. The calculation of diffuse scattering for monoatomic solid is not inaccessible at the present level, but it is rather lengthy. Here, the main steps and present the final result, so that its structure can be analysed:

- Only phonons with the same wave vector k and the same branch index contribute coherently to first-order TDS. The scattering from one such phonon is not difficult to calculate: it generate satellite peaks at ±k around any given Bragg peaks.
- One needs to sum the scattered intensity of phonons with different wave vectors, taking into account the fact that their amplitude is related to the thermal population.
- The final result is:

$$\left(\frac{d\sigma}{d\Omega}\right)_{1^{st}\,TDS} = Nr_o^2 \mathcal{P}(\gamma) \frac{\hbar}{2M} \sum_j \frac{1}{\omega_j(q)} \coth\left(\frac{\hbar\omega_j(q)}{2k_B T}\right) |F_j(q)|^2$$

  where the sum is over the different phonon branches and,

$$F_j(q) = f(q) e^{-W} \left[q \cdot e_j(q) \right]$$

  is the so-called one-phonon structure factor and $e_j(q)$ is the polarisation vector of phonon branch j at the wavevector q. In order to evaluate the quantities $e_j(q)$ and $\omega_j(q)$, one needs to reduce q to the first Brillouin zone in the usual way.
- The function $\frac{1}{\omega_j(q)} \coth\left(\frac{\hbar\omega_j(q)}{2k_B T}\right)$ decreases rapidly with increasing frequency. Therefore low energy, acoustic phonons provide the largest contribution to first-order TDS. For the

same reason, the TDS diverges at the Bragg peak positions and is minimum between the Bragg peaks.

- When the same reduced wave vector is considered in different Brillouin zones, one finds that the cross section is proportional to q^2, similar to the diffuse scattering due to uncorrelated displacements (in above equation).

Scattering from Non-Crystalline Solids and Liquids

Our discussion of the scattering cross sections for "defective" crystal would not be complete without a mention of the most extreme form of defects — the complete loss of crystalline order, as in the case of non-crystalline solids (glasses) or liquids. In this case, it is no longer possible to define a crystal lattice; nevertheless, the formalism to calculate the scattering cross section is very similar to the one we have already encountered in equation $\frac{d\sigma}{d\Omega} = r_0^2 \mathcal{P}(\gamma) \sum_{R_{ij}} e^{-iq\cdot R_{ij}} \sum_{\text{P airs with } R_{ij}} \left[f_i(q) f_j^*(q) e^{-iq.u_{ij}} \right]$. The main difference is that, for obvious reasons, decomposing position vectors in R_{ij} and u_{ij} will no longer make sense, and we will therefore employ the original notation r_{ij},

$$\frac{d\sigma}{d\Omega} = r_0^2 \mathcal{P}(\gamma) \sum_{m=1}^{N} \sum_{r_{ij}} f_i(q) f_j^*(q) e^{-iq\cdot r_{ij}}$$

where the first sum runs over all the atoms in the sample. For simplicity, we will proceed from here assuming that all the atoms in the sample are the same, as appropriate, for example, for a monoatomic liquid like mercury. The extension to polyatomic liquids or glasses is rather straightforward but requires a heavier notation. With our assumption, we can extract $N|f(q)|^2$ from all the summations and replace the sums with averages; we will also isolate the autocorrelation term ($r_{ij} = 0$) which yields 1:

$$\frac{d\sigma}{d\Omega} = N r_0^2 \mathcal{P}(\gamma) |f(q)|^2 \left[1 + \sum_{r_{ij} \neq 0} \left\langle e^{-iq\cdot r_{ij}} \right\rangle \right]$$

We now want to take advantage of the fact that, in a disordered sample, if a vector r_{ij} connecting two atoms is present, then all other vectors with the same modulus r_{ij} will also be present with equal probability somewhere in the sample. This will enable us average over the angular variables and replace the summation in above equation with a sum over r_{ij}. In doing so, we have to account for the fact that any given atom may be surrounded by more than one atom at the same distance r_{ij}, and we will do so by introducing the average coordination number $n(r_{ij})$.

$$\frac{d\sigma}{d\Omega} = N r_0^2 \mathcal{P}(\gamma) |f(q)|^2 \left[1 + \sum_{r_{ij} \neq 0} n(r_{ij}) \frac{1}{4\pi} \int_0^{\pi} e^{-iqr_{ij}\cos\theta} \sin\theta \, d\theta \, d\phi \right]$$

The integral is straightforward and yields $4\pi \sin qr_{ij}/qr_{ij}$.

$$\frac{d\sigma}{d\Omega} = N r_0^2 \mathcal{P}(\gamma) |f(q)|^2 \left[1 + \sum_{r_{ij} \neq 0} n(r_{ij}) \frac{\sin qr_{ij}}{qr_{ij}} \right]$$

The structure of above equation clearly illustrates the main feature of the scattering form non-crystalline materials: the cross section only depends on q, so the scattering will occur as uniform spheres in reciprocal space. Analysis of the scattering will provide information about the average coordination number at every distance r_{ij} around a given atom.

References

- Crystallography: files.wordpress.com, Retrieved 17, May 2020
- Solid-state: unf.edu, Retrieved 03, January 2020
- Crystal-Structure: physics.ox.ac.uk, Retrieved 26, August 2020

3

Thermal Properties

Some of the different aspects studied in relation to the thermal properties of solids are atomic bonding, heat capacity of solids, lattice vibrations, specific heat capacities and the Dulong-Petit Law and Einstein's and Debye's theory. The diverse applications of these thermal properties of solids have been thoroughly discussed in this chapter.

Atomic Bonding

In a gaseous state of matter, the distance of separation between molecules is large. The interatomic force between the atoms or molecules is very weak. The molecules or atoms exist in a continuous motion colliding with each other and the walls of the container in which they are present. When they are brought closer and closer, they first condense into liquid in which the translational motion of the molecules is totally restricted. The energy is then lowered compared to that of free atoms. If the liquid is then condensed to form a solid, the atomic motion is restricted. The atoms will have only vibrational motion, and the rotatory motion, and translational motion are totally absent. The atoms will be held by the bonding forces. The energy of the solid is much lower than that of the liquid state or gaseous state. The atoms in a solid are held by an energy called the bond energy.

The liquid, like the melt of a solid, consists of a short range order that is not stable. There is a continuous agitation of molecules or atoms. The short range order is broken and restored again. This process takes place continuously with a relaxation time. When a solid is formed out of such a state, there is a heat of crystallization that is evolved. The atoms are brought closer to an equilibrium distance at a temperature that is the freezing temperature. In solids, the forces that operate between the atoms are the attractive forces and repulsive forces, and these forces determine the potential energy associated with the atoms in the solid.

Let us examine how does the resultant force between the atoms changes when the atoms are brought closer to each other. At relatively large distances, the attractive force between the atoms dominates. It grows as the interatomic distance diminishes. This variation is shown in figure below as curve 1. When the atoms come sufficiently closer, the repulsive forces come into play ($F_{repul} \propto 1/r^n$) and they grow faster than attractive forces. When the atoms are brought closer, their charge distributions overlap. This overlapping causes electrostatic energy to change. The overlap energy is repulsive at close distances. The repulsion occurs because Pauli Exclusion Principle prevents multiple occupancy. When atoms reach a distance of separation $r = r_o$ the attractive forces just balance the repulsive forces and the resultant force becomes zero (curve 3). The distance of separation, r_o is called 'the equilibrium distance'.

The interaction energy U also changes as $F = -\partial U/\partial r$. The interaction energy at $r = r_o$ is called binding energy of pair of atoms. The state defined by $(U)_{r=r_0}$ is a stable equilibrium state. What is said above is valid for all types of primary bonds and the general character of the forces is the same for all types of solids.

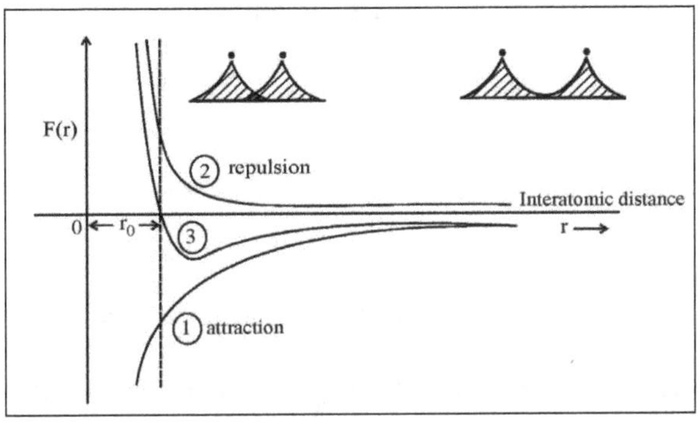

The variation of force between atoms as they are brought closer to each other curve 1 attractive force 2 repulsive force 3 the resultant force between the atoms. The solid circles and the shaded peaks represent charge distribution overlap.

Types of Bonding

Ionic Bond

Ionic bond is formed when the atoms with small number of electrons on the outer shells, have a tendency to give up electrons to the atoms with an almost filled outer shell. Ionic bond is the simplest type of interatomic bond. By losing electrons, metallic atoms like alkali-metals, alkaline-earth metals etc., and go into the nearest noble gas configuration, while the atoms like halogens take electrons to gain the inert gas configuration. Alkali halides are the simplest type of ionic bonded crystals. The formation of Na Cl by ionic bonding is shown in figure.

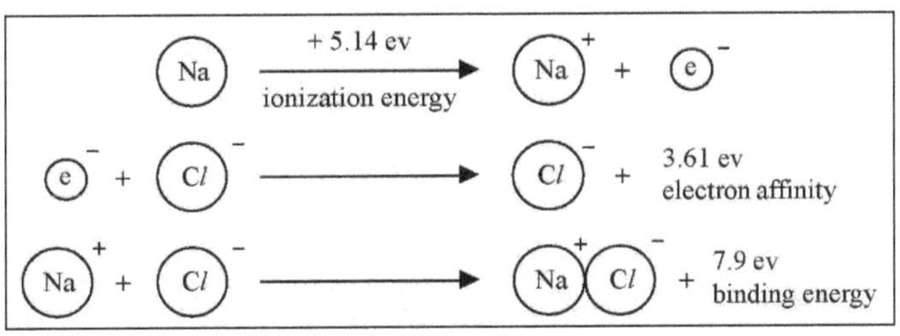

Ionic bonding in Na Cl.

The ionic crystals consist of positive ions and negative ions. Sodium chloride is the best example for ionic bonding and it has the highest degree of ionic bonding, the sodium atom has one $3S^1$ electron and upon ionization process gives up electron to become sodium ion. The chlorine atom which has half-filled p-orbital, takes the electron from sodium, forming chloride ion. The two ions exist as a single unit. The sodium atom with 0.19 nm size is reduced in size to sodium ion with size of 0.095 nm.

Consider that the two ions, sodium ion and chloride ion are initially far apart. They approach each other as they are attracted to each other by Coulomb forces. The Coulomb force is given as,

$$F_{Coul} = \frac{-(Z_1 q)(+Z_2 q)}{4\pi\epsilon_0 r^{m_1}} = \frac{-Z_1 Z_2 q^2}{4\pi\epsilon_0 r^{m_1}} = \frac{+a}{\gamma^{m_1}}$$

where m_1 is an exponent that determines the nature of the force. Z_1 and Z_2 are the charges of the ions, q is the charge, r is the intertonic distance and ϵ_o is the permittivity of free space (ϵ_o = 8.85 × 10^{-12} C/N–m²).

The repulsive force becomes significant when the ions are brought very close to each other, this repulsive force is expressed as,

$$F_{rep} = \frac{b}{r^{\eta}}$$

where η and b are constants.

The change of energy as the ions come closer to each other can be computed easily.

The potential energy associated with attraction (U_{attr}) and the repulsion (U_{rep}) are given as,

$$U_{attr} = -\int \frac{a}{r^{m_1}} dr = \frac{a}{m_1 - 1} \frac{1}{r^{\eta_1 - 1}} + C_1$$

$$U_{rep} = -\int \frac{b}{r^{\eta}} dr = \frac{b}{\eta - 1}\left(\frac{1}{r^{\eta-1}}\right) + C_2$$

∴ The binding energy of the ions is,

$$U_b = \frac{-A}{r^m} + \frac{B}{r^{\eta}} + C$$

where $A = \frac{-a}{m}$; $m = m_1 - 1$; $B = \frac{b}{\eta - 1}$; $\eta - 1 = \eta$;

$$C = C_1 + C_2 \text{ (the constants of integration)}$$

when r = infinity, $U_b(\infty) = 0$ i.e., $C = 0$

$$\therefore\ U_b(r) = \frac{-A}{r^m} + \frac{B}{r^n}$$

At equilibrium distance, the first derivative of the total-energy should vanish i.e., the forces of attraction and repulsion should balance each other i.e., at r = r_o, the equilibrium distance,

$$\left(\frac{dU_b}{dr}\right)_{r=r_o} = 0.$$

$$\left[+\frac{A_m}{r_o^{m+1}} - \frac{B_n}{r_o^{n+1}}\right] = 0$$

$$\therefore\quad \frac{r_o^{m+1}}{r_o^{n+1}} = +\left(\frac{A}{B}\right)\frac{m}{n}$$

$$\text{i.e.,}\quad r_o^{m-n} = +\left(\frac{A}{B}\right)\left(\frac{m}{n}\right)$$

Then, $U_{b(r=r_o)} = -\frac{A}{r_o^m} + A\left(\frac{m}{n}\right)\frac{1}{r_o^m}$

i.e., $\left[U_b\right]_{r=r_o} = -\frac{A}{r_o^m}\left(1-\frac{m}{n}\right)$

In the case of ionic crystals, $m_1 = 2$; $m = 1$.

$$\therefore \quad \left[U_b\right]_{r=r_o} = \frac{-A}{r_o}\left(1-\frac{1}{n}\right) = -\frac{Z_1Z_2q^2}{4\pi\epsilon_o r_o}\left(1-\frac{1}{n}\right)$$

Lattice Energy in Ionic Solids

The above condition that $(du_b/dr)_{r=r_o} = 0$ is the criterion for thermal equilibrium. The stable configuration is possible only when the energy is minimum (curve 3 of figure). Let us see the variation of forces with the distance of separation. The net force is zero at the equilbrium position. When like charges approach each other, they require energy and when unlike charges approach each other they release energy. When the ions are brought closer and closer, there exists repulsion due to electrons of the adjacent atoms. The Pauli Exclusion Principle also comes into operation and the electron-repulsion energy becomes significant as the atoms approach closer and closer. The energy (curve 3 in figure), possesses a sharp minimum as inter atomic distance becomes r_o. This is a stable inter atomic distance. Energy is required to reduce or increase this distance.

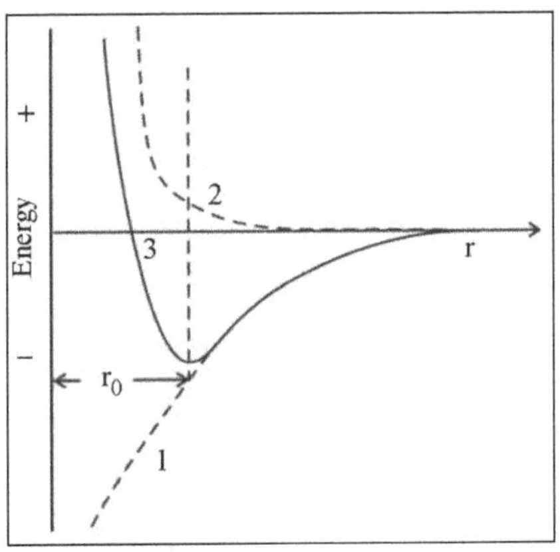

Potential energy variation: The solid line (3) gives the energy of the solid, the resultant of the two energies. The dotted lines (2 and 1) correspond to the repulsive and attractive energies.

Suppose the lattice is built of N_p ion pairs, then the energy of the lattice is obtained as follows:

$$\frac{nB}{r_o^{n+1}} = \frac{-A}{r_o^2}$$

i.e., $B = \frac{Z_1Z_2q^2r_o^{n-1}}{4\pi\epsilon_o n}$

for the U_b equation $[U_b]r = r_o = \frac{-A}{r_o}\left(1-\frac{1}{n}\right) = \frac{Z_1Z_2q^2}{4\pi\epsilon_o r_o}\left(1-\frac{1}{n}\right)$ gives,

$$U_b = -\left(\frac{q^2 Z_1 Z_2}{4\pi \in_o r_o}\right)\left(1-\frac{1}{n}\right)$$

As the lattice is built up of 'N_p' ion pairs, the energy of the lattice is,

$$U_{Lat} = -\frac{N_p Z_1 Z_2 q^2}{4\pi\in_o r_o}\left(1-\frac{1}{n}\right) = \frac{-AN_p}{V_o^{1/3}}\left(1-\frac{1}{n}\right)$$

where V_o represents the volume.

The energy trough will be very narrow for strongly bonded atoms. As a result, the slope, dF/dr, of the force curve at $r = r_o$ in above figure becomes very steep. This produces an interatomic distance which is constant within 0.001 Å at a given temperature. In such a case, we can take the ions as hard balls with definite radii.

We can determine the value of B in the expression in equation $F_{rep} = \frac{b}{r^{\eta}}$ for a crystal like Na Cl. The interionic distance is 2.76×10^{-10}m. The attractive force is,

$$F_{att} = \frac{-Z_1 Z_1 e^2}{4\pi\in_o r^2} = \frac{(1.6\times10^{-19}C)^2}{4\pi(8.55\times10^{-12}C^2/N.m^2)(2.76\times10^{-10}m)^2}$$

$$= -3.02 \times 10^{-9} N.$$

Near $r = r_o$, this is equal to the repulsive force. Then, from equation $\therefore U_b(r) = \frac{-A}{r^m} + \frac{B}{r^n}$ we get,

$$B = \frac{r_o^{n+1}}{n} \times F_{rep}$$

$$\text{i.e., } B = \frac{3.01\times10^{-9}\times(2.76\times10^{-10})^{n+1}}{n}$$

n = 9 for Na Cl.

$$\therefore B = \frac{3.02\times10^{-9}\times(2.76\times10^{-10})^{10}}{9} = 8.6\times10^{-106} N\cdot m^{10}$$

Then, the binding energy for Na^+- Cl^- pair is,

$$U_b = \frac{-q^2}{4p\in_o r_o} + \frac{8.6\times10^{-106}N\cdot m^{10}}{(2.76\times10^{-10}m)^9}$$

$$= \frac{-(1.6\times10^{-19}C)^2}{4\pi(8.55\times10^{-12})(2.76\times10^{-10})} + \frac{8.6\times10^{-106}}{(2.76\times10^{-10})^9}$$

$$\sim -8.34\times10^{-19} J + 0.92\times10^{-19} J$$

$$= -7.42\times10^{-19} J.$$

The bond energy due to attraction is – 5.23 eV and the contribution of the repulsive force to the energy is 0.575 eV.

Ionic Radii and Coordination Number

A crystal has a certain number of ions per unit volume that are distributed in three dimensions. They can be considered as having spherical shape with a characteristic value for radius in crystals. There are a large number of ionic crystals, namely, alkali halides, alkaline earth halides, oxides, transition metal oxides etc. The ionic radius of ions in these crystals depends on the coordination number. Since ions can be treated as hard spheres, carrying a charge of normal parent atom of the crystal. They interact through central electrostatic forces. The crystal structures of ionic compounds to a large extent, are determined by packing of ions together as closely as possible under the strong electrostatic forces. The ratio of the ionic radii in the crystal structure is very important aspect of ionic solids.

A crystal of sodium chloride consists of positive ions (cation, Na^+) and negative ions (anion, Cl^-). The energy of attraction between the two ions spaced at a distance r_o apart,

$$U_o = -\frac{Z_1 Z_2 q^2}{4\pi \in_o r_o} = \frac{-A}{V_o^{1/3}}$$

where q is the charge of the ion. The reduction of energy on attachment of an electron to an anion, is higher than the increase of energy on detachment of an electron from a cation. The energy of an aggregate composed of positive and negative ions is reduced further due to the fact that each ion of one sign has ions of opposite sign as its nearest neighbours. This does occur as a result of coulomb attraction of the oppositely charged ions. The number of neighbours surrounding an ion and the type of crystal lattice are determined by the ratio of atomic radii of cation and anion. If this is R, then,

$$R = r_c / r_a$$

where r_c and r_a are the radii of the cation and anion respectively. The lattice of a crystal is stable only when the cores of the ions of one sign are in direct contact with the cores of opposite sign.

The coordination number z for the ionic lattices depends on the value of R. The following table gives the R values and the corresponding Z values:

R	0.73 – 1.00	0.41 – 0.73	0.22 – 0.41	0.15 – 0.22
Z	8	6	4	3

The example for Z = 8 is cesium chloride (CsCl) Figure (c), for Z = 6, it is NaCl figure (b), and Z = 4, it is CaF_2 or LiF. The Figure (a) gives the coordination polyhedron of an ionic bond that has Z = 6. When this polyhedron is repeated to fill space, the arrangement of atoms in a crystal is shown in figure (b). The alkali metal halides and hydrides, the alkaline earth oxides and silver halides (except AgI) belong to this type of ionic bonding.

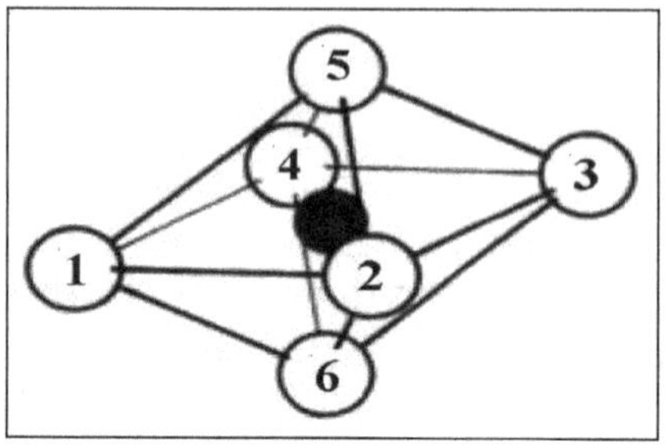

Coordination polyhedron for coordination number 6. • cation o anion. The anions 1, 2, 3, 4 are at the apices of the polyhedron.

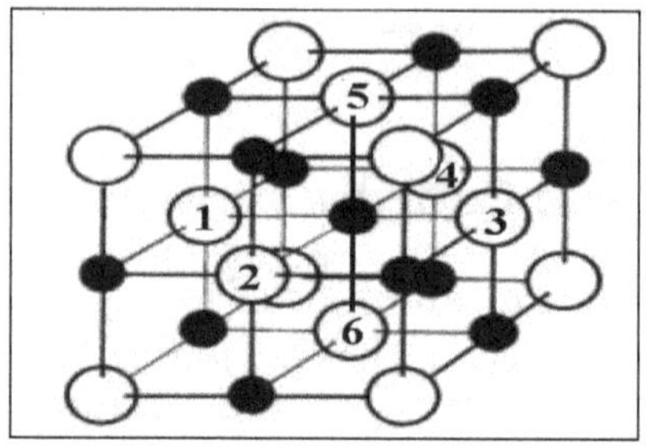

The unit cell of a crystal in which Z = 6. The structure is known as rocksalt structure.

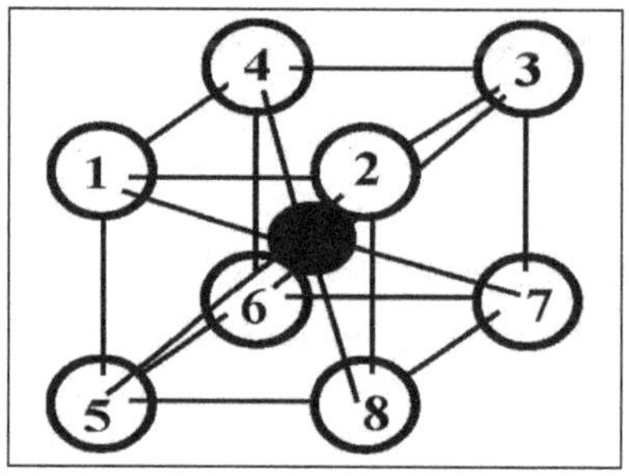

CsCl, with coordination number 8.

The structure called the rocksalt structure becomes unstable when the anions come into contact along the coordination polyhedron edges. Then, if a further decrease in r_c/r_a is caused, the coordination number falls to 4. In a fourfold structure, tetrahedral structure, the anions are at the apices of a tetrahedron. Decreasing r_c/r_a produces instability on the hard sphere model of f.c.c when the anions touch along the edges of the tetrahedron. From geometrical considerations, the tetrahedron coordination is possible only for $0.414 > r_c / r_a > 0.225$.

Tetrahedral coordination polyhedra may be arranged or stacked to fill space in two ways. They are 1) the structure showing cubic structure and known as zinc blende 2) wurtzite. ZnS is the example and it exists in both the forms with a coordination number '4'.

Properties of Ionic Crystals

A large class crystals like alkali halides, alkaline earth halides, metal oxides, sulphides, selenides etc., belong to the group of ionic crystals. Alkali halides are the most basic crystals for study. We shall briefly give the general properties of the ionic crystals.

The ionic bond is a strong bond. The properties of crystals depend on the lattice energy. Some basic properties like heat of sublimation, melting point, boiling point, solubility etc., show a regular dependence on the inter ionic distance. For example if we take MF, MCl, MBr, MI, the halogenides, of alkali metals (Li, Na, K, Rb, Cs), the boiling points and melting points decrease as one goes from fluorides to iodies. In general, the ionic solids possess high melting points.

The ionic compounds exhibit large band gaps and therefore come under the class of insulators. The resistivities of alkali halides will be of the order of 10^{12} Ω–cm or more at room temperature. The oxides exhibit much higher resistivities. For example, the resistivity of Al_2O_3 is of the order of 10^{16} Ω–cm at room temperature.

The resistivity of all ionic solids decreases with the rise of temperature. That is, they show negative temperature coefficient of resistivity. The resistivity is a strong function of temperature in ionic crystals. The conduction is basically electrolytic in nature. The ions move through the lattice via the point defects, when electric field is applied to a crystal. The study of electrical conductivity helps to determine the characteristics of defects in ionic crystals.

The defects like point defects and line defects are charged defects. The addition of impurities strongly affects the conductivity of ionic crystals. This is made use of, in making the electrolytic solid state battery cells. Super ionic conductors are useful materials for solid state batteries. Many ionic crystals are transparent to radiation over a wide wavelength range. Normally the alkali halides have absorption bands in the infrared region. Therefore, they are useful as infrared filters and windows. When crystals like alkali halides are subjected to high energy radiation like X - rays or γ rays, they get colored i.e., they absorb light in visible radiation at certain wave lengths. The absorption is due to 'Colour centres'. The formation of color centers is very important in energy storage utility. Many oxides, sulphides and alkali halides exhibit luminescence properties if they are subjected to external stimuli like high energy radiation or electric field. Ionic crystals exhibit high strength. They are also brittle. The fracture takes place along the cleavage planes. The ionic solids are hard. Ionic compounds dissolve in polar solvents.

Covalent Bond and Covalent Crystals

Covalent Bonds in Crystals

It is known that the word 'covalent' describes the sharing of valence electrons by adjacent atoms. Hydrogen molecule is a good example of covalent bonding. The two protons share two electrons of opposite spins. The electrons lie between the hydrogen atoms for major part of the time. To a first approximation, the covalent bond in a hydrogen molecule can be considered to arise from the attraction of positive ions to the intervening pair of electrons with opposite spins.

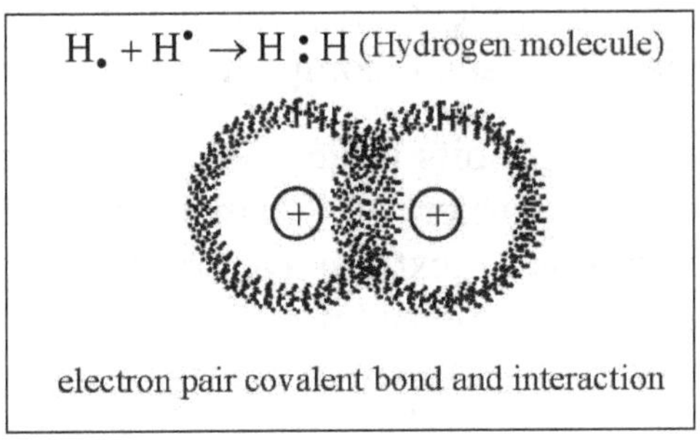

Covalent bonding.

In the bonding process of forming the hydrogen molecule, the potential energy of the hydrogen atom is lowered as shown in figure. The molecules of fluorine, chlorine, oxygen, nitrogen etc., (the diatomic molecules) are formed on the basis of electron-pair covalent bonds. In these cases, two p-electrons are shared between the atoms. The above examples have single bonds i.e., their coordination number is one. But many elements form multiple bonds, Carbon is the best example. Carbon can form four covalent bonds even though it should form only two-covalent bonds. Sulphur, selenium, tellurium etc., form two bonds; Phosphorous, nitrogen, arsenic, antimony can form three bonds.

The covalent bond is pictured as an overlap of the electron orbitals of the two participating atoms.

Pauli's exclusion principle explains why only two electrons can take part in the bond formation. According to it, only two electrons that have the same quantum numbers but opposite spins can pair up. If a third electron has to take part in the bonding, it would have its spin vector in the same sense as either of the earlier electrons. Therefore, it will be repelled. Hund's rule states that, for equivalent orbitals, electrons tend not to share the same orbitals and the electrons occupying different but equivalent orbitals tend to have their spins parallel.

Nitrogen (outer valence electrons $2S^2\ 2p^3$) has three unpaired p-electrons. It can attain the neon noble gas electron configuration ($2S^2\ 2p^6$) by sharing three 2p-electrons with another nitrogen atom to form the diatomic molecule. The most important covalent bonded material is carbon. The carbon is the basic element for many organic materials like polymeric materials. The carbon atom has the ground state electron configuration $1s^2\ 2s^2\ 2p^2$. Therefore, it should form two covalent bonds with its two half-filled 2p-orbitals as per Hund's rule.

However, carbon shows a covalency of four and form four covalent bonds of equal strength. To explain this, let us consider the electron configuration of carbon. There are 2s states and 2p states. The 2s states and 2p states are closely spaced and the energy difference between them is small. Therefore electrons can redistribute themselves as follows:

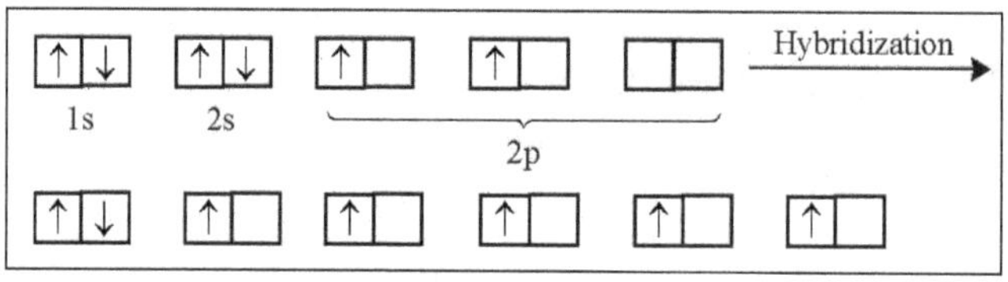

Hybridization of carbon.

The carbon promotes one of its 2s electrons to a 2p orbital, the 2s and three 2p orbitals merge to form sp^3 hybrid orbitals. These are elliptical orbitals which are disposed toward the four corners of a tetra hadron. Each orbital contains one electron. This type of arrangement is known as hybridization. The process of hybridization enables carbon to form four equivalent covalent bonds instead of only two. This process leads to the formation of very stable compounds, or elemental compounds like diamond.

The carbon in the form of diamond exhibits the covalent bonding. The four sp^3 hybrid orbitals are directed symmetrically toward the corners of a regular tetrahedron. Each carbon is covalently bonded to four adjacent atoms. Diamond is extremely hard. It has high bond strength and melting point (3550 °C). Its bond energy is about 711 kJ/mol.

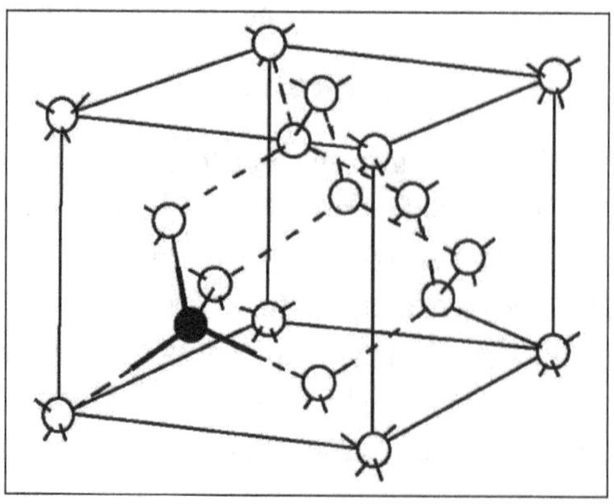

In a Diamond structure, each carbon atom is covalently bonded to four adjacent atoms. The bond angle is 109.5 °C.

The covalent bonds are highly directional i.e., stereospecific. The coordination numbers for covalently bonded atoms are mostly controlled by radii ratios. The covalent bonds follow the following characteristics:

- Multiple bonds have shorter bond lengths.
- Double bonds are stronger than single bonds. This aspect is very important in polymers.

Properties of Covalent Crystals

The covalent crystals because of their directionality in bonding have different properties compared to ionic crystals. The co-ordination numbers for covalently bonded crystals are not controlled by radii ratios. There is always a strong covalent attraction only between first neighbours. As with ionic bonds, the lattice energy is associated with covalent crystals. Shortest bond lengths exist only in covalent crystals. The melting and boiling points of covalent crystals are usually low as compared to those of ionic crystals.

These compound solids exhibit smaller energy band gaps. Pure covalent solids are good insulators at lower temperatures. However, as temperature increases, the conductivity increases. The conduction in most of the covalent solids is electronic in nature. Covalent solids are very hard, for example, diamond is hardest substances known. These solids have high yield strength and are not brittle. Covalent substances are soluble in non-polar solvents. Hydrogen peroxide, propylene etc., are other examples for a covalent bonding. Silicon and Germanium are excellent examples for covalent bonding. Silicon and Germanium are semiconductors. Silicon and Germanium have the same structure and have a finite number of electrons which are able to leave the covalent bond locations.

Metallic Bonding

Nature of Metallic Bonding

This is the third type of chemical bonding that is important in materials. It is somewhat intermediate between covalent and ionic bonding. Metals are the major class of materials that are used in engineering. From the scientific point of view also, they place themselves as the most interesting species among different materials. The atoms are packed relatively close together in a systematic pattern in metallic solids. Metallic atoms i.e., those which possess loosely held valence electrons may be bonded together into a stable solid.

We shall consider the case of a simple metal like sodium. When two sodium atoms are brought closer, the unpaired 3s electrons of the atoms form an electron-pair bond with the spins of the 3s electrons opposed. As a third sodium atom approaches this pair, its 3s electron will be repelled by the 3s electrons of the other two atoms. This is because of Pauli's exclusion principle. However, the energy of the unoccupied 3p states is very nearly the same as the energy of a 3s state. Consequently, the third electron can go over into this state without violating the Pauli Exclusion Principle. In a similar fashion, a large number of sodium atoms can thus surround any sodium atom. The first sodium atom or the central atom that has only one unpaired electron takes turns forming electron-pair bonds with each of the surrounding atoms. Therefore, the sodium forms less than a whole electron-pair bond with each neighbour. Solid sodium solidifies into a body-centered cubic structure in which each sodium atom is surrounded by eight other atoms. Therefore, each sodium atom can form one-eighth of an electron-pair bond, or one-quarter of an electron bond, with each of its nearest neighbour.

Similarly, copper atoms in crystalline copper are so close together that their outer valence electrons are attracted to the nuclei of their numerous neighbours. Each atom is surrounded by twelve neighbours (closest). The valence electrons are associated with all nuclei and are spread out among all atoms. Now, electrons exist as 'electron gas' in which ion cores are present. The metallic bond is essentially a covalent bond without saturation. The metal atoms are held together by a mutual sharing of valence electrons. The density of electrons between the atoms is much lower than that is allowed by Pauli Exclusion Principle. Therefore, the electrons move freely as gas molecules without expending energy. The metallic bond is non-directional. The coordination number is very high. The metallic lattice is shown in Figure.

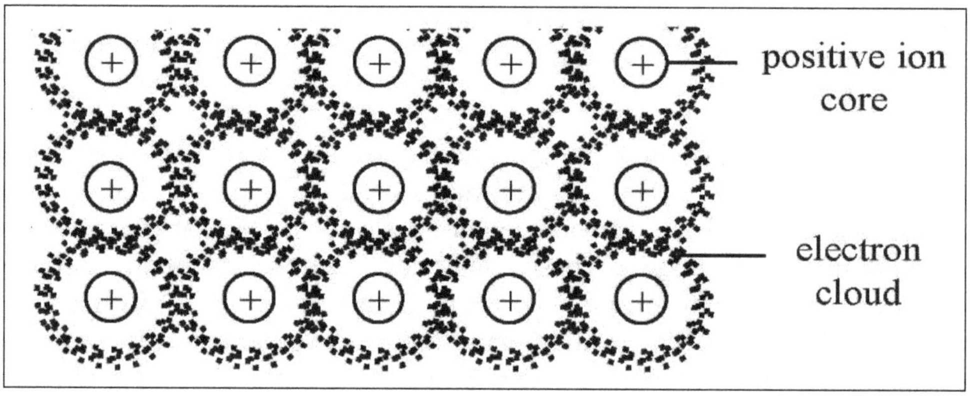

Metallic lattice.

A large number of metals like Al, Cl Mg, Ti, Ni, Zn, etc., have 12 neighbors for each atom. Other metals like Li, Na, K, V, Mn, W and Fe (at room temperature) have eight neighbours for each atom. The binding energy for potassium is about 89.6 KJ/mol, for Nickel it is 423 KJ/mol, Zinc, it is 131 KJ/mol.

Properties of Metallic Bonding

Metallic bonding is weaker than ionic and covalent bonding, but stronger than Van der waals' bonding. Metals possess high thermal and electrical conductivities. This is because of the free electron movement in the lattice. Metals exhibit ductility. They are brittle below a certain temperature. Metals exhibit brittle to ductile fracture transition at a certain temperature. Many metals like tungsten are tough. But some metal like pure silver, aluminium, gold are soft and malleable. Their melting points vary from 64 °C (Potassium) to 2000 °C [Titanium (1812 °C), Molybdenum, Tungsten]. Metallic alloys are extremely useful for all engineering applications. Their high conducting values (gold, platinum, silver, copper etc.,) make them useful as electrodes in scientific and technical devices. Metals are opaque to light and they reflect light energy very efficiently.

Hydrogen Bonding

A chemical bond, in which a hydrogen atom of one molecule is attracted to an electronegative atom, of the same or another molecule is called 'Hydrogen bond'. Hydrogen forms bonding with electronegative nitrogen, oxygen, or fluorine atom, usually of another molecule. The interaction which occurs when a hydrogen atom, covalently bonded to an electronegative atom (as in AH), interacts with another atom Y to form the aggregate AH ··· Y.

The shortest and strongest bond is indicated as AH, while the secondary and weaker interaction is written as H ··· Y. Thus, AH is a proton donor, while (Y) is a proton acceptor which often contains lone pair electrons and can act as a base. Hydrogen atom acts as a bridge between two atoms, holding one atom by a covalent bond and the other atom by a hydrogen bond. The hydrogen bond is represented by dotted line (.....), while covalent bond is represented by the solid line(—).

$$\ldots\ldots H^{\delta+} - A^{\delta-} \ldots\ldots\ldots\ldots\ldots\ldots H^{\delta+} - A^{\delta-} \ldots\ldots\ldots\ldots\ldots\ldots H^{\delta+} - A^{\delta-}$$

Example: Hydrogen fluoride (HF)—The hydrogen atom while remaining bonded with fluorine atom, forms another weak bond with fluorine atom of adjacent bonding molecule. Hence, the formation of a hydrogen bonding results in the formation of a cluster of hydrogen fluoride (HF) molecules as $(HF)_n$, as shown below:

......HF......HF.....HF....HF.....HF...

Hydrogen fluoride can be solidified at low temperatures. In solid form, it contains long zig-zag chains of molecules as shown below.

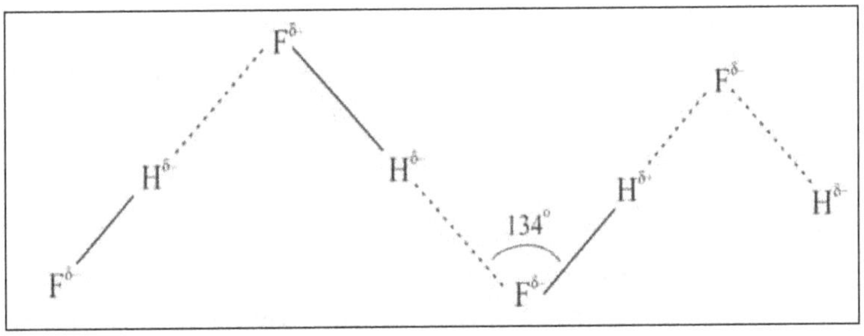

Hydrogen bonding $(HF)_n$ in HF solid state.

Covalent bonding is a strong bond. Compared to it, hydrogen bond is much weaker. The strength of a hydrogen bond is of the order of 10^{-40} KJ-mol^{-1} as compared to the normal covalent bond that would have 40 KJ mol^{-1}. The bond length is longer in hydrogen bonding than that in covalent bonding. In the case of HF molecule, the covalent bond length is 109 pm while the bond length between F and H is 155 pm.

It should be noted that strong electronegativity of the other atom or molecule is a must for the formation of hydrogen bonding. Two good examples we can cite here are: ammonia (NH_3) and water (H_2O). In the case of ammonia, (NH_3), the electronegative nitrogen atom forms a bond with three hydrogen atoms. As nitrogen is more electronegative and hydrogen acquires a positive charge, the bond takes place with good strength. Because of the above reason, i.e. the existence of positive and negative sites, the molecules of ammonia is associated through hydrogen bonding in long chains. The bonding is shown in the Figure.

One of the conditions for hydrogen bonding is the size of electronegative atoms or molecule should be small. For example, in the case of NH_3 and HCl, only NH_3 shows hydrogen bonding and not HCl. This is because of the fact that N is smaller than Cl.

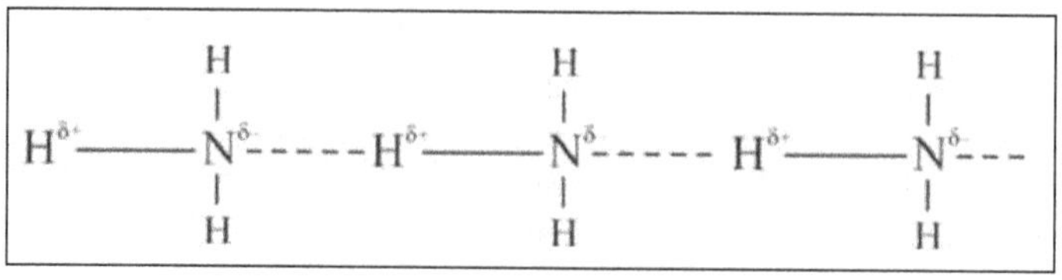

Ammonia bonding, NH_3.

Water is a good example of polar molecule that exhibits hydrogen bonding. Apart from two (O-H) bonds, water molecule has oxygen atoms forming two hydrogen bonds as shown in the figure.

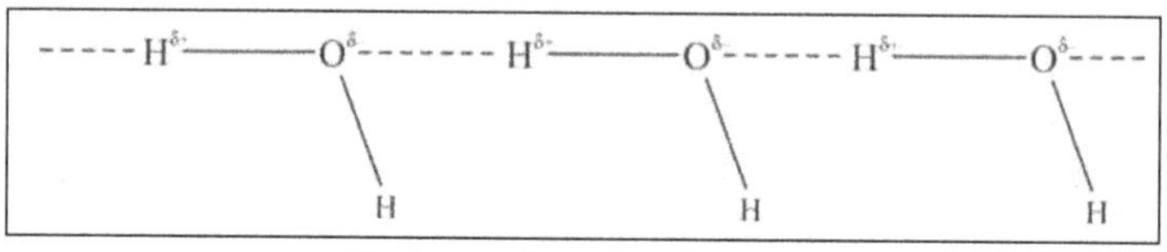

Hydrogen bonding in H_2O.

In H_2O, four hydrogen atoms, two by covalent bonding and two hydrogen bonds tetrahedrally surround each oxygen atom. The molecular formula is $(H_2O)_n$. In the solid form, ice is a very interesting crystal. The crystal lattice consists of linked hexagonal rings of water molecules with considerable open space in the center of the ring. All naturally occurring ice crystals are hexagonal in shape and all snowflakes reflect this basic six-sided crystal habit.

Good examples of crystals in which hydrogen bonding plays very important role in determining their properties are Potassium dihydrogen phosphate ($KH_2 PO_4$, KDP) and Ammonium di hydrogen phosphate ($NH_4H_2 PO_4$, ADP). KDP is a ferroelectric below 120K and piezoelectric even at room temperature. KDP exhibits electro-optic effect. ADP is anti-ferroelectric and piezoelectric.

The strongest hydrogen bonds are formed between the most electronegative (A) atoms such as fluorine, nitrogen, and oxygen which interact with (Y) atoms having electronegativity greater than that of hydrogen (C, N, O, S, Se, F, Cl, Br, I). The weaker the hydrogen bond, the shorter the lifetime of the complex it forms. An important aspect of weak hydrogen bond formation is that the different molecular aggregates which do form can be easily and reversibly transformed. Thus the small energy changes resulting in the rapid making and breaking of hydrogen bonds in biological systems are of great importance; for example, hydrogen bonding determines the configuration of the famous α-helix of DNA, and the structures of most proteins, thereby serving an important function in determining the nature of all living things.

Secondary (Van Der Waals) Bonding

Primary bonding (ionic, covalent and metallic) is strong and the energies involved range from about 100 to 1000 kJ/mole. In contrast, secondary bonding is weak, involving energies ranging from about 0.1 to 10 kJ/mole. While this type of bonding, also referred to as "residual", is weak, it is essential in the functioning of our environment.

The energy difference between the liquid and vapor states of a given system is given by the heat of vaporization, i.e. the heat required to convert a given liquid into a vapor (normally) at the boiling point temperature at 1 atm pressure. The energy difference is due to intermolecular attraction between molecules at close distance of separation. This phenomenon of attraction through secondary bonding can best be considered between a single pair of molecules, but recognizing that the forces are of longer range. The following intermolecular forces are identified:

1. Dipole-Dipole Interaction: Molecules with permanent dipoles (such as water, alcohol and other organic compounds with functional groups) exert a net attractive force on each other as a result of varying degrees of alignment of oppositely charged portions of the molecules. For two polar molecules with a dipole moment of (μ) separated by a distance of (r), the energy of attraction can be quantified as:

$$\mu = L.\ \partial q$$

where L is the distance of separation of the charges. In molecules, the dipole moment is given by the vector sum of electronic charges of the polar bonds. Thus, because of molecular geometry (symmetry), molecules such as CCl_4 exhibit no dipole-moment.

The Dipole-Dipole interaction is given by,

$$E_{\text{Dipole-Dipole}} = -\frac{2}{3}\frac{\mu^4}{Dr^6} \times \frac{1}{kT}$$

where μ is the dipole moment, r is the distance of approach of the oppositely charged molecular portions, k is the Boltzmann constant and T is the absolute temperature in K.

The molar energies of attraction associated with dipole-dipole interaction range from 0 to about 10 kJ/mole. These forces are primarily responsible for the liquid state (at room temperature) of most polar organic molecules; they are a contributing factor for H_2O to be liquid at room temperature, and are responsible for alcohol being a liquid.

2. Dipole-Induced Dipole Interaction: A dipole in one molecule can interact with and polarize the electrons of a neighboring non-polar molecule, thus generating an induced dipole which will experience an attractive force with the polarizing, polar dipole. Debye showed that in a molecule with a "polarizability" of (α) the attractive potential arising from dipole-induced dipole interaction is given as:

$$E_{\text{Dipole-Induced dipole}} = -\frac{2\alpha\mu^2}{Dr^6}$$

Induced dipole interaction is important in aqueous solutions and very effective during adsorption of inert molecules on active solid substrates.

3. London Dispersion Forces: It is a well-known fact that all substances, including rare gases and hydrogen, assume liquid state at finite temperature, an indication of the existence of attractive interatomic and intermolecular forces, even in the absence of permanent dipole systems. The origin of this force has been proposed by F. London in 1930. Accordingly, orbiting electrons will at any instant generate a "temporary dipole", the configuration of which changes as the electrons move. Since all atoms of a given system similarly experience temporary instantaneous dipole moments, their effect is expected to be cancelled because of the statistically random orientation of dipoles. It is evident that, should the dipoles be synchronized in a given assembly of atoms, then, a net

attractive force would result. But, such an attractive force constitutes a lowering of the energy of a given system, synchronization can and will take place because all systems will attempt to assume minimum energy configuration. The London dispersion force can be formulated as:

$$E_{London} = -K\frac{\alpha^2}{r^6}$$

where k and α are constants, r is the distance between molecules.

The attractive London forces are small, as manifested by the very low boiling points of some of the rare gases, hydrogen and nitrogen. The dispersion forces are much weaker then covalent bonds within molecules. Dispersion forces occur in addition to dipole-dipole interactions. The molecules is liquids, which have temporary fluctuating dipoles, have low boiling points.

Cohesive Energy

The energy required to bind the atoms or ions in a molecule is called cohesive energy. It is also the energy necessary to dissociate a molecule into atoms.

Madelung constant 'A' represents the effect of the interaction of all the ions on any ion in the ionic crystal. All the ions in the lattice will be contributing to the energy of cohesion. We shall discuss a simple way of evaluating the Madelung constant for a crystal like NaCl. In ionic crystal, individual atoms do not exist. Every anion is surrounded by a certain number of cations and every cation is surrounded by as many anions. The coordination number is determined by the ratio of cation and anion radii.

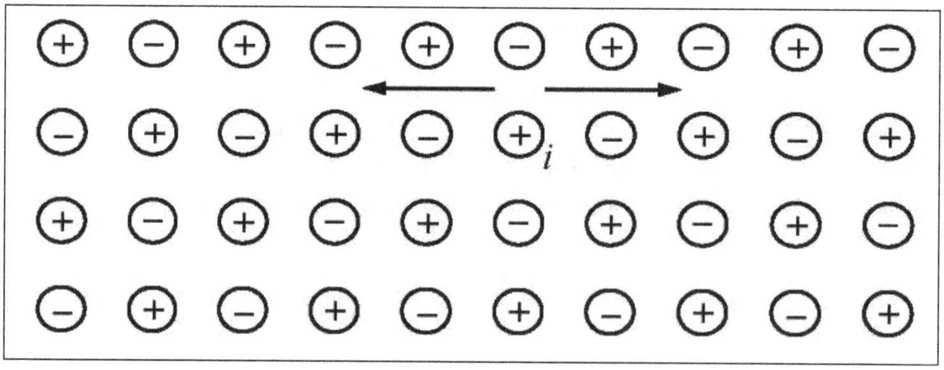

Array of ion pairs in an ionic crystal.

Consider a crystal as arrays of ions (NaCl), where a two dimensional array is shown. Let u_{ij} be the interaction energy between any two ions i and j. The interactions are pair-wise additive, the total interaction energy between i^{th} ion taken as a reference ion, and all the other ions in the crystals is given as a summation over all pairs of ions as given below:

$$u_i = \sum_j{}' u_{ij}(r_{ij})$$

where $\sum_j'$ indicates that the summation is taken to include all ions except j = i (i.e., the same ion). This interaction energy does not depend whether the reference ion 'i' is positive or negative or the position of the ion in the array or crystal.

The term u_{ij} consists of first the long-range coulombic interaction. It is a central force as has already been stated, the second force is the short range interaction. We have defined the coulomb energy as,

$$u_{ij} = -\frac{Z_i Z_j e^2}{4\pi \in_0 r_{ij}}$$

Suppose $r_{ij} = p_{ij} r_o$ and r_o is the shortest distance between unlike ions in the crystal. Then the total electrostatic energy between the ith ion and other ions in the crystal is given as,

$$\sum_j{}' + \frac{Z_i Z_j e^2}{p_{ij} 4\pi \in_0 r_{ij}}$$

$$= -\frac{Z_i Z_j A}{4\pi \in_0 r_o}$$

Where, $A = \sum_j{}' \mp \frac{1}{p_{ij}}$

The constant A called Madelung constant is introduced in the equation $u_i = \sum_j{}' u_{ij}\left(r_{ij}\right)$ already. Madelung constant is the property of a particular crystal structure. It can be calculated.

We consider one row of ions for NaCl. Consider the ions on either side of the reference ion 'i'.

$$u_{i+1} = \frac{e^2}{4\pi \in_0 r_o}$$

$$u_{i,\,i-1} = \frac{e^2}{4\pi \in_0 r_o}$$

∴ The net energy is $\frac{-2e^2}{4\pi \in_0 r_o}$.

The next two ions on either side contribute,

$$\frac{2e^2}{4\pi \in_0 2r_o}$$

The next-two ions on the either side contribute,

$$\frac{-2e^2}{4\pi \in_0 \cdot 3r_o}$$

This can be done upto the end of the chain on either side which can be taken as $- \propto$ to $+ \propto$. Then,

$$\frac{A}{r_o} = 2\left[\frac{1}{r_o} - \frac{1}{2r_o} + \frac{1}{3r_o} - \frac{1}{4r_o} + \ldots\right]$$

Thus, the factor two comes in, because there are two ions, one left side and the other on the right side at equal distance.

$$\text{or } A = 2\left[1 - \frac{1}{2} + \frac{1}{3} - \frac{1}{4} + \ldots\right]$$

$$\text{i.e., } A = 2 \ln 2$$

Thus, the Madelung constant for one dimensional chain is A = 2 ln 2.

The cohesive energy decreases monotonously in alkali halides with the generic formula, M^+X^-, where M^+ can be Li^+, Na^+, K^+ or Rb^+ and X^- can be F^-, Cl^-, Br^-, and I^- for same M^+ ion. Thus, the following table gives the cohesive energies for a series of compounds. 'A' depends only on the type of the lattice and is independent of the lattice dimensions and the kind of ions. The attractive potential energy U of the lattice energy is proportional to $\frac{1}{r_o}$. The calculated Madelung constants for a few crystals are given below:

NaC*l*	1.74756
CsC*l*	1.76267
$M^{++}X_2^-$	5.03878
Rutile	4.816

For isolated Na^+Cl^- molecule, the Madelung constant has a value equal to one. But in a crystal of NaCl, it is 1.74756 i.e., about 75% more.

If there are 'N' ion pairs, 'Z' is the ion valence, e the elementary charge, 'r_o' is the distance of separation. Then,

$$U_{coul} = -A\frac{NZ^2e^2}{4\pi \epsilon_o} \times \frac{1}{r_o}$$

If $Z_{cation} \neq Z$ anion, and Z is taken as the largest common factor.

Born Repulsive Exponent

The repulsive energy is given as $\frac{B}{r^n}$. B is evaluated in the equation $\therefore\ U_b(r) = \frac{-A}{r^m} + \frac{B}{r^n}$. 'n' is called the Born exponent. It is evaluated on the basis of compressibility studies in ionic crystals. During the compression of a crystal, the ions are pushed to closer distances and hence their electron outer shells interact vigorously. Therefore, the study of compressibility gives the information about repulsive forces. The study of compressibility helps is in the estimation of born exponent. If N_p is the number of ion pairs per unit volume, x is the compressibility of the crystal, then,

$$\frac{1}{\chi} = \frac{n-1}{9} A \frac{Z^2e^2}{4\pi \epsilon_o} \cdot \frac{N_p}{r_o^4}$$

The value of n is nearly 10 for many ionic crystals. The value of n and the lattice energies for a number of alkali halides is given in the table.

Table: The values of Born's exponent and lattice energies calculated and measured. The energies mentioned are in KCal/mole.

	LiF	LiC*l*	LiBr	LiI	NaF	NaC*l*	NaBr	NaICsF	CsF	CsC*l*	CsBr	CsI
n	6.2	7.3	7.7	7.0	6.4	8.4	8.3	8.0	10.2	10.6	10.5	11.1
U(calculated) KCal/mole	240	194	183	165.7	210	180.9	170	157	176	152	146	138.6
measured in KCa*l*/mole	242	199	188.8	177.7	214	182	173.6	163	172	155.4	149	142.4

The compressibility and repulsion exponent are inversely proportional to one another. Many cases, the value of n is high confirming the fact that the ions can be regarded as hard spheres ($n \to \propto$) which are arranged in the lattice in such a way that the nearest neighbours touch each other. The repulsion force is principally due to Pauli exclusion principle Knowing the exponent value allows us to calculate the lattice energy, the main part of which is due to coulomb energy. The values of Madelung constants for various crystals are,

$$NaCl - 1.748; \quad CsCl - 1.763; \quad CaF_2 - 5.039; \quad ZnS - 1.641.$$

Heat Capacity of Solids

The ratio of the amount of heat, ΔQ absorbed by a body to the corresponding change in its temperature, ΔT is known as heat capacity. It is usually denoted by the letter C. Expressed mathematically, we can write,

$$C = \frac{\Delta Q}{\Delta T}.$$

For example, when a body is heated at constant volume, we define heat capacity at constant volume (C_v). Similarly, when pressure on a body is kept constant during heating process, we define heat capacity at constant pressure (C_p). For gases, the values of heat capacities at constant volume and at constant pressure differ significantly even at ordinary temperatures. This is because gases can be compressed due to large interatomic separation and weak interatomic forces. On the other hand, for liquids and solids, which are almost incompressible, the values of C_p and C_v are almost the same upto about 300K. At higher temperatures, measurable difference has been reported in some cases. Since it is more difficult to keep the volume of a solid constant when thermal energy is supplied to it, the experimental observations enable us to measure C_p. But, theoretically it is simpler to calculate C_v.

The thermodynamic relation connecting C_p and C_v:

$$C_p - C_v = \frac{TV\alpha^2}{k_T},$$

where α is the coefficient of volume expansion and K_T is isothermal compressibility.

Since the coefficient of is volume expansion is three in times the coefficient of linear expansion α_l, equation above can be rewritten as:

$$C_p - C_v = \frac{9TV\alpha_L^2}{k_T},$$

What information do experiments provide us about heat capacity of solids? What is the current status of our theoretical understanding, and so on? You will discover answers to all such questions as you read through this unit. We begin our discussion with the classical findings of Dulong and Petit.

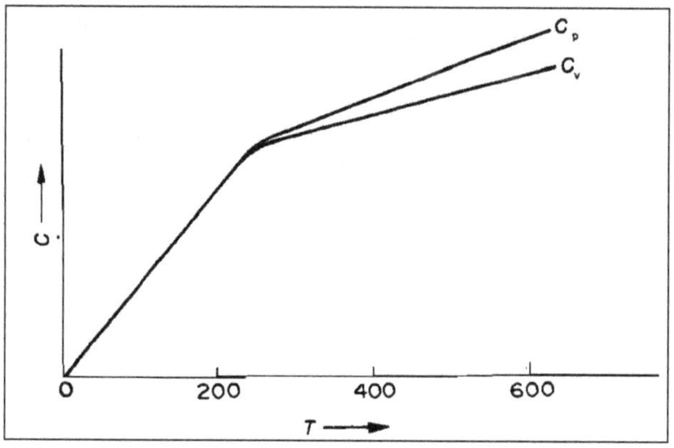

Temperature variation of C_v and C_p for copper.

Dulong-Petit Law

In early 19th century, Dulong and Petit conducted a series of classical experiments to measure heat capacities of solids at different temperatures. Figure (a) below shows a typical temperature versus heat capacity curve obtained by Dulong and Petit. Figure (b) shows the temperature dependence of heat capacity of silver and germanium, respectively a metal and a semiconductor. Note that:

- At high temperatures, heat capacity is independent of temperature and is equal to 3 R for all solids, where R is the universal gas constant. Its value is 8.3 JK^{-1}. This is known as Dulong-Petit law.
- As temperature decreases, heat capacity decreases almost exponentially and approaches zero as T→ OK.

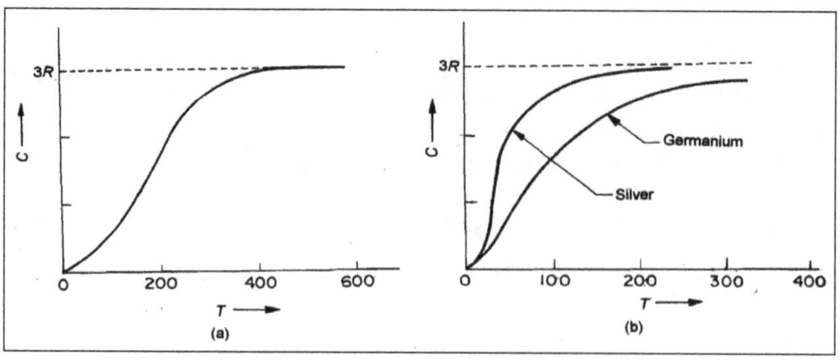

(a) Typical temperature dependence of heat capacity of solids; and (b) heat capacity versus temperature curves for silver and germanium.

Now, you may like to know: What is the theoretical basis for the observed temperature dependence of heat capacity? Let us first understand the observations qualitatively. You know that a solid is a collection of atoms arranged along a periodic lattice. Since we confine our discussion to heat capacity at constant volume, it is logical to think that the (heat) energy supplied to a solid will be completely used in increasing its internal energy. This can happen via two mechanisms. Firstly, atoms can oscillate about their equilibrium positions much more vigorously by absorbing heat energy. Secondly, in metals and semiconductors, the heat energy can also be absorbed by electrons in outer orbits and be excited to higher energy states. Corresponding to these mechanisms of absorption of heat energy, we have lattice and electronic contributions to the heat capacity.

Early attempts to explain Dulong-Petit law were made on the basis of classical notions about the energy of vibrating atoms based on Maxwell-Boltzmann (M-B) distribution law. It gave rise to the classical theory of heat capacity.

Classical Theory of Heat Capacity

This theory is based on the assumption that atoms in a periodic lattice vibrate about + P their equilibrium sites and obey Hookes' law. That is, each atom is bound to its equilibrium site by a harmonic force and when thermal energy is supplied, it vibrates like a harmonic oscillator. It is further assumed that each atom is equivalent to a classical three-dimensional (3-D) harmonic oscillator and vibrates independently. This means that a solid consisting of N atoms can be treated as a collection of as many three-dimensional, independent harmonic oscillators. Equivalently, N vibrating atoms can be considered as a collection of 3N, one-dimensional (I-D) independent harmonic oscillators. And, to obtain an expression for the heat capacity of a solid, we need to know the average energy of a 1-D harmonic oscillator and multiply it by 3N. Therefore, our problem reduces to obtaining an expression for the average energy of a I-D harmonic oscillator.

The equivalence of N, 3-D harmonic oscillators and 3N, 1-D harmonic oscillators follows from the principle of equipartition of energy. According to this principle, vibrational energy of an atom is distributed equally among its three degrees of freedom one along each coordinate axes. Therefore, an atom (or a 3-D harmonic oscillator) is equivalent to three 1-D harmonic oscillators.

The total energy of a 1-D harmonic oscillator is written as,

$$\varepsilon = \text{Kinetic energy} + \text{Potential energy}$$

$$= \frac{p^2}{2m} + \frac{1}{2} m\omega^2 x^2,$$

where m, p, ω and x respectively denote mass, momentum, angular frequency and displacement from equilibrium position of an atom. According to classical mechanics, energy of an oscillator can take any value between zero and infinity. To calculate the average energy: we note that energy distribution of a classical system is governed by M-B distribution. The M-B distribution function, which gives the probability that an oscillator will have energy ε, is given by,

$$f(\varepsilon) = A \exp\left[-\frac{\varepsilon}{k_B T}\right],$$

where A is an arbitrary constant and k_B is Boltzmann constant. Thus, the average energy of a 1-D oscillator will be given by,

$$<\varepsilon>=\frac{\int_0^\infty \varepsilon\, f(\varepsilon)\, d\varepsilon}{\int_0^\infty f(\varepsilon)\, d\varepsilon}=k_B T.$$

Since a solid having N atoms is equivalent to 3N one-dimensional harmonic oscillators, its total energy is given by,

$$E = 3N\, k_B T$$

and the heat capacity is given by,

$$C_v = \left(\frac{\partial E}{\partial T}\right)_v = 3Nk_B.$$

If we consider one mole of a substance, the total number of atoms will be equal to the Avogadro number, N_A, and the heat capacity per mole or the molar heat capacity will be equal to,

$$C_V = 3N_A k_B = 3R \text{ mole}^{-1},$$

where $R = N_A k_B$ is the universal gas constant. Let us pause for a while and try to understand the significance of this result. It shows that heat capacity of solids is constant, equal to 3R and independent of temperature. This explains Dulong-Petit law, which holds rather well at room temperature and above. However, as it became possible to carry out measurements at lower temperatures, it was observed that heat capacity decreases with temperature for almost all solids. Physically, it implies that as temperature decreases, the atomic motion is constrained and at absolute zero, all motion should cease. The classical theory, which predicts temperature independent heat capacity, fails to explain why heat capacity tends to zero as temperature approaches absolute zero and is not consistent with the third Iaw of thermodynamics. This posed an interesting challenge to the physicists in the twentieth century. First successful attempt to explain this behavior was made by Einstein on the basis of quantum ideas.

Einstein Theory

Einstein developed the quantum theory of heat capacity and proposed that atoms in a solid should be treated as Planck oscillators characterized by a constant frequency. This implies that energy (ε_n) of a quantum oscillator is nhv, where n is a positive integer, h is Planck constant and v is frequency of oscillation. Note that Einstein oscillate, an oscillator must assumed that all oscillators vibrate with the same frequency.

According to Einstein's theory, a solid made of N atoms is equivalent to 3N, 1-D independent harmonic oscillators.

To obtain an expression for the heat capacity, Einstein assumed that the probability of an oscillator possessing energy nhv, when the system is at temperature T, is proportional to $\exp\left[-\frac{nhv}{k_B T}\right]$. That is, Einstein also used M-B distribution law probably because the oscillators are distinguishable from heat each other due to their location at different lattice sites. Thus, in Einstein's theory, we analyses the statistical behavior of quantum oscillators using classical (M-B) distribution law. Since the energy is quantized, we obtain the average energy of an oscillator by summation, rather than by integration. Under these conditions, we may write the average energy of a quantum harmonic oscillator as:

$$<\varepsilon>=\frac{\sum_{n=0}^{\infty}\varepsilon_n \exp\left[-\frac{\varepsilon_n}{k_B T}\right]}{\sum_{n=0}^{\infty}\exp\left[-\frac{\varepsilon_n}{k_B T}\right]}.$$

Substituting $\varepsilon_n = nhv$ in this expression, we get,

$$<\varepsilon>=\frac{hv\sum_{n=0}^{\infty} n\exp\left[-\frac{nhv}{k_B T}\right]}{\sum_{n=0}^{\infty}\exp\left[-\frac{nhv}{k_B T}\right]}.$$

To evaluate the RHS of above equation, we introduce a change of variable and write $\frac{hv}{k_B T}=x.$ Then, expression for average energy simplifies to,

$$<\varepsilon>=\frac{hv\sum_{n=0}^{\infty} ne^{-nx}}{\sum_{n=0}^{\infty} e^{-nx}}=\frac{hv\left[e^{-x}+2e^{-2x}+3e^{-3x}+\ldots\right]}{\left[1+e^{-x}+e^{-2x}+e^{-3x}+\ldots\right]}$$

$$=\frac{hvN}{D}.$$

Now let us first consider the expression in the numerator:

$$N=e^{-x}+2e^{-2x}+3e^{-3x}+\ldots$$

Multiply by e^x on both sides. This gives,

$$e^x N = 1+2e^{-x}+3e^{-2x}+\ldots$$

On subtracting the expression in equation $N=e^{-x}+2e^{-2x}+3e^{-3x}+\ldots$ from that in equation $e^x N=1+2e^{-x}+3e^{-2x}+\ldots$ we get,

$$\left(e^x-1\right)N=1+e^{-x}+e^{-2x}+\ldots$$

The right hand side in this expression is an infinite geometric series with common ratio e^{-x} and is simply identical to the denominator in equation $<\varepsilon>=\frac{hv\sum_{n=0}^{\infty} ne^{-nx}}{\sum_{n=0}^{\infty} e^{-nx}}=\frac{hv\left[e^{-x}+2e^{-2x}+3e^{-3x}+\ldots\right]}{\left[1+e^{-x}+e^{-2x}+e^{-3x}+\ldots\right]}=\frac{hvN}{D}.$

Hence,

$$\frac{N}{D}=\frac{1}{e^x-1}$$

and the expression for average energy of an Einsteinian harmonic oscillator simplifies to,

$$<\varepsilon>=\frac{hv}{e^x-1}=\frac{hv}{\exp\left[\frac{hv}{k_BT}\right]-1}.$$

If a solid consists of 3N one-dimensional oscillators, the expression for its total energy is obtained by multiplying above equation by 3N:

$$E=\frac{3Nhv}{\exp\left[\frac{hv}{k_BT}\right]-1}$$

so that the heat capacity is given by,

$$C_v=3Nk_B\left(\frac{hv}{k_BT}\right)^2\frac{\exp\left[\frac{hv}{k_BT}\right]}{\left(\exp\left[\frac{hv}{k_BT}\right]-1\right)^2}.$$

If we take one mole of the solid, N is replaced by N_A in this expression to obtain molar heat capacity:

$$C_v=3R\left(\frac{hv}{k_BT}\right)^2\frac{\exp\left[\frac{hv}{k_BT}\right]}{\left(\exp\left[\frac{hv}{k_BT}\right]-1\right)^2}.$$

This result shows that heat capacity is a function of temperature - a feature missing in the classical theory. Note that the key to success of Einstein's theory is the concept of quantisation of energy.

To compare the predictions of this theory with experimental results, we need to know the frequency of the oscillators so that the theoretical value of C_v at a given temperature can be obtained using above equation. The question is: How to estimate the value of frequency? This difficulty is overcome by choosing such a value of frequency for which above equation fits the experimental curve. That is, it is assumed that each solid has a characteristic frequency, called Einstein frequency, V_E. It is customary to write,

$$hv_E=k_BT_E,$$

where T_E denotes the Einstein temperature. Like the characteristic frequency, T_E will have a fixed value for a given solid. In terms of Einstein temperature, the expression in equation,

$$C_v=3R\left(\frac{hv}{k_BT}\right)^2\frac{\exp\left[\frac{hv}{k_BT}\right]}{\left(\exp\left[\frac{hv}{k_BT}\right]-1\right)^2}$$ for heat capacity takes the form,

$$C_v = 3R\left(\frac{T_E}{T}\right)^2 \frac{\exp\left[\frac{T_E}{T}\right]}{\left(\exp\left[\frac{T_E}{T}\right]-1\right)^2}.$$

C_v is a function of (T_E/T). You may now like to investigate the nature of temperature dependence. For copper, a plot of above equation is shown in figure. This relation reproduces all the general features of the observed curve. Now, let us consider the nature of temperature dependence in two limiting cases: at high temperatures as high temperatures ($T >> T_E$) and at low temperatures ($T << T_E$).

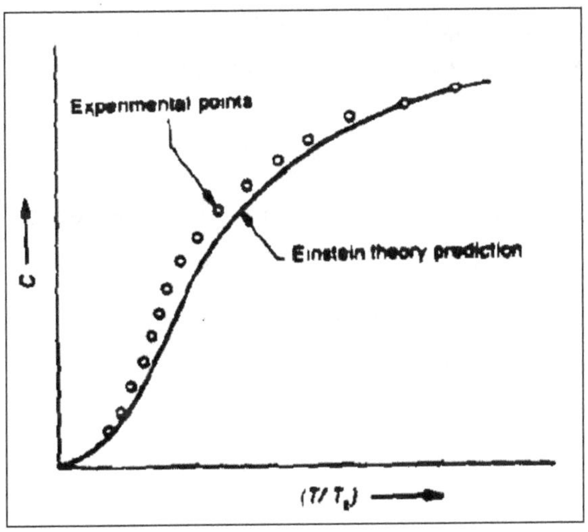

Heat capacity versus temperature plot for copper at low temperatures as predicted by Einstein's theory and as obtained experimentally,

$$\exp(x) = 1 + x + \frac{x^2}{2!} + \frac{x^2}{3!} + \ldots$$

For $x << 1$, so that $\exp(x) - 1 = x + \frac{x^2}{2!} + \frac{x^2}{3!} + \ldots$

High Temperature Limit ($T>>T_E$)

In this case, $(T_E/T) << 1$ and we can write the exponential term as,

$$\exp\left[\frac{T_E}{T}\right] = 1 + \frac{T_E}{T} + \frac{1}{2!}\left(\frac{T_E}{T}\right)^2 + \ldots$$

Using this expansion in equation $C_v = 3R\left(\frac{T_E}{T}\right)^2 \frac{\exp\left[\frac{T_E}{T}\right]}{\left(\exp\left[\frac{T_E}{T}\right]-1\right)^2}$ we get,

$$C_v = 3R\left(\frac{T_E}{T}\right)^2 \frac{\left(1+\frac{T_E}{T}+\frac{1}{2!}\left(\frac{T_E}{T}\right)^2+\ldots\right)}{\left[\frac{T_E}{T}+\frac{1}{2!}\left(\frac{T_E}{T}\right)^2+\frac{1}{3!}\left(\frac{T_E}{T}\right)^3+\ldots\right]^2}$$

You can readily simplify this expression to obtain,

$$C_v = 3R\left[1-\frac{1}{12}\left(\frac{T_E}{T}\right)^2+\ldots\right]$$

For many materials, molar heat capacity at $T_E \sim 200$ K arid high temperatures will approach 3R. From this we can conclude that Einstein's theory provides sound basis for validity of Dulong-Petit law.

Low Temperature Limit ($T << T_E$)

For $(T_E/T) >> I$, we can ignore unity in comparison to the exponential term in the denominator of equation $C_v = 3R\left(\frac{T_E}{T}\right)^2 \frac{\exp\left[\frac{T_E}{T}\right]}{\left(\exp\left[\frac{T_E}{T}\right]-1\right)^2}$ and write:

$$C_v = 3R\left(\frac{T_E}{T}\right)^2 \exp\left[-\frac{T_E}{T}\right].$$

We know that exponential function changes more rapidly than any power of x. Therefore, Einstein theory predicts that heat capacity at low temperatures will decay exponentially and approach zero as temperature tends to absolute zero. This prediction of Einstein led to a lot of experimental activity on measurement of heat capacity at low temperatures and results for several solids were reported. Figure depicts the temperature variation of heat capacity of copper as predicted by Einstein's theory along with experimentally observed curve. Note that, at low temperatures, the exponential decrease predicted by Einstein is much faster compared to experimental result. We may, therefore, conclude that Einstein's theory is qualitative, particularly at low temperatures, as it fails to explain experimental observations near absolute zero for most solids.

The discrepancy between experimental curve and Einstein's predication could possibly be due to Einstein's assumption that all the atomic oscillators in a solid are identical, independent of each other and vibrate with the same characteristic frequency. Indeed, arguing along these lines, Debye modified Einstein's oversimplified model by assigning a range of frequencies to atomic oscillators.

Debye Theory

To overcome the limitations of Einstein theory, Debye suggested that it is unrealistic to assume that all atoms in a solid vibrate with one frequency and independent of each other. According to him, motion of one atom influences the motion of all other atom. Debye, therefore, proposed that a solid should be treated as a collection of coupled harmonic oscillators having a spectrum of frequencies. However, Debye assumed that each mode (way of oscillation) of vibration of a crystal lattice is equivalent to a harmonic oscillator whose average energy is given by equation,

$$<\varepsilon>=\frac{hv}{e^x-1}=\frac{hv}{\exp\left[\frac{hv}{k_BT}\right]-1}.$$

You have seen that in Einstein's theory, the total energy of a solid is obtained by multiplying average energy of an oscillator by the number of modes (that is, 3N). Bu writing expression for total energy of a solid for Debye model is not straightforward because:

- Enumeration of number of modes is quite difficult for a system of coupled oscillators.
- Exact knowledge of the distribution of frequencies among different modes is not easy to obtain.

Thus, to write an expression for total energy, Debye introduced a parameter called density of modes or density of states, which gives the number of modes per unit frequency range. We denote it by the symbol g(v) and write the total vibrational energy of a solid as:

$$E=\int\langle\varepsilon(v)\rangle g(v)dv,$$

where integral is over all the allowed values of frequencies and g(v) dv gives the number of modes whose frequencies are in the range v and v + dv.

Debye assumed that all waves (generated by the system of coupled oscillators constituting a solid) with frequencies in a certain range, $0 < v < v_{max}$, could propagate through the solid. And for a 3-D solid, he showed that the number of allowed vibrational modes between frequencies v and v + dv is given by,

$$g(v)dv=\frac{9N}{V_{max}^3}v^2dv \qquad 0<v<v_{max}$$
$$=0. \qquad v>v_{max}$$

The total number of modes is defined by,

$$\int_0^{v_{max}} g(v)dv=3N,$$

where v_{max} is the maximum allowed frequency, also called Debye frequency. On substituting for average energy and density of states from equations $<\varepsilon>=\frac{hv}{e^x-1}=\frac{hv}{\exp\left[\frac{hv}{k_BT}\right]-1}$ and $g(v)dv=\frac{9N}{V_{max}^3}v^2dv \ldots v>v_{max}$ respectively in equation $E=\int\langle\varepsilon(v)\rangle g(v)dv$, we can write the total energy of the solid as,

$$E=\frac{9Nh}{v_{max}^3}\int_0^{v_{max}}\frac{v^3dv}{\exp\left[\frac{hv}{k_BT}\right]-1}.$$

To simplify, we rewrite above equation in terms of the energy of an elastic wave by defining,

$$\xi = hv$$

so that, $d\xi = hdv$ and $v^3dv = \frac{\xi^3 d\xi}{h^4}$.

On inserting these results in equation $E = \frac{9Nh}{v_{max}^3}\int_0^{v_{max}} \frac{v^3 dv}{\exp\left[\frac{hv}{k_B T}\right]-1}$ we get,

$$E = \frac{9N}{\xi_{max}^3}\int_0^{\xi_{max}} \frac{\xi^3 d\xi}{\exp\left[\frac{\xi}{k_B T}\right]-1}$$

where $_{max}$ is the maximum energy.

The heat capacity is therefore given by,

$$C_v = \left(\frac{dE}{dT}\right)_v = \frac{9N}{\xi_{max}^3}\int_0^{\xi_{max}} \frac{\xi^4 d\xi \exp\left[\frac{\xi}{k_B T}\right]}{k_B T^2\left[\exp\left[\frac{\xi}{k_B T}\right]-1\right]^2}$$

$$= \frac{9N}{(k_B T_D)^3}\int_0^{k_B T_D} \frac{\xi^4 \exp\left[\frac{\xi}{k_B T}\right] d\xi}{k_B T^2\left[\exp\left[\frac{\xi}{k_B T}\right]-1\right]^2}.$$

where $T_D = \frac{\xi_{max}}{k_B}$ is Debye temperature.

Integrating by parts, we get,

$$C_v = \frac{9N}{(k_B T_D)^3}\frac{1}{T}\left[-\frac{(k_B T_D)^4}{\left[\exp\left[{}^{T_D}/_{T}\right]-1\right]} + \int_0^{k_B T_D} \frac{4\xi^3 d\xi}{\left[\exp\left[\frac{\xi}{k_B T}\right]-1\right]}\right].$$

The expression for integration by parts is $\int f(x)g'(x)dx = f(x)g(x) - \int g(x)f'(x)dx$ where $f'(x) = \frac{df(x)}{dx}$ and $g'(x) = \frac{dg(x)}{dx}$.

To simplify this expression, we introduce a dimensionless variable through the $x = \frac{\xi}{k_B T}$. As $\xi \to 0, x \to 0$ and the lower limit of integration will remain zero. However, as $\xi \to \xi_{max}$, $x \to \frac{T_D}{T}$ so that $x_{max} = \left(\frac{T_D}{T}\right)$ gives the upper limit of integration. Thus, we can write,

$$C_v = 3Nk_B f_D\left(\frac{T_D}{T}\right),$$

where

$$f_D\left(\frac{T_D}{T}\right) = \frac{1}{x_{max}^3}\left[12\int_o^{x_{max}} \frac{x^3\,dx}{[\exp[x]-1]} - \frac{3x_{max}^4}{[\exp[x_{max}]-1]}\right].$$

Equation $C_v = 3Nk_B f_D\left(\frac{T_D}{T}\right)$, is the Debye formula for heat capacity of solids. Here f_D is known as Debye function and cannot be evaluated analytically for general values of x. Therefore, the value of heat capacity has to be obtained numerically at different temperatures. However, the expression for f_D can be evaluated in the limiting cases.

High Temperature Limit ($T >> T_D$ or $x_{max} << 1$)

In high temperature limit, x (= T_D/T) is very small and we may write $\exp[x] \cong 1+x$, so that $\exp[x]-1 = x$, if we ignore powers of x higher than one. Thus, in the first order approximation, Equation $C_v = 3Nk_B f_D\left(\frac{T_D}{T}\right)$, reduces to,

$$C_v = 9Nk_B \frac{1}{x_{max}^3}\left[12.\frac{x_{max}^3}{3} - 3x_{max}^3\right]$$
$$= 3Nk_B$$

This shows that Debye theory conforms to Dulong-Petit law at high temperatures. This convergence can be understood physically. At high temperatures, there is sufficient thermal energy available in the solid so that each and every mode receives energy equal to classical value k_BT. And since the density of states is 3N, we write total energy of the solid as,

$$E = 3N k_B T$$

which gives $C_v = 3Nk_B$. However, from the above, you may be justified in concluding that it is not a special achievement of the Debye's theory because high temperature heat capacity is pretty well explained by classical as well as Einstein's theories.

Low Temperature Limit ($T << T_D$ or $x_{max} >> I$)

For $\frac{T}{T_D} << 1$, the exponential term in equation $C_v = 3Nk_B f_D\left(\frac{T_D}{T}\right)$, increases faster than any power of x. Therefore, the quantity $\frac{x_{max}^n}{\exp[x_{max}]-1}$ will tend to zero as $x_{max} \to \infty$. That is, the second term in the Debye function approaches zero. Further, in the integral occurring in Debye function, we may replace x_{max} by infinity without introducing much error. Then equation $f_D\left(\frac{T_D}{T}\right) = \frac{1}{x_{max}^3}\left[12\int_o^{x_{max}} \frac{x^3\,dx}{[\exp[x]-1]} - \frac{3x_{max}^4}{[\exp[x_{max}]-1]}\right]$ simplifies to,

$$f_D(x_{max}) = \frac{12}{x_{max}^3} \int_0^\infty \frac{x^3\,dx}{\exp[x]-1}.$$

Integral in this expression for $f_D\left(\frac{T_D}{T}\right)$ is of the form of a standard integral given as,

$$\int_0^\infty \frac{x^{p-1}\,dx}{\exp[x]-1} = (p-1)! \sum_{n=1}^{\infty} \frac{1}{n^p}.$$

So, we can write,

$$f_D\left(\frac{T_D}{T}\right) = \frac{12}{x_{max}^3} \int_0^\infty \frac{x^{4-1}\,dx}{\exp[x]-1}$$

$$= \frac{12}{x_{max}^3}(3)! \sum_{n=1}^{\infty} \frac{1}{n^4}$$

$$= \frac{72}{x_{max}^3} \sum_{n=1}^{\infty} \frac{1}{n^4}.$$

The summation term in the above expression represents Riemann Zeta function of order 4, that is,

$$\sum_{n=1}^{\infty} \frac{1}{n^4} = \zeta(4)$$

and its value is equal to $\frac{\pi^4}{90}$. Thus, the expression for Debye function simplifies to,

$$f_D\left(\frac{T_D}{T}\right) = \frac{72}{x_{max}^3} \times \frac{\pi^4}{90} = \frac{4}{5}\frac{\pi^4}{x_{max}^3}.$$

Substituting this value of $f_D\left(\frac{T_D}{T}\right)$ in equation $C_v = 3Nk_B f_D\left(\frac{T_D}{T}\right)$, we get,

$$C_v = \frac{12\pi^4}{5} N k_B \left(\frac{T}{T_D}\right)^3$$

This result shows that at low temperatures, heat capacity varies as T^3. This is known as Debye T^3-law. Debye's theory successfully explains heat capacity at low temperatures for most solids.

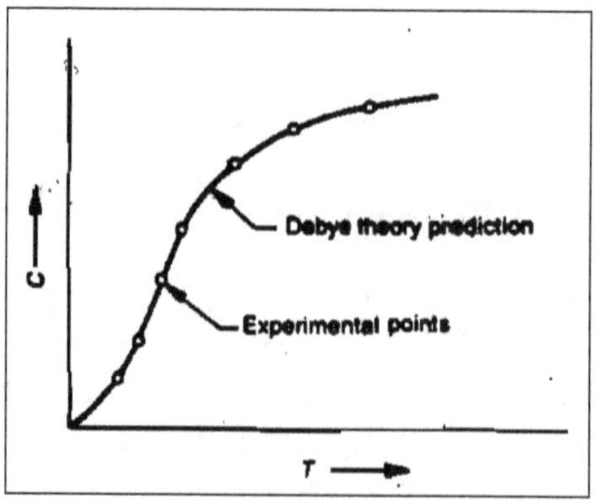

Comparison of temperature variation of heat capacity as per Debye's theory and experimental observation.

Debye's theory uses phenomenological model for frequency distribution function. It is based on the approximation that solids can be treated as isotropic homogeneous media whose dispersion relations are linear. The dispersion curves for monatomic and diatomic crystal lattices are not linear for the entire frequency range. At some stage, you may like to consider such refinements in Debye model by modifying the expression for density of states.

In metals, thermal energy is absorbed by crystal lattice as well as by free electrons. Therefore, we should also expect an electronic contribution to the heat capacity. At normal temperatures, this contribution is negligible compared to the contribution due to lattice vibrations and may be ignored. At low temperatures, however, the lattice heat capacity decreases very rapidly (Debye T^3-law), whereas the heat capacity due to electrons decreases linearly with T. As a consequence, electronic heat capacity dominates. Therefore, heat capacity of a metal at low temperatures will be given by the expression,

$$C = aT + bT^3,$$

where a and b are constants.

We have consistently assumed that vibrating atoms in a solid are analogous to harmonic oscillators. This is known as harmonic oscillator approximation and it has served us well in case of heat capacity. There are, however, many other thermal properties of solids, such as thermal expansion, which cannot be explained on the basis of harmonic approximation. To explain those thermal properties, we must take into account what is known as a harmonic interactions between the atoms of a solid.

Lattice Vibrations

When the lattice is at equilibrium each atom is positioned exactly at its lattice site. Now suppose that an atom displaced from its equilibrium site by a small amount. Due to force acting on this atom, it will tend to return to its equilibrium position. This results in lattice vibrations. Due to interactions between atoms, various atoms move simultaneously, so we have to consider the motion of the entire lattice.

One-Dimensional Lattice

For simplicity we consider, first, a one-dimensional crystal lattice and assume that the forces between the atoms in this lattice are proportional to relative displacements from the equilibrium positions.

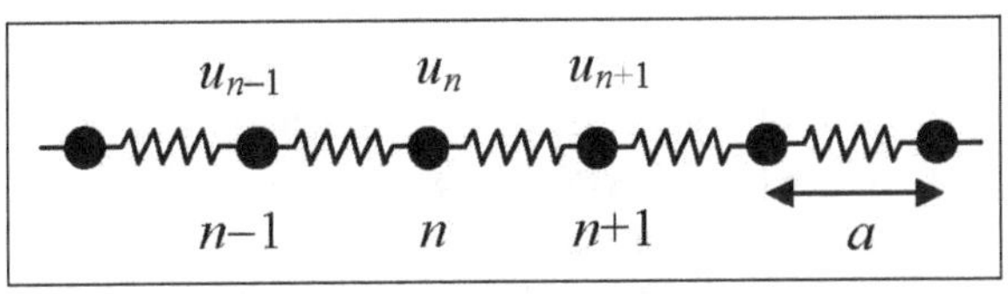

This is known as the harmonic approximation, which holds well provided that the displacements are small. One might think about the atoms in the lattice as interconnected by elastic springs. Therefore, the force exerted on n-the atom in the lattice is given by,

$$F_n = C(u_{n+1} - u_n) + C(u_{n-1} - u_n),$$

where C is the interatomic force (elastic) constant. Applying Newton's second law to the motion of the n-th atom we obtain,

$$M\frac{d^2u_n}{dt^2} = F_n = C(u_{n+1} - u_n) + C(u_{n-1} - u_n) = -C(2u_n - u_{n+1} - u_{n-1}),$$

where M is the mass of the atom. We neglected here by the interaction of the n-th atom with all but its nearest neighbors. A similar equation should be written for each atom in the lattice, resulting in N coupled differential equations, which should be solved simultaneously (N is the total number of atoms in the lattice). In addition the boundary conditions applied to the end atom in the lattice should be taken into account.

Now let us attempt a solution of the form,

$$u_n = Ae^{i(qx_n - \omega t)}$$

where x_n is the equilibrium position of the n-th atom so that x_n=na. This equation represents a traveling wave, in which all the atoms oscillate with the same frequency ω and the same amplitude A and have wavevector q. A solution of above equation is only possible because of the transnational symmetry of the lattice.

Now substituting above equation into equation $M\frac{d^2u_n}{dt^2} = F_n = C(u_{n+1} - u_n) + C(u_{n-1} - u_n) = -C(2u_n - u_{n+1} - u_{n-1})$, and canceling the common quantities (the amplitude and the time-dependent factor) we obtain,

$$M(-\omega^2)e^{iqna} = -C\left[2e^{iqna} - e^{iq(n+1)a} - e^{iq(n-1)a}\right]$$

This equation can be further simplified by canceling the common factor e^{iqna}, which leads to,

$$M\omega^2 = C\left[2 - e^{iqa} - e^{-iqa}\right] = 2C(1 - \cos qa) = 4C\sin^2\frac{qa}{2}.$$

We find therefore the dispersion relation for the frequency,

$$\omega = \sqrt{\frac{4C}{M}}\left|\sin\frac{qa}{2}\right|$$

which is the relationship between the frequency of vibrations and the wavevector q. This dispersion relation have a number of important properties.

1. Reducing to the first Brillouin zone: The frequency of above equation and the displacement of the atoms (equation $u_n = Ae^{i(qx_n - \omega t)}$) do not change when we change q by q+2π/a. This means that

these solutions are physically identical. This allows us to set the range of independent values of q within the first Brillouin zone, i.e.

$$-\frac{\pi}{a} \le q \le \frac{\pi}{a}.$$

Within this range of q the ω versus q is shown in figure.

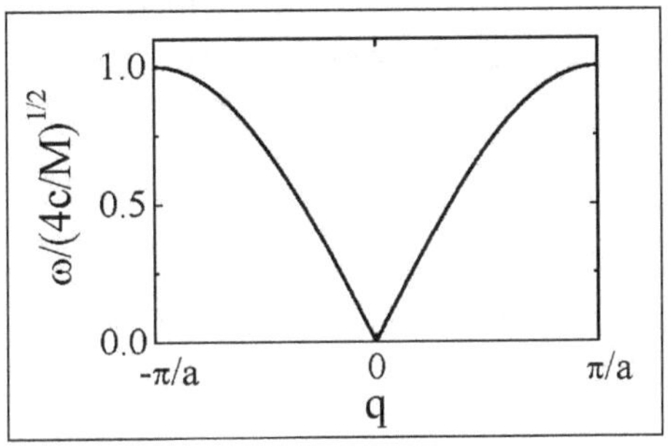

The maximum frequency is $\sqrt{4C/M}$. The frequency is symmetric with respect to the sign change in q, i.e. ω(q) = ω(-q). This is not surprising because a mode with positive q corresponds to the wave traveling in the lattice from the left to the right and a mode with a negative q corresponds to the wave traveling from the right to the left. Since these two directions are equivalent in the lattice the frequency does not change with the sign change in q.

At the boundaries of the Brillouin zone q=±π/a the solution represents a standing wave $u_n = A(-1)^n e^{i-\omega t}$: atoms oscillate in the opposite phases depending on whether n is even or odd. The wave moves neither right nor left.

2. Phase and group velocity: The phase velocity is defined by,

$$v_p = \frac{\omega}{q}$$

and the group velocity by,

$$v_g = \frac{d\omega}{dq}.$$

The physical distinction between the two velocities is that v_p is the velocity of the propagation of the plane wave, whereas the vg is the velocity of the propagation of the wave packet. The latter is the velocity for the propagation of energy in the medium.

For the particular dispersion relation (equation $\omega = \sqrt{\frac{4C}{M}} \left|\sin\frac{qa}{2}\right|$) the group velocity is given by,

$$v_g = \sqrt{\frac{Ca^2}{M}} \cos\frac{qa}{2}.$$

As is seen from above equation the group velocity is zero at the edge of the zone where q=±π/a. Here the wave is standing and therefore the transmission velocity for the energy is zero.

3. Long wavelength limit: The long wavelength limit implies that λ>>a. In this limit qa<<1. We can then expand the sine in equation $\omega = \sqrt{\frac{4C}{M}}\left|\sin\frac{qa}{2}\right|$ and obtain for the positive frequencies:

$$\omega = \sqrt{\frac{C}{M}}qa.$$

We see that the frequency of vibration is proportional to the wavevector. This is equivalent to the statement that velocity is independent of frequency. In this case,

$$v = \frac{}{q} = \sqrt{\frac{}{M}}a.$$

This is the velocity of sound for the one dimensional lattice which is consistent with the expression we obtained earlier for elastic waves.

Diatomic 1D Lattice

Now we consider a one-dimensional lattice with two non-equivalent atoms in a unit cell. It appears that the diatomic lattice exhibit important features different from the monoatomic case. Figure below shows a diatomic lattice with the unit cell composed of two atoms of masses M_1 and M_2 with the distance between two neighboring atoms a.

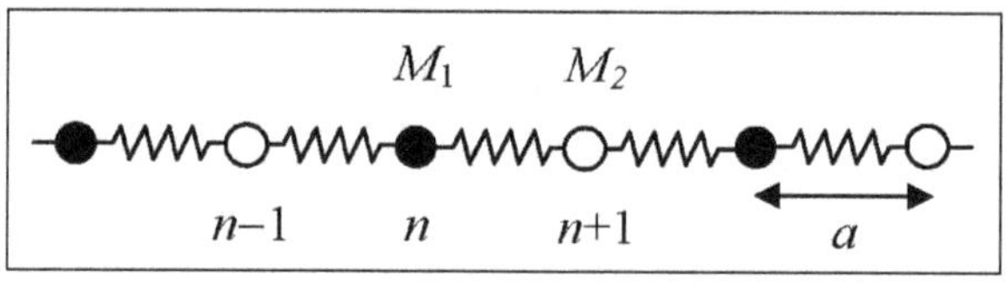

We can treat the motion of this lattice in a similar fashion as for monoatomic lattice. However, in this case because we have two different kinds of atoms, we should write two equations of motion:

$$M_1\frac{d^2u_n}{dt^2} = -C(2u_n - u_{n+1} - u_{n-1})$$

$$M_2\frac{d^2u_{n+1}}{dt^2} = -C(2u_{n+1} - u_{n+2} - u_n)$$

In analogy with the monoatomic lattice we are looking for the solution in the form of traveling mode for the two atoms:

$$\begin{bmatrix} u_n \\ u_{n+1} \end{bmatrix} = \begin{bmatrix} A_1e^{iqna} \\ A_2e^{iq(n+1)a} \end{bmatrix} e^{-i\omega t},$$

which is written in the obvious matrix form. Substituting this solution to equation $M_1\frac{d^2u_n}{dt^2} = -C(2u_n - u_{n+1} - u_{n-1})$ $M_2\frac{d^2u_{n+1}}{dt^2} = -C(2u_{n+1} - u_{n+2} - u_n)$ we obtain,

$$\begin{bmatrix} 2C - M_1\omega^2 & -2C\cos qa \\ -2C\cos qa & 2C - M_2\omega^2 \end{bmatrix}\begin{bmatrix} A_1 \\ A_2 \end{bmatrix} = 0.$$

This is a system of linear homogeneous equations for the unknowns A_1 and A_2. A nontrivial solution exists only if the determinant of the matrix is zero. This leads to the secular equation,

$$\left(2C - M_1\omega^2\right)\left(2C - M_2\omega^2\right) - 4C^2\cos^2 qa = 0.$$

This is a quadratic equation, which can be readily solved:

$$\omega^2 = C\left(\frac{1}{M_1} + \frac{1}{M_2}\right) \pm C\sqrt{\left(\frac{1}{M_1} + \frac{1}{M_2}\right)^2 - \frac{4\sin^2 qa}{M_1 M_2}}$$

Depending on sign in this formula there are two different solutions corresponding to two different dispersion curves.

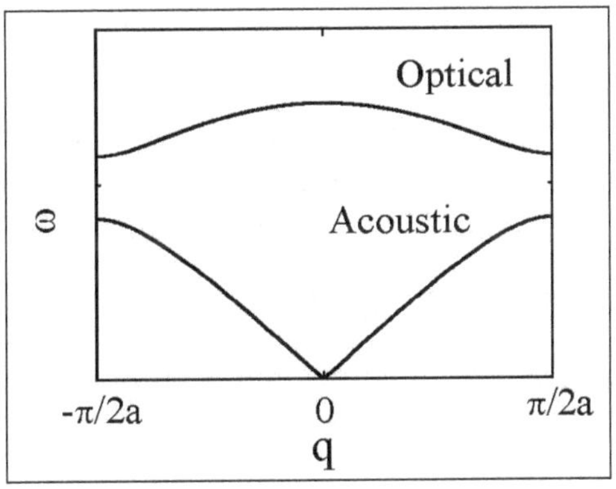

The lower curve is called the acoustic branch, while the upper curve is called the optical branch. The optical branch begins at q=0 and ω=0. Then with increasing q the frequency increases in a linear fashion. This is why this branch is called acoustic: it corresponds to elastic waves or sound. Eventually this curve saturates at the edge of the Brillouin zone. On the other hand, the optical branch has a nonzero frequency at zero q,

$$\omega_0 = \sqrt{2C\left(\frac{1}{M_1} + \frac{1}{M_2}\right)}$$

and it does not change much with q.

The distinction between the acoustic and optical branches of lattice vibrations can be seen most clearly by comparing them at q=0 (infinite wavelength). As follows from Equation $\begin{bmatrix} 2C - M_1\omega^2 & -2C\cos qa \\ -2C\cos qa & 2C - M_2\omega^2 \end{bmatrix}\begin{bmatrix} A_1 \\ A_2 \end{bmatrix} = 0$ for the acoustic branch ω=0 and A_1=A_2. So in this limit the two atoms in the cell have the same amplitude and the phase. Therefore, the molecule oscillates as a rigid body, as shown in figure below for the acoustic mode.

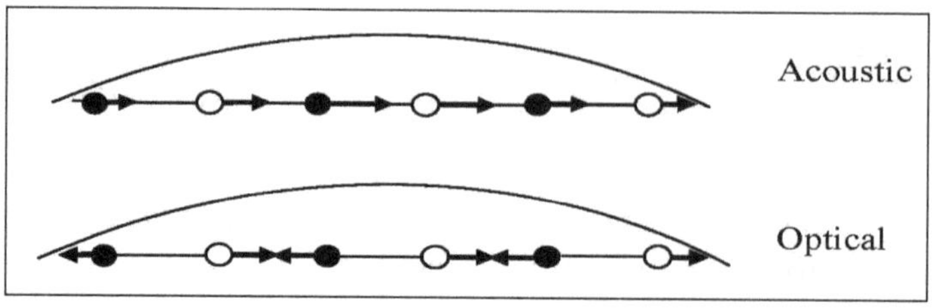

On the other hand, for the optical vibrations, substituting above equation to equation $\begin{bmatrix} 2C-M_1\omega^2 & -2C\cos qa \\ -2C\cos qa & 2C-M_2\omega^2 \end{bmatrix}\begin{bmatrix} A_1 \\ A_2 \end{bmatrix}=0$ we obtain for q=0:

$$M_1A_1 + M_2A_2 = 0.$$

It implies that the optical oscillation takes place in such a way that the center of mass of a molecule remains fixed. The two atoms move in out of phase as shown in above figure. The frequency of these vibrations lies in infrared region which is the reason for referring to this branch as optical.

Three Dimensional Lattice

These considerations can be extended to the three-dimensional lattice. To avoid mathematical details we shall present only a qualitative discussion. Consider, first, the monatomic Bravais lattice, in which each unit cell has a single atom. The equation of motion of each atom can be written in a manner similar to that of equation $M\frac{d^2u_n}{dt^2} = F_n = C(u_{n+1}-u_n)+C(u_{n-1}-u_n) = -C(2u_n - u_{n+1} - u_{n-1})$. The solution of this equation in three dimensions can be represented in terms of normal modes,

$$u = Ae^{i(qr-\omega t)}$$

where the wave vector q specifies both the wavelength and direction of propagation. The vector A determines the amplitude as well as the direction of vibration of the atoms. Thus this vector specifies the polarization of the wave, i.e., whether the wave is longitudinal (A parallel to q) or transverse (A perpendicular to q).

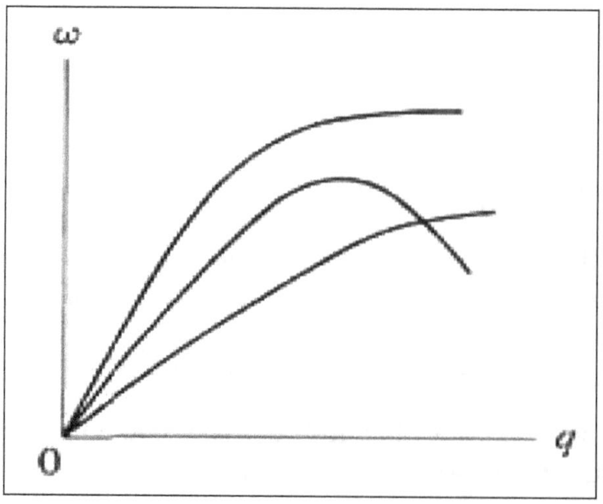

When we substitute above equation into the equation of motion, we obtain three simultaneous equations involving A_x, A_y, and A_z, the components of A. These equations are coupled together and are equivalent to a 3 x 3 matrix equation. The roots of this equation lead to three different dispersion relations, or three dispersion curves. All three branches pass through the origin, which means all the branches are acoustic. This is of course to be expected, since we are dealing with a monatomic Bravais lattice.

The three branches in figure above differ in their polarization. When q lies along a direction of high symmetry - for example, the or directions – these waves may be classified as either pure longitudinal or pure transverse waves. In that case, two of the branches are transverse and one is longitudinal. One usually refers to these as the TA - transverse acoustic and LA – longitudinal acoustic branches, respectively. However, along non-symmetry directions the waves may not be pure longitudinal or pure transverse, but have a mixed character.

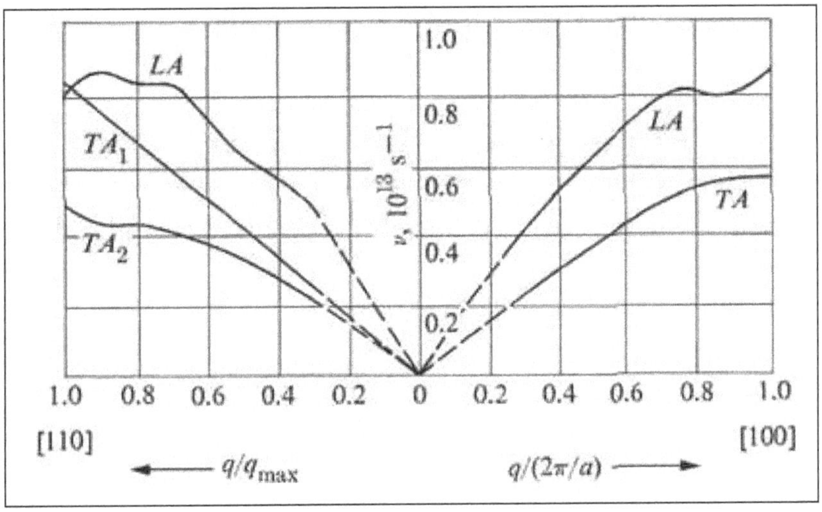

Figure shows the dispersion curves for Al in the and directions. Note that in certain high-symmetry directions, such as the in Al, the two transverse branches coincide. The branches are then said to be degenerate.

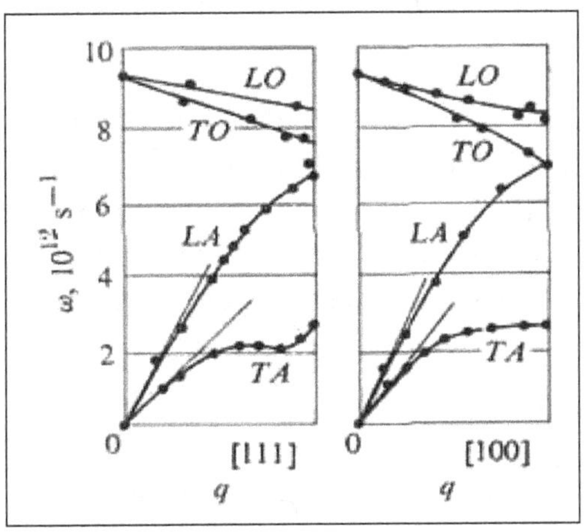

We turn our attention now to the non-Bravais three-dimensional lattice. Here the unit cell contains two or more atoms. If there are s atoms per cell, then on the basis of our previous experience we conclude that there are 3s dispersion curves. Of these, three branches are acoustic, and the remaining (3s −3) are optical. The mathematical justification for this assertion is as follows: We write the equation of motion for each atom in the cell, which results in s equations. Since these are vector equations, they are equivalent to 3s scalar equations, which have 3s roots. Three of these roots always vanish at q = 0, which results in three acoustic branches. The remaining (3s −3) roots, therefore, belong to the optical branches.

The acoustic branches may be classified, as before, by their polarizations as TA_1, TA_2, and LA. The optical branches can also be classified as longitudinal or transverse when q lies along a high-symmetry direction, and one speaks of LO and TO branches. As in the one-dimensional case, one can also show that, for an optical branch, the atoms in the unit cell vibrate out of phase relative to each other. As an example of a non-Bravais lattice, the dispersion curves for Ge are shown in figure. Since there are two atoms per unit cell in germanium, there are six branches: three acoustic and three optical.

Specific Heat Capacity and the Dulong-Petit Law

Since heat is not a thermodynamic quantity, however it becomes so under the constraints of constant volume and constant pressure i.e,

$$dQ|_p = dH$$

and

$$dQ|_v = dE$$

where H and E are enthalpy and internal energy respectively.

The specific heat is defined as the amount of heat energy required to raise the temperature of a unit mass of solid by one degree. Also it is the heat capacity per unit mass. Consequently $(\partial H/\partial T)_p = C_p$ and $(\partial E/\partial T)_v = C$ are then the expressions for heat capacity at constant pressure and heat capacity at constant volume respectively.

For solids and liquids, $C_p \sim C_v$, especially at low temperatures (~0-20K), but even at higher temperatures (~300K), the difference is not more than 2%. With the supply of heat energy to a solid, there is an increase in its internal energy. The increase in the internal energy is manifested as an increase in the vibrations of the atoms about their mean positions and also as an increase in the kinetic energies of the free electrons. The specific heat corresponding to lattice energy is called lattice specific heat.

Table: Specific heats of a few materials.

Material	Specific Heat (J/kg/K)
Lead	126
Copper	381
Iron	438

Aluminium	878
Ethanol	2410
Water	4200
Hydrogen	14150

There are various theories of lattice specific heat. In all the theories the vibrational energy of a crystal containing N number of atoms is considered equivalent to the energy of a 3N harmonic oscillator. The distinction in various theories is the difference in proposed frequency spectrum of the oscillators and the problem regarding calculations of wavelengths and frequencies of the possible modes of vibration in the crystal remains the centric concern.

In order to understand the lattice dynamics and the phenomenon that occur when heat or thermal energy is supplied to a solid and is consequently raised to a temperature (it occurs even at absolute zero), it will be easier to begin with the first law of thermodynamics which states that whenever some amount of heat dQ is given to a system, it results in the increase in energy dE of the system plus the amount of work done, i.e ,

$$dQ = dE + pdV$$

Considering that the work done by the system is of a mechanical nature only. Now, E is determined uniquely by the temperature and volume of the system. Hence,

$$dE = \left(\frac{\partial E}{\partial T}\right)_V dT + \left(\frac{\partial E}{\partial V}\right)_T dV$$

Equation above may be rewritten in the form,

$$dQ = \left(\frac{\partial E}{\partial T}\right)_V dT + \left[\left(\frac{\partial E}{\partial V}\right)_T T + p\right] dV$$

dQ/dT is the general expression for specific heat. Our interest remains in Specific heat at constant volume C_v and specific heat at constant pressure C_p.

From equation above,

$$C_v = \left(\frac{\partial Q}{\partial T}\right)_V = \left(\frac{\partial E}{\partial T}\right)_V$$

The second law of thermodynamics is stated as relation between C_p and C_v in the form,

$$C_p \cdot C_v = -T\left(\frac{\partial V}{\partial T}\right)_P^2 \left(\frac{\partial P}{\partial V}\right)_T$$

Rewriting above equation in terms of the volume expansion coefficient α_v and the compressibility K, defined by,

$$\alpha_v = (1/V)\left(\frac{\partial V}{\partial T}\right)_P \text{ and}$$

$$K = -(1/V)\left(\frac{\partial V}{\partial P}\right)_T$$

$$C_p - C_v = \alpha_v^2 TV / K$$

Since both α_v and K are positive quantities,

$$C_p - C_v \geq 0$$

As depicted from figure below at low temperatures their difference becomes very small and both C_p & C_v go to zero at T = 0. In general, the variation of C_v with time is studied. We assume that a change in volume is not much with a small increase in temperature.

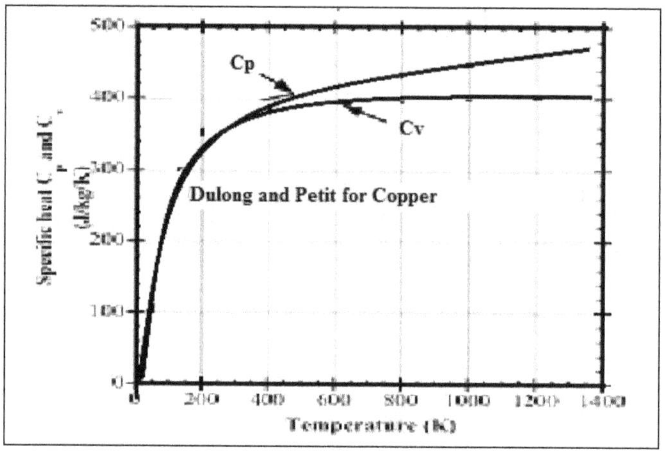

Temperature variation of C_p and C_V.

As the temperature increases from absolute zero, there is a rapid increase in specific heat and it finally levels off at a nearly constant value (6 cal/mole) at high temperature. This is the classical approach put forward as in Dulong Petit's Law.

Dulong and Petit Law

The physical properties of the solids are roughly assumed to be:

- Due to the contributions resulting from the atomic vibrations (as in crystals).
- Due to additional contribution to specific heat from electronic system (as in metals and semiconductors). Although this contribution is relatively small to that of lattice vibrations.

Taking these assumptions into consideration and assuming the behaviour of the atoms of the crystal as to be independent classical harmonic oscillator, the heat capacity of the solid crystal can be calculated simply by finding average thermal energy of a single oscillator and then multiplying it by total number of oscillators, N.

In comparison to the translational motion of the ideal gas molecules, the constituent atoms in a crystal, have almost fixed positions which vibrate about their mean positions executing simple harmonic motion. The total energy of the oscillator at an instant is a composition of its instantaneous potential energy and kinetic energy. Classically a harmonic oscillator vibrating with its natural frequency ω has energy expression as,

$$E = P^2 / 2m + ½\, m\omega^2 \left(x^2 + y^2 + z^2\right)$$

$$E = P^2 / 2m + m\omega^2 q^2 / 2$$

where p is the momentum in kinetic energy term and q is the displacement from equilibrium position in potential energy term.

The average energy of each harmonic oscillator as given by Plank's distribution law is,

$$\langle\varepsilon\rangle = \frac{\int E \exp\left(\frac{-E}{kT}\right) dE}{\int \exp\left(\frac{-E}{kT}\right) dE}$$

here k is Boltzmann constant.

On substituting equation $E = P^2 / 2m + m\omega^2 q^2 / 2$ in above equation we get,

$$\langle\varepsilon\rangle = \frac{\int\int (P2/2m + m\omega 2q2/2) \exp\left(\frac{-P2}{2mkT}\right) \exp\left(\frac{-m\omega 2q2}{2kT}\right) dpdq}{\int\int \exp\left(\frac{-P2}{2mkT}\right) \exp\left(\frac{m\omega 2q2}{2kT}\right) dpdq}$$

$$= \frac{\int\int (P2/2m) \exp\left(\frac{-P2}{2mkT}\right) \exp\left(\frac{-m\omega 2q2}{2kT}\right) dpdq}{\int\int \exp\left(\frac{-P2}{2mkT}\right) \exp\left(\frac{-m\omega 2q2}{2kT}\right) dpdq} + \frac{\int\int (m\omega 2q2/2) \exp\left(\frac{-P2}{2mkT}\right) \exp\left(\frac{-m\omega 2q2}{2kT}\right) dpdq}{\int\int \exp\left(\frac{-P2}{2mkT}\right) \exp\left(\frac{-m\omega 2q2}{2kT}\right) dpdq}$$

$$= \frac{\int_{-\infty}^{\infty} (P2/2m) \exp\left(\frac{-P2}{2mkT}\right) dp}{\int_{-\infty}^{\infty} \exp\left(\frac{-P2}{2mkT}\right) dp} + \frac{\int_{-\infty}^{\infty} (m\omega 2q2/2) \exp\left(\frac{-m\omega 2q2}{2kT}\right) dq}{\int_{-\infty}^{\infty} \exp\left(\frac{-m\omega 2q2}{2kT}\right) dq}$$

Substituting A^2 for $P^2/2mkT$ and B^2 for $m\omega^2q^2/2kT$, on further simplification we arrive at,

$$\langle\varepsilon\rangle = \frac{kT\int_{-\infty}^{\infty} A^2 \exp\left(-A^2\right) dA}{\int_{-\infty}^{\infty} \exp\left(-A^2\right) dA} + \frac{kT\int_{-\infty}^{\infty} B^2 \exp\left(-B^2\right) dB}{\int_{-\infty}^{\infty} \exp\left(-B^2\right) dB}$$

Making use of standard integrals as under,

$$\int_{-\infty}^{\infty} A^2 \exp\left(-A^2\right) dA = \sqrt{\pi}/2$$

And

$$\int_{-\infty}^{\infty} \exp\left(-A\ \right) dA = \sqrt{\pi}$$

We arrive at the following result,

$$\langle \varepsilon \rangle = \frac{kT\sqrt{\pi}/2}{\sqrt{\pi}} + \frac{kT\sqrt{\pi}/2}{\sqrt{\pi}} = kT$$

This gives the vibrational energy of one harmonic oscillator. To obtain total vibrational energy of the crystal which has N independent harmonic oscillators in three dimensions, we have,

$$\begin{aligned} &\epsilon = 3N\langle \epsilon \rangle \\ \text{or } &\epsilon = 3NkT \end{aligned}$$

therefore,

$$C_v = \left(\frac{\partial E}{\partial T}\right)_V = 3Nk = 3R \text{ per mole} = 5.96 \text{ cal/ mole.}$$

This is called Dulong Petit's classical law of specific heat. This turns out to be that the heat capacity is independent of temperature and has a constant value at room temperature.

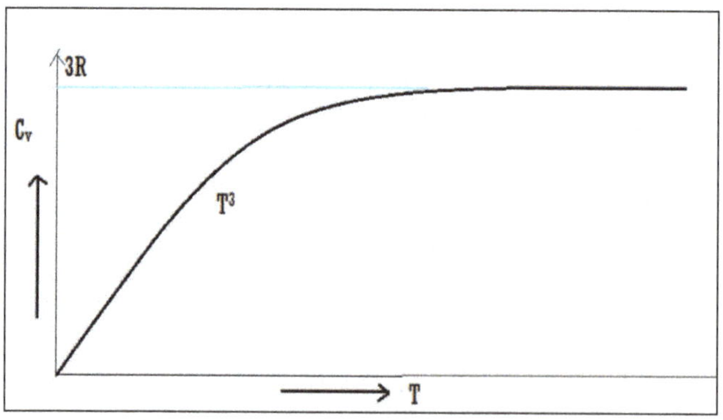

Variation of specific heat with temperature.

Shortfall of Classical Theory

The classical theory gives a good idea of energy contributions to the total energy of the system as long as the volume remains constant, which is given in above equation. The specific heat at constant volume as obtained for a solid containing N number of atoms is given as,

$$C_v = \left(\frac{\partial E}{\partial T}\right)_V = 3Nk = 3R = 5.96 \text{ cal / mole.}$$

This is the level off value of the experimental results at high temperature.

Equation $\varepsilon = 3N\langle\varepsilon\rangle$ or $\varepsilon = 3NkT$ is a frequency independent expression and also it depends only on temperature. The results obtained here are in quantitative agreement with those obtained from experiments but at higher temperatures only. This does not explain the decrease of specific heat at low temperatures as observed for all solids. The discrepancy to a certain extent is removed by using quantum theory but before that it is important to understand the concept of quantization of elastic waves and the concept of phonons.

Quantization of Elastic Waves

The atoms in solids vibrate about their equilibrium positions and these lattice vibrations can be expressed in the form of waves, like repetitive and systematic sequence of atomic displacements that can be longitudinal or transverse or even a combination of both. The energy of a lattice vibration is quantized and the quantum of energy is called 'Phonon', analogous to quanta of electromagnetic energy – photon.

Vibrations of a linear lattice connected through springs and their particle motion can be quantized in a similar way as that for a harmonic oscillator or a combination of coupled harmonic oscillators. The Hamiltonian for a harmonic oscillator is,

$$H = (1/2M)\, p^2 + ½\, Cx^2$$

With energy eigen values for n = 1, 2, 3,....

Hence average thermal energy of the oscillator is,

$$\epsilon_n = (n + ½)\hbar\omega$$

Where ω is the angular frequency.

The term ½ ħω is called zero point energy of the mode. Thus thermal lattice vibrations are thermally excited phonons. Thermal conduction in non-metallic crystals is a consequence of annihilation or creation of a phonon. The energy of phonons is ~0.1 ev.

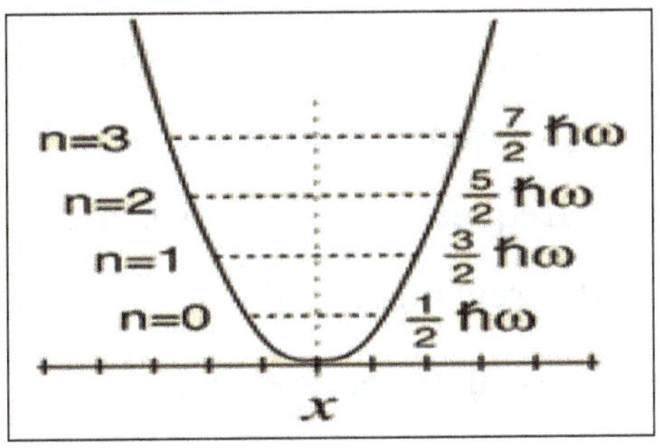

Discrete energy levels of harmonic oscillator.

If the particles have zero spin then these follow Bose – Einstein statistics and the probability of such particles having energy E is given as,

$$f_{BE} = \frac{1}{\exp\left\{\frac{E-\alpha}{kBT}\right\} - 1}$$

if the particles are not constant with time then α = 0.

$$f_{BE} = \frac{1}{\exp\left\{\frac{E}{kBT}\right\} - 1}$$

The following observations confirm the experimental evidence for the quantization of lattice vibrational energy:

- At absolute zero the lattice contribution of heat energy always approaches zero.
- The neutrons and x rays are scattered in elastically by crystals and the changes in energy and momentum correspond to creation or absorption of phonons.

Einstein's and Debye's Theory

C_v falls below the limiting value of 3NR, where R is universal gas constant. It was assumed that some of the oscillators stop vibrating as the temperature is lowered, and Einstein supported the logic on the basis of Plank's hypothesis on quantization of energy. As a consequence, the there is a decrease in the number of oscillators oscillating at lower temperatures (below room temperatures). An oscillator must possess at least one quantum of energy (phonon) to oscillate at all. But the oscillators which do not have even single quantum will not oscillate and have zero energy according to Plank's hypothesis. But Einstein visualized some ingenuity in this and rather assumed a rigid lattice made of identical oscillators and all of them to be oscillating with same frequency.

Einstein's Theory

To understand the dip in specific heat curve at low temperatures Einstein employed a physical model (although oversimplified to arrive at desired results), but it indicated that the problem had its solution in quantum approach. The suggested model assumes the following:

- Lattice containing N atoms is equivalent to 3N harmonic oscillators.
- Atoms vibrate independently of each other.
- All atoms vibrate with same frequency (as all have same environment).
- Harmonic oscillators have discrete quantized energy levels represented as $\epsilon_n = nh\nu$.

The average energy of the harmonic oscillator is given as,

$$\langle \varepsilon \rangle = \frac{\int_0^\infty \varepsilon e^{\varepsilon/kT}}{\int_0^\infty e^{\varepsilon/kT}} d\varepsilon$$

Replacing the integral with summation above equation takes the form,

$$\langle \varepsilon \rangle = \frac{\sum_{n=0}^{\infty} nh\nu \, e^{-nh\nu/kT}}{\sum_{n=0}^{\infty} e^{-nh\nu/kT}}$$

Simplifying the denominator,

$$\sum_{n=0}^{\infty} e^{-nh\nu/kT} = \left(1-e^{-nh\nu/kT}\right)^{-1}$$

Differentiating above equation w.r.t 1/KT,

$$\frac{\partial S}{\partial\left(\frac{1}{kT}\right)} = -\sum_{n=0}^{\infty} nh\nu\, e^{-nh\nu/kT} = \frac{-h\nu\, e^{-nh\nu/kT}}{\left(1-e^{-nh\nu/kT}\right)^{2}}$$

From above three equations, the expression obtained is same as given in Plank's relation for average energy of an oscillator at a temperature viz.

$$\langle\varepsilon\rangle = \frac{h\nu}{e^{h\nu/kT}}$$

Equation above is frequency dependent expression unlike equation $\langle\varepsilon\rangle = \frac{\int_0^{\infty} \varepsilon e^{a/kT}}{\int_0^{\infty} e^{\varepsilon/kT}} d\varepsilon$. Thus energy of solid containing N atoms is,

$$E = 3N\langle\varepsilon\rangle = 3N\frac{h\nu}{e^{h\nu/kT}}$$

And specific heat at constant volume is,

$$C_v = \frac{\partial E}{\partial T} = 3R\left(h\nu/kT\right)^{2} \frac{e^{h\nu/kT}}{\left(e^{-nh\nu/kT}-1\right)^{2}}$$

The possible energy levels of a harmonic oscillator according to quantum mechanics is given as,

$$\varepsilon_n = \left(n+1/2\right) h\nu \qquad n = 0,1,2..$$

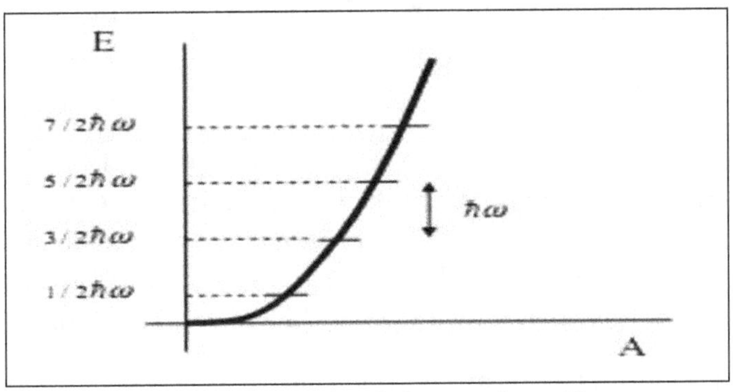

Discrete energy levels of a harmonic oscillator.

So, $\langle\varepsilon\rangle = h\nu/2 + h\nu/e^{h\nu/kT} - 1$

Now for T=0, $\langle\varepsilon\rangle = h\nu/2$ and not zero, so the first term in above equation is referred to as 'zero point energy' and according to quantum mechanics the atoms possess vibrational energy even at absolute zero.

The energy hν of the oscillator indicates the difference between the allowed energy states of the given oscillator. In case the difference is small, ie hν/kT is negligible in comparison with unity then Cv will be equal to 3R at all finite temperatures. However at low temperatures, unity is negligible as compared to hν/kT and Cv becomes proportional to $e^{-h\nu/kT}$.

Heat capacity decreases exponentially at low temperatures. This is in fact an indirect evidence of quantization of energy as proposed by Plank's hypothesis. At kT>>hν, above equation reduces to classical result value, however at lower temperatures specific heat decreases. It is better understood by rewriting equation $E = 3N\langle\varepsilon\rangle = 3N\frac{h\nu}{e^{h\nu/kT}}$ in terms of θ_E, (Einstein temperature) as,

$$C_v / 3R = (\theta_E / T)^2 \frac{e^{\theta E/T}}{\left(e\theta E^{/T} - 1\right)^2} = F_E(\theta_E / T)$$

Where F_E is Einstein's function determining the ratio of specific heat at temperature T and classical value 3R. For T >> θ_E,

$$C_v = 3R,$$

ie Dulong Petit law is obeyed.

For T << θ_E,

$$C_v = 3R(\theta_E /)^2 e^{-\theta_E/T}$$

the specific heat is proportional to θ_E/T.

This is in good agreement with experimental results at higher temperatures but fails to give exact values at very low temperatures. The reason for this discrepancy must be sought in the oversimplification of the model assumed by Einstein considering the vibrations to be independent of each other. A modification suggesting a solution is made by Debye known as Debye's approximation theory.

Debye's Theory

Although Einstein's assumption that the atoms vibrate independently of each other in the lattice with a constant frequency had been very successful in explaining the exponential decrease in heat capacity at low temperatures but the predicted decrease was much sharper than the experimentally observed values. Further Einstein's model failed to explain and fit T^3 dependence. To explain the variation of specific heat of solids with temperature, Debye relied on the following assumptions:

- The vibrations of all atoms are coupled together.
- Solid acts as an isotropic elastic continuum.
- Atomic vibrations produce a continuous spectrum of frequencies.
- The frequency is same for longitudinal and transverse vibrations produced in solids.
- Beyond a certain frequency V_D (Debye frequency), no vibrations are produced.

Consider an elastic wave u(x, y, z, t) propagating through a medium. Writing its wave equation as,

$$\frac{\partial^2 u}{\partial x^2}+\frac{\partial^2 u}{\partial y^2}+\frac{\partial^2 u}{\partial z^2}=\frac{1}{v^2}\frac{\partial^2 u}{\partial t^2}$$

where u is the displacement and v is the wave velocity. It is assumed that the medium is continuous and in the form of a cube of dimension L such that it's all faces are fixed having a solution of the form,

$$u = A\sin\left(\frac{n_x \pi x}{L}\right)+\sin\left(\frac{n_y \pi y}{L}\right)+\sin\left(\frac{n_z \pi z}{L}\right)\cos 2\pi \nu t$$

n_x, n_y, n_z are integers and is the frequency of the wave.

Double partial differentiations of above equation w.r.t x,y,z and t and substituting in equation $\frac{\partial^2 u}{\partial x^2}+\frac{\partial^2 u}{\partial y^2}+\frac{\partial^2 u}{\partial z^2}=\frac{1}{v^2}\frac{\partial^2 u}{\partial t^2}$ gives,

$$n_x^2+n_y^2+n_z^2=\left[\frac{2\nu L}{v}\right]^2$$

$$\text{or, } R^2=n_x^2+n_y^2+n_z^2=\left[\frac{2\nu L}{v}\right]^2$$

$$R=\frac{2\nu L}{v} \quad \text{and, } dR=(2L/v)d\nu$$

Equation above shows that the different values of frequencies are quantized. Different no. of modes of vibration can be determined using this equation. Let Z(ν) is the number of modes of vibration in frequency interval ν and ν+ dν Therefore, the number of modes of vibration is,

$$Z(\nu)d(\nu)=\frac{1}{8}\left[\frac{4}{3}\pi(R+dR)^3\frac{4}{3}\pi R^3\right]$$

$$Z(\nu)d(\nu)=\frac{4\pi V}{v^3}\nu^2\,d\nu$$

Where V = L³.

Equation above gives the possible modes of vibration also known as Density of states of elastic continuous medium. In deriving equation $R=\frac{2\nu L}{v}$ and, $dR=(2L/v)d\nu$ it has been assumed that the velocity v is irrespective of longitudinal or transverse nature of the wave propagating in the medium. But actually the frequency is associated with one longitudinal mode and two transverse modes; hence equation above gets modified as,

$$Z(\nu)d(\nu)=\left[\frac{2}{v_t^3}+\frac{1}{v_l^3}\right]4\pi V\nu^2\,d\nu$$

Equation above gives the total number of modes of vibration in elastic continuum medium. Debye proposed that a solid is a continuously vibrating medium giving rise to a spectrum of frequencies (wavelengths comparable to inter-atomic separations) called Debye's continuum. Then the total energy expression becomes,

$$\varepsilon = \int \langle \varepsilon \rangle \; Z(v)d(v)$$

As there are N atoms therefore the frequency spectrum has to comply with 3N modes of vibration i.e there must be a maximum frequency V_D beyond which no frequencies are possible such that the internal energy expression becomes,

$$\varepsilon = \frac{9N}{v_D^3} \int_0^{v_D} \frac{hv^3 \, dv}{e^{(hv/kT)} - 1}$$

Making use of the following assumptions:

- $x = hv / kT, \quad dv = kTdx / h$
- $x_D = hv_D / kT$

Therefore above equation becomes,

$$\varepsilon = \frac{9N}{v_D^3} (kT / hv_D)^3 \int_0^{x_D} \frac{x^3 \, dx}{e^{(x)} - 1}$$

For $hv_D = k\theta_D$, where θ_D is called Debye temperature, we arrive at result,

$$\varepsilon = 9NkT(T / \theta_D)^3 \int_0^{x_D} \frac{x^3 \, dx}{e^{(x)} - 1}$$

For high temperature range $T >> \theta_D$, equation above reduces to classical limit i.e, $\epsilon = 3NkT$ confirming the outcomes of Einstein's model.

At low temperature range, the internal energy is ,

$$\varepsilon = \frac{3}{5} \pi^4 NkT \left[\frac{T}{\theta_D} \right]^3$$

And the expression for specific heat becomes,

$$C_v = \frac{12}{5} \pi^4 Nk \left[\frac{T}{\theta_D} \right]^3$$

This is Debye's T^3 law which holds for sufficiently low temperatures for long wavelength excitations, i.e the modes that may be treated in elastic continuum with macroscopic elastic constants. Except for the extreme situation of temperatures, Debye's approximation falls back to classical model.

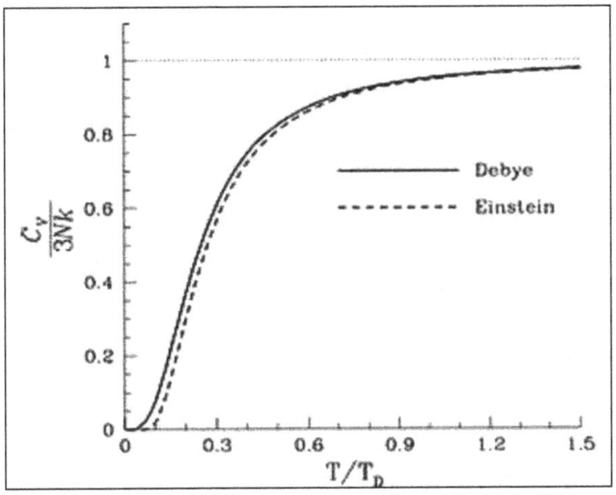

Debye's T³ law which holds good for sufficiently low temperatures.

Neutron Scattering

The neutrons have a great advantage to be used as a probe in condensed matter physics. Scattering of free neutrons with the matter is used to measure details which are of prime importance in material research. Elastic scattering (neutron diffraction) is used in determining structures whereas inelastic scattering provides valuable information on atomic vibrations and excitations. The interaction of phonons is assumed as scattering collisions.

A neutron sees the crystal lattice by interacting with the nuclei of the atoms and exchange energy which is used to measure details of atomic and molecular motions by the use of inelastic neutron scattering (INS). When a crystal gets scattered by a crystal, it results either in absorption or emission of energy of an amount equal to quantum of phonon (hv). The wave vectors before and after the interaction gives information about the phonon energies.

Elastic Scattering

In elastic scattering of X-rays in a crystal (Bragg's diffraction) the magnitude of wave vector (or the frequency) remains unchanged i.e,

$$\omega = \omega' \text{ and } k + G = k'$$

where k and k' are wave vectors of incident and scattered photons, which obey selection rules. G is one of the reciprocal lattice vectors.

Also the energy and momenta are conserved i.e,

$$\hbar\omega = \hbar\omega' \text{ and } \hbar k + \hbar G = \hbar k'$$

As the frequencies of incident and scattered photons are unchanged the crystal as a whole recoils with momentum -ħG. This process is called N-Process.

Inelastic Scattering

But in case the scattering is inelastic, then the process leads to either emission or absorption of a phonon of wave vector K. This phonon will interact with photons, neutrons and electrons as if having a momentum ħk, however a phonon carries no physical momentum on lattices as the phonon coordinates (other than k = 0) are linked with the relative coordinates of the constituents of the lattice. But for the sake of practical purpose, a phonon is considered to have its momentum, more precisely referred to as crystal momentum.

The selection rule so obeyed is as under,

$$\omega = \dot{\omega} + \Omega$$

Here Ω is the frequency of the emitted phonon.

And energy and momentum conservation laws become,

$$\hbar\omega = \hbar\dot{\omega} + \hbar\Omega \text{ and } \hbar k + \hbar G = \hbar k' + \hbar k$$

The conservation of the selection rules require that,

$$K + G = k' \pm K$$

It is found that the frequency of incident photon and scattered photon is not same i.e. $\omega \neq$ þ and the momentum is transferred to the whole crystal. This process is known as U-Process. In neutron elastic scattering process a neutron interacts with the crystal by interacting with the nuclei of the atom. It is used for determination of phonon dispersion relation.

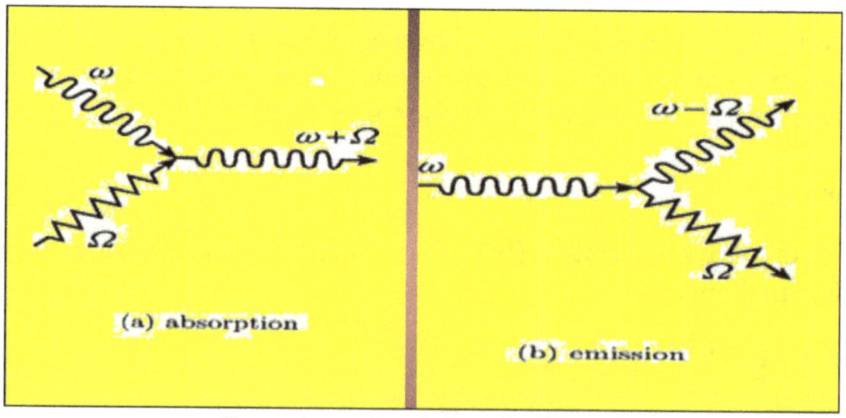

Absorption and emission of a phonon.

The K.E. of incident neutron is $\frac{\hbar^2 k^2}{2M_n}$ and that of scattered neutron is $\frac{\hbar^2 k'^2}{2M_n}$. Therefore energy conservation expression becomes,

$$\frac{\hbar^2 k^2}{2M_n} = \frac{\hbar^2 k'^2}{2M_n} \pm \hbar\Omega$$

$\hbar\Omega$ is the energy of the phonon created (+) or absorbed (-).

Anharmonic Interactions in Crystals

The lattice vibrations and atomic motions are studied within the harmonic approximations, under which the consequences are that the lattice waves never interact and the wave form never changes or decays with time. To deal with real time phenomenon of thermal expansion, thermal conductivity, temperature dependence of elastic constants and the problem of increase in heat capacity at T > θ, the anharmonic effects are to be taken into account.

Let for an atom in any position obeying Hook's law, the energy expression is,

$$\varepsilon_r = \varepsilon_{r_o} + A(r - r_o)^2$$

In such a case, the phonon interaction becomes impossible. One cannot neglect anharmonicity which is a source of coupling between lattice waves and makes scattering of waves possible. Therefore energy expression has to be of the form,

$$\varepsilon_r = \varepsilon_{r_o} + A(r - r_o)^2 + B(r - r_o)^3 + \_\ \_\ \_ + \_\ \_$$

The higher order anharmonic terms makes phonons to collide with perfect crystal by making mean free path of finite dimensions.

Thermal Expansion in Solids

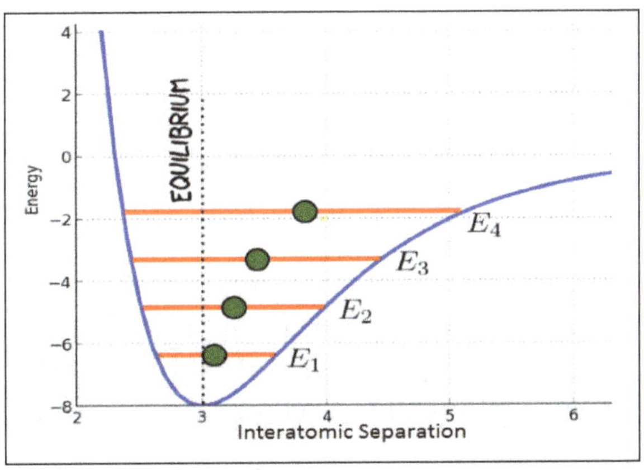

Figure showing the change in equilibrium position of vibrations with respect to energy.

The general answer to why do solids expand on increasing temperature is that an increase in energy results in an increase in equilibrium spacing of atomic bonds. Thermal expansion is measured by the coefficient of linear thermal expansion defined as the increase in length per unit rise in temperature. The thermal expansion in solids is a direct consequence of anharmonicity of the atomic interactions. With the increase in temperature the amplitude of lattice vibrations increases and the equilibrium position shifts as the atoms spend time at greater than original spacing due as the repulsion at short distances is greater than the corresponding attraction at farther distances.

In case of a one dimensional solid if the vibrations are considered harmonic, the potential energy will be,

$$V(x) = \frac{1}{2} f x^2$$

And it will be a parabola with x_o as the equilibrium position, then the harmonic potential is a perfectly parabola function and the mean position does not shift with rise in temperature.

The mean displacement,

$$\overline{X} = \frac{\int_{-\infty}^{\infty} x e^{\left(\frac{1/2 f x^2}{k_B T}\right)} dx}{e^{\left(\frac{1/2 f x^2}{k_B T}\right)} dx} = 0$$

But the atomic oscillators differ from perfectly harmonic oscillators and essentially have harmonicity in the potential with comes into play with higher order terms in 'x'.

Therefore for anharmonic potential,

$$V(x) = \frac{1}{2} f x^2 - g x^3 - h x^4$$

And the expression for mean displacement becomes,

$$\overline{X} = \frac{3g}{f x^2} k_B T$$

$$\overline{X} = \frac{\int_{-\infty}^{\infty} x e^{\left[-\left(\frac{\frac{1}{2 f x^2} - g x^3 - h x^4}{k_B T}\right)\right]} dx}{e^{\left[-\left(\frac{\frac{1}{2 f x^2} - g x^3 - h x^4}{k_B T}\right)\right]} dx}$$

The above equation suggests a prominent anharmonicity or a departure from harmonic behavior and the curve starts to deviate. The mean positions start shifting with increasing temperature corresponding to thermal expansion in physical behavior as demonstrated in figure.

For small anharmonicity the first term of the numerator takes the form as under,

$$\exp\left[\frac{g x^3 + h x^4}{k_B T}\right] = I + \frac{g x^3}{h x^4} + \frac{h x^4}{k_B T}$$

Therefore the expression for mean displacement reduces to,

$$\overline{X} = \frac{\int_{-\infty}^{\infty} \frac{g x^3}{k_B T} e^{-\left(\frac{1 f x^2}{2 k_B T}\right)} dx}{e^{-\left(\frac{1 f x^2}{2 k_B T}\right)} dx}$$

We find that there appears a non-zero mean displacement in the presence of the anharmonicity which contributes to thermal expansion.

Thermal Conductivity

An attempt is made to understand the heat conductivity and energy transmission in solids taking phonon factor into account. For a long rod let a steady heat flows along its length with temperature gradient dT/dx. It is assumed that the temperature variation is very small such that the average phonon number is definite. As the neighboring regions have slightly varying temperatures, the phonon number behaves as a function of position. The energy transmitted across unit area per unit time called the flux of thermal energy is given as,

$$E_{th} = -K\, dT/dx$$

Where K is the coefficient of thermal conductivity in solids.

Consider phonon gas analogous to molecular gas then on the basis of kinetic theory of gases,

$$K = 1/3\, C_v v\lambda$$

C_v is the lattice specific heat which is a measure of phonon density.

v is the average phonon velocity and λ is the mean free path of the phonons.

Hence the expression for thermal flux reduces to,

$$E_{th} = -1/3 C_v v\lambda\, dT/dx$$

Actually total thermal conductivity receives contribution from conductivity due to electrons and phonons each, hence,

$$K_{total} = K_{electron} + K_{phonon}$$

The expression for thermal conductivity due to electrons is deduced and is,

$$K_{electron} = \left[\pi^2 N k^2 \tau / 3m\right] T$$

Here N is the number of free electrons per mole and τ is the mean free time of the electron before collision with the positive ion where it gives its entire thermal energy.

It is observed that in the expression, $K_{electron}$ is almost independent of temperature as the mean free time τ varies as T^{-1} above Debye's temperature θ_D.

So the phonons carry most of the heat energy while electrons stay immobile. The lattice specific heat as given by Debye's law as,

$$C_v = \alpha T^3 \text{ for } T < \theta_D.$$

And $C_v = 3Nk$ for $T > \theta_D$.

Hence the corresponding thermal conductivity expressions are,

$$K_{phonon} = 1/3\, v\lambda\, \alpha T^3 \text{ for } T < \theta_D$$

$$K_{phonon} = 3\, v\lambda\, Nk \text{ for } T < \theta_D.$$

The above expressions clearly indicate that thermal conductivity of insulators is constant at high temperatures but is proportional to T^3 at lower temperatures and this variation in K arises due to phonon – phonon interaction or as we said anharmonocity. The phonon mean free path λ is found to be inversely proportional to absolute temperature so consequently the expression for thermal conduction due to phonons becomes,

$$K_{phonon} \alpha T^{-1}$$

So thermal conductivity of an insulator is proportional to T^{-1} at high temperatures.

References

- Bonding-in-Solids: bspublications.net, Retrieved 11, February 2020
- Thermal-properties: ac.in, Retrieved 29, June 2020
- Lattice-Vibrations: unl.edu, Retrieved 06, April 2020

4

Magnetic Properties

The magnetic properties of solids are due to the electrons present in the molecules or atoms. The materials which are used and studied for their magnetic properties are known as magnetic materials. The topics elaborated in this chapter will help in gaining a better perspective about the magnetic properties of solids as well as magnetic materials and theories of magnetism.

Magnetic Materials

All materials can be classified in terms of their magnetic behaviour falling into one of five categories depending on their bulk magnetic susceptibility. The two most common types of magnetism are diamagnetism and paramagnetism, which account for the magnetic properties of most of the periodic table of elements at room temperature.

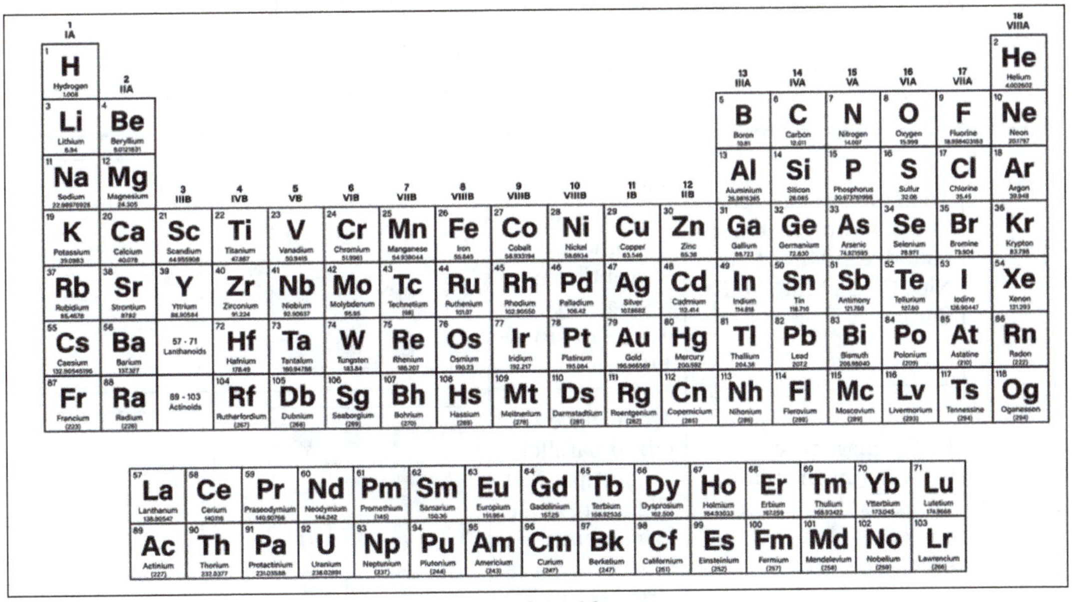

Periodic table.

The two most common types of magnetism are diamagnetism and paramagnetism, which account for most of the periodic table of elements at room temperature. These elements are usually referred to as non-magnetic, whereas those which are referred to as magnetic are actually classified as ferromagnetic. The only other type of magnetism observed in pure elements at room temperature is antiferromagnetism. Finally, magnetic materials can also be classified as ferrimagnetic although this is not observed in any pure element but can only be found in compounds, such as the mixed oxides, known as ferrites, from which ferrimagnetism derives its name. The value of magnetic susceptible falls into a particular range for each type of material and this is shown in table.

Table: Different types of magnetic behaviour.

Type	Example	Atomic/Magnetic Behaviour		
Diamagnetism	Inert gases; many metals e.g. Au, Cu, Hg; non-metallic elements e.g. B, Si, P, S; many ions e.g. Na^+, Cl^- & their salts; diatomic molecules e.g. H_2, N_2; H_2O; most organic compounds.	Atoms have no magnetic moment. Susceptibility is small & negative, -10^{-6} to -10^{-5}.	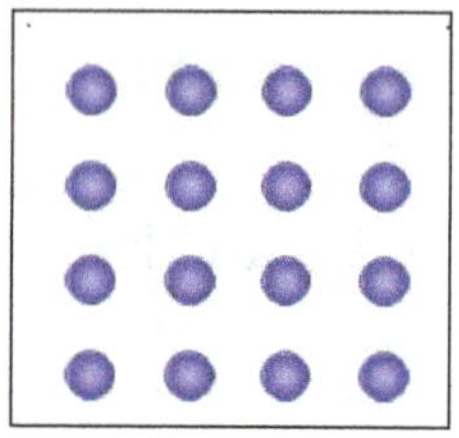	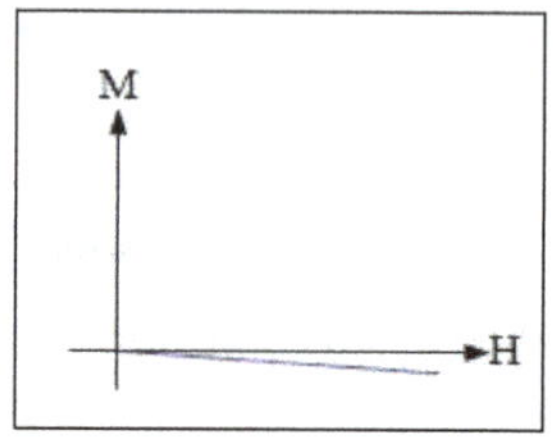
Paramagnetism	Some metals, e.g. Al; some diatomic gases, e.g. O_2, NO; ions of transition metals and rare earth metals, and their salts; rare earth oxides.	Atoms have randomly oriented magnetic moments. Susceptibility is small & positive, $+10^{-5}$ to $+10^{-3}$.	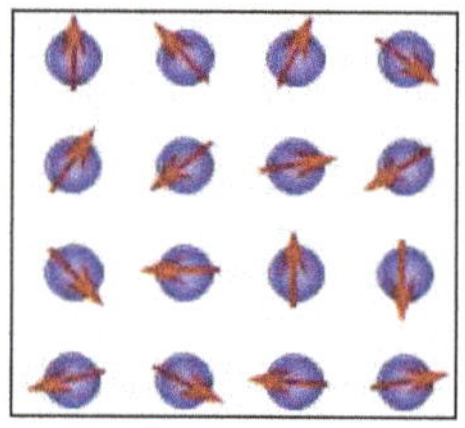	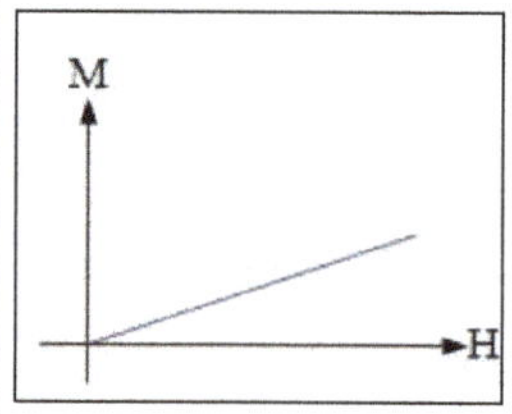
Ferromagnetism	Transition metals Fe, H, Co, Ni; rare earths with $64 \le Z \le 69$; alloys of ferromagnetic elements; some alloys of Mn, e.g. MnBi, Cu_2MnAl.	Atoms have parallel aligned magnetic moments. Susceptibility is large (below T_C).	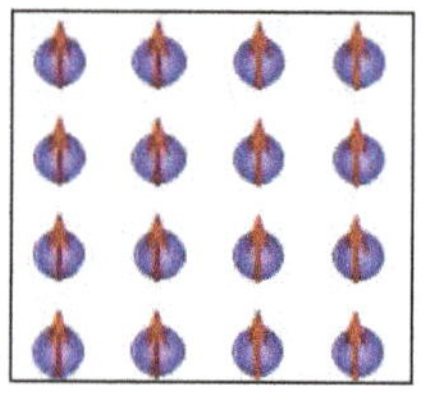	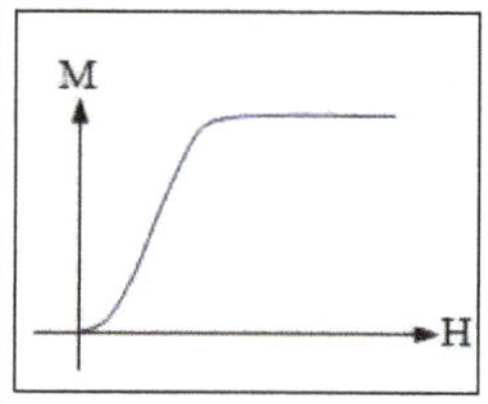
Antiferromagnetism	Transition metals Mn, Cr & many of their compound, e.g. MnO, CoO, NiO, Cr_2O_3, MnS, MnSe, $CuCl_2$.	Atoms have antiparallel aligned magnetic moments. Susceptibility is small & positive, $+10^{-5}$ to $+10^{-3}$.	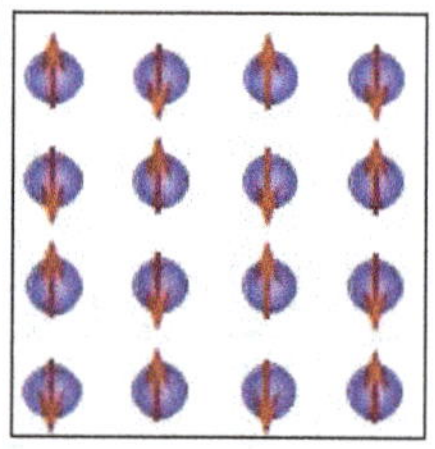	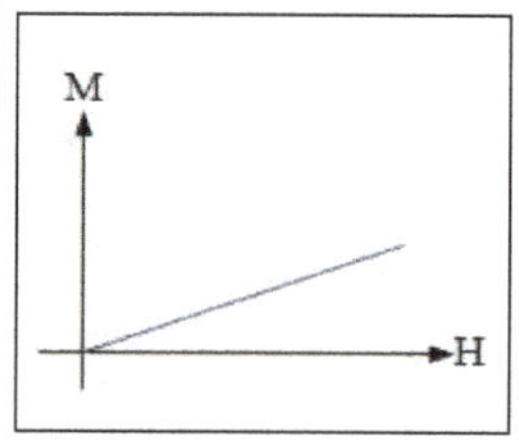
Ferrimagnetism	Fe_3O_4 (magnetite); γ-Fe_2O_3 (maghemite); mixed oxides of iron and other elements such as Sr ferrite.	Atoms have mixed parallel and antiparallel aligned magnetic moments. Susceptibility is large (below T_C)	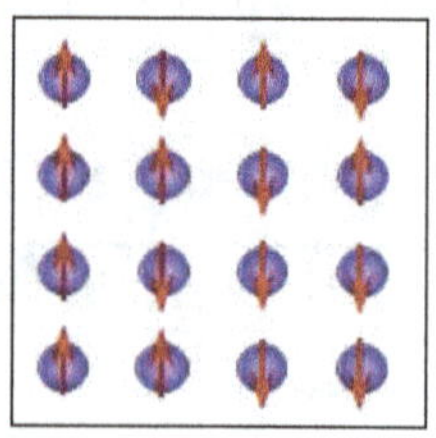	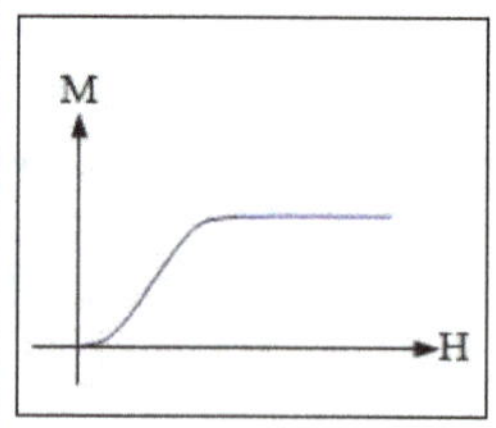

Diamagnetism

In a diamagnetic material the atoms have no net magnetic moment when there is no applied field. Under the influence of an applied field (H) the spinning electrons process and this motion, which is a type of electric current, produces a magnetisation (M) in the opposite direction to that of the applied field. All materials have a diamagnetic effect; however, it is often the case that the diamagnetic effect is masked by the larger paramagnetic or ferromagnetic term. The value of susceptibility is independent of temperature.

Paramagnetism

There are several theories of paramagnetism, which are valid for specific types of material. The Langevin model, which is true for materials with non-interacting localized electrons, states that each atom has a magnetic moment which is randomly oriented as a result of thermal agitation. The application of a magnetic field creates a slight alignment of these moments and hence a low magnetisation in the same direction as the applied field. As the temperature increases, then the thermal agitation will increase and it will become harder to align the atomic magnetic moments and hence the susceptibility will decrease. This behaviour is known as the Curie law and is shown below in equation below, where C is a material constant called the Curie constant.

$$x = \frac{C}{T}$$

Materials which obey this law are materials in which the magnetic moments are localized at the atomic or ionic sites and where there is no interaction between neighbouring magnetic moments. The hydrated salts of the transition metals, e.g. $CuSO_4{\cdot}5H_2O$, are examples of this type of behaviour as the transition metal ions, which have a magnetic moment, are surrounded by a number of non-magnetic ions/atoms, which prevent interaction between neighbouring magnetic moments.

In fact the Curie law is a special case of the more general Curie-Weiss law (equation below), which incorporates a temperature constant (θ) and derives from Weiss theory, proposed for ferromagnetic materials, that incorporates the interaction between magnetic moments.

$$x = \frac{C}{T - \theta}$$

In this equation q can either be positive, negative or zero. Clearly when $\theta = 0$ then the Curie-Weiss law equates to the Curie law. When θ is non-zero then there is an interaction between neighbouring magnetic moments and the material is only paramagnetic above a certain transition temperature. If θ is positive then the material is ferromagnetic below the transition temperature and the value of θ corresponds to the transition temperature (Curie temperature, T_C). If θ is negative then the material is antiferromagnetic below the transition temperature (Néel temperature, TN), however the value of θ does not relate to TN. It is important to note that this equation is only valid when the material is in a paramagnetic state. It is also not valid for many metals as the electrons contributing to the magnetic moment are not localized. However, the law does apply to some metals, e.g. the rare earths, where the 4f electrons, that create the magnetic moment, are closely bound.

The Pauli model of paramagnetism is true for a material where the electrons are free and interact to form a conduction band this is valid for most paramagnetic metals. In this model the conduction electrons are considered essentially to be free and under an applied field an imbalance between electrons with opposite spin is set up leading to a low magnetisation in the same direction as the applied field. The susceptibility is independent of temperature, although the electronic band structure may be affected, which will then have an effect on the susceptibility.

Ferromagnetism

Ferromagnetism is only possible when atoms are arranged in a lattice and the atomic magnetic moments can interact to align parallel to each other. This effect is explained in classical theory by the presence of a molecular field within the ferromagnetic material, which was first postulated by Weiss in 1907. This field is sufficient to magnetize the material to saturation. In quantum mechanics, the Heisenberg model of ferromagnetism describes the parallel alignment of magnetic moments in terms of an exchange interaction between neighbouring moments.

Weiss postulated the presence of magnetic domains within the material, which are regions where the atomic magnetic moments are aligned. The movement of these domains determines how the material responds to a magnetic field and as a consequence the susceptible is a function of applied magnetic field. Therefore, ferromagnetic materials are usually compared in terms of saturation magnetisation (magnetisation when all domains are aligned) rather than susceptibility.

In the periodic table of elements only Fe, Co and Ni are ferromagnetic at and above room temperature. As ferromagnetic materials are heated then the thermal agitation of the atoms means that the degree of alignment of the atomic magnetic moments decreases and hence the saturation magnetisation also decreases. Eventually the thermal agitation becomes so great that the material becomes paramagnetic; the temperature of this transition is the Curie temperature, TC (Fe: $T_C = 770$ °C, Co: $T_C = 1131$ °C and Ni: $T_C = 358$ °C). Above T_C then the susceptibility varies according to the Curie-Weiss law.

Antiferromagnetism

In the periodic table the only element exhibiting antiferromagnetism at room temperature is chromium. Antiferromagnetic materials are very similar to ferromagnetic materials but the exchange interaction between neighbouring atoms leads to the anti-parallel alignment of the atomic magnetic moments. Therefore, the magnetic field cancels out and the material appears to behave in the same way as a paramagnetic material. Like ferromagnetic materials these materials become paramagnetic above a transition temperature, known as the Néel temperature, T_N. (Cr: $T_N = 37$°C).

Ferrimagnetism

Ferrimagnetism is only observed in compounds, which have more complex crystal structures than pure elements. Within these materials the exchange interactions lead to parallel alignment of atoms in some of the crystal sites and anti-parallel alignment of others. The material breaks down into magnetic domains, just like a ferromagnetic material and the magnetic behaviour is also very similar, although ferrimagnetic materials usually have lower saturation magnetisations. For example in Barium ferrite ($BaO.6Fe_2O_3$) the unit cell contains 64 ions of which the barium and oxygen ions have no magnetic moment, 16 Fe^{3+} ions have moments aligned parallel and 8 Fe^{3+} aligned antiparallel giving a net magnetisation parallel to the applied field, but with a relatively low magnitude as only 1/8 of the ions contribute to the magnetisation of the material.

Theories of Magnetism

Diamagnetism can be regarded as originating from shielding currents induced by an applied field in the filled electron shells of ions. Classically, diamagnetism is related to changes in the orbital motion of electrons known as Larmor precession that occurs when atomic systems are placed in a magnetic field. Atoms or ions in which the orbital and spin angular momenta cancel in pairs have J=0 and have no permanent magnetic moment. It may be recalled that the current induced in closed electrical circuit by a magnetic field is always in such a direction so as to keep the total flux unchanged. Thus, the circuit has a negative susceptibility. The effect is retained even in systems of charges that must be treated by quantum mechanics and is responsible for diamagnetism. Since all atoms or ions produced a diamagnetic contribution to the total susceptibility, although it may be marked by the other types; it is a consequence of magnetic moment induced in the atoms by an external field. In this respect diamagnetism may be compared with the electron polarization in an electric field. Both are essentially independent of temperatures. There exists, however, an essential difference; in the electrical case the induced moment lies along the direction of an applied field leading to the positive electrical susceptibility; in the magnetic case the induced moment produce is a negative susceptibility.

Classical Theory of Diamagnetism

Let us consider the Bohr model of the atom in which a central nucleus is surrounded by revolving electrons with some frequency, say ω_0 as shown in figure. If and external field is applied, the current changes so as to oppose the change in flux. This appears as a change in the frequency of revolution. In the absence of external field, the force acting on the electron will be:

$$F = m\omega_0^2 r = \frac{ze^2}{4\pi \epsilon_0 r^2}$$

$$\omega_0 = \sqrt{\frac{ze^2}{4\pi \epsilon_0 r^3}}$$

where r is the radius of the orbit, Ze is the nuclear charge and m is mass of the electron.

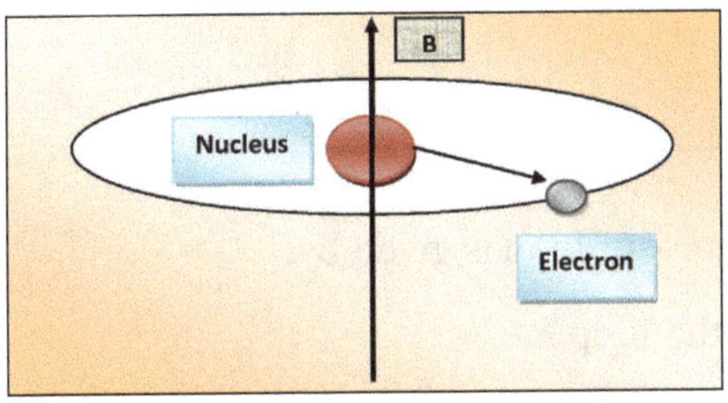

An atom under the influence of external magnetic field.

In the presence of magnetic field, the Lorentz force acting on the electron will be:

$$\vec{F} = e\left(\vec{V} \times \vec{B}\right)$$

If B is perpendicular to r, then, $F = er\omega B$

where ω is the frequency in the presence of field. Now the force equation is,

$$m\omega^2 r = \frac{ze^2}{4\pi \epsilon_0 r^2} - er\omega B$$

$$m\omega^2 r = m\omega_0^2 r - er\omega B$$

The negative sign with Lorentz force represents negative charge. Now above equation can be modified as,

$$\omega^2 + \frac{eB}{m} - \omega_o^2 = 0$$

On solving above equation, we get,

$$\omega = \frac{\frac{-eB}{m} \pm \sqrt{\left(\frac{eB}{m}\right)^2 + 4\omega_o^2}}{2}$$

$$\omega = \frac{-eB}{2m} \pm \sqrt{\left(\frac{eB}{2m}\right)^2 + \omega_o^2}$$

If $\frac{eB}{2m} << \omega_o$ then $\omega = \pm\, \omega_o - \frac{eB}{2m}$

The ± sign on ω_o means that those electrons whose orbital moments were parallel to the field and those whose moments were antiparallel are speeded up by an amount $\frac{eB}{2m}$. This result is called Larmor theorem. This frequency change gives rise to magnetization. The reason for this is that the frequency change is equivalent to an additional current and this current component in every atom is in the same direction, whereas the original circulating currents were in random directions and cancel each other. In the absence of the field, the electron motions produce no net currents or flux. In the presence of field we can write a current for each electron due to its frequency change. This additional current is given by:

$$I = \text{Charge} \times \text{revolutions per unit time}$$

$$I = (-Ze)\left(\frac{1}{2\pi} \cdot \frac{eB}{2m}\right) = \frac{-ze^2B}{4\pi m}$$

The magnetic moment μ of a current loop is given by,

$$\mu = \text{Current} \times \text{area of the loop}$$

$$\mu = \frac{-ze^2B}{4m}\langle \rho^2 \rangle$$

where $\langle \rho^2 \rangle$ is the average radius of the electron from the field axis. If the field is in the z direction, then $\langle \rho^2 \rangle = \langle x^2 \rangle + \langle y^2 \rangle$. If $\langle r^2 \rangle$ is the average radius from the nucleus then $\langle r^2 \rangle = \langle x^2 \rangle + \langle y^2 \rangle + \langle z^2 \rangle$. If the atom has spherical symmetry, then $\langle x^2 \rangle = \langle y^2 \rangle = \langle z^2 \rangle$. Therefore, $\langle \rho^2 \rangle = \frac{2}{3}\langle r^2 \rangle$. This result expresses the fact that the projected area of a great circle titled w.r.t. the equatorial plane is less than $\pi\langle r^2 \rangle$. Thus,

$$\mu = \frac{-ze^2B}{4m} \cdot \frac{2}{3}\langle r^2 \rangle = \frac{-ze^2B}{6m}\langle r^2 \rangle$$

If N is the number of atoms per unit volume, then $M = N\mu$ and the diamagnetic susceptibility per unit volume is,

$$\chi_{dia} = -\frac{\mu_0 N\mu}{B} = -\frac{\mu_0 N Ze^2}{6m}\langle r^2 \rangle$$

It is evident to above equation that the diamagnetic susceptibility is independent of temperature and this result is the classical Langevin result.

Quantum Theory of Diamagnetism

In the quantum-mechanical treatment, we have to consider that the electrons are described by wave functions φ, where φ^2 at every point is the probability of finding the electron. Alternatively, we may consider the electron as a charge cloud of intensity φ^2 at each point in space. It can be shown that the quantum-mechanical result is correctly given by above equation, provided we use the expectation value for the squared electron position parameter r.

$$\langle r^2 \rangle = \frac{\int r^2 |\varphi(r)|^2 \, dr}{\int |\varphi(r)|^2 \, dr}$$

where the integration extends over the whole space.

The Hamiltonian for a mononuclear system under the influence of magnetic field can be expressed as,

$$H' = \frac{ie\hbar}{2mc}(\nabla.A + A.\nabla) + \frac{e^2}{2mc^2}A^2$$

These terms can be considered as a small perturbation for an atomic electron. If we consider that a uniform magnetic field is applied in the z direction then we can write as below:

$$A_x = -\frac{1}{2}yB$$

$$A_y = \frac{1}{2}xB$$

$$A_z = 0$$

Therefore, equation $H' = \frac{ie\hbar}{2mc}(\nabla.A + A.\nabla) + \frac{e^2}{2mc^2}A^2$ becomes,

$$H' = \frac{ie\hbar B}{2mc}\left(x\frac{\partial}{\partial y} - y\frac{\partial}{\partial x}\right) + \frac{e^2B^2}{8mc^2}(x^2 + y^2)$$

For a spherical symmetric system and by first order perturbation theory, the second term on the RHS gives a contribution as below,

$$E' = \frac{e^2B^2}{12mc^2}\langle r^2 \rangle$$

Therefore, the associated magnetic moment is diamagnetic and is as below,

$$\mu = -\frac{\partial E'}{\partial B} = \frac{-e^2B}{6mc^2}\langle r^2 \rangle$$

This result is in agreement with the classical result.

Paramagnetism

Paramagnetism occurs in those atoms which have permanent magnetic moments. Paramagnetic materials have positive magnetic susceptibility. The direction of magnetization is parallel to the applied magnetic field. The permanent magnetic moments of atoms result from the following contributions:

- The intrinsic or spin moment of the electrons.
- The orbital motion of the electrons.
- The nuclear magnetic moment.

Paramagnetism is found in:

- All atoms having an odd number of electrons: The atoms of groups I, III, V and VII of the periodic table satisfy this condition. If the atom forms a diatomic molecule then it will be out of consideration as in a diatomic molecule the two electrons have a chance to pair up.
- Atoms with unfilled inner shells: The rare-earth elements and the transition elements have atoms with unfilled inner shells.
- Free radicals: Some organic compounds have a single unpaired valence electron. These can be made stable under certain circumstances.
- Metals: The electrons in metals behave in many respects as if they are free to move throughout the lattice like molecules in a gas. They tend to pair up but there are always a few unpaired electrons to produce a weak temperature independent paramagnetism known as Pauli paramagnetism.

Classical Theory of Paramagnetism

The classical theory of paramagnetism was developed by Paul Langevin. He considered a paramagnetic gas in which each atom or molecule possesses a permanent magnetic moment. He neglected the mutual magnetic interaction between the gas particles. The dipoles tend to align in the direction of magnetic field when an external magnetic field is applied and this alignment is hampered due to thermal agitations. Thus equilibrium is established and the dipoles orient themselves with respect to the applied field as shown in figure.

The probability that a dipole is oriented at an angle θ to the direction of applied field is proportional to Boltzmann factor exp $(-\mu_B\cos\theta/k_BT)$. According to statistical mechanics, the number of molecules (dn) having inclination θ and $\theta + d\theta$ and which fall in solid angle $2\pi\sin\theta d\theta$ is given as,

$$dn = C\exp(-\mu B\cos\theta / k_BT)2\pi\sin\theta d\theta$$

where C is a constant. Further each particle out of dn contributes a component of magnetic moment $\mu\cos\theta$ to the magnetization, while by symmetry the components which are perpendicular to the direction of the field cancel each other.

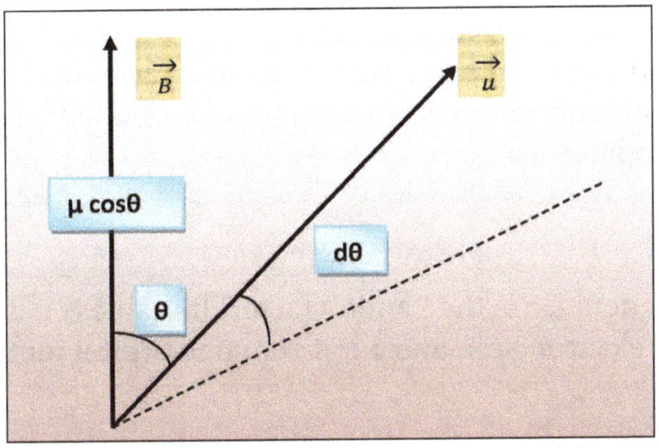

Magnetic dipole of moment μ oriented at angle θ to B.

Therefore, the average component of the magnetic moment of each atom along the direction of the field multiplied by the number of atoms per unit volume, N gives magnetization M as below:

$$M = \frac{N\mu\int_0^{\pi} \exp\left(-\frac{\mu B\cos\theta}{k_B T}\right)\cos\theta \sin\theta\, d\theta}{\int_0^{\pi} \exp\left(-\frac{\mu B\cos\theta}{k_B T}\right)\sin\theta\, d\theta}$$

which on solving gives,

$$M = N\mu\left[\cot hx - \frac{1}{x}\right];\ x = \frac{\mu B}{k_B T}$$

$$M = N\mu L(x)$$

where L(x) is the Langevin function.

Let us plot a Langevin curve i.e. L(x) versus $\frac{\mu B}{k_B T}$.

It can be seen that for large value of x i.e. for high field strength and for low temperature L(x) approaches unity.'

When $x << 1$, $L(x) = x/3$. Thus, $M = \frac{N\mu^2 B}{3k_B T}$

Since $\chi_{para} = \frac{\mu_0 M}{B}$ therefore, $\chi_{para} = \frac{\mu_0 N\mu^2}{3k_B T} = \frac{C}{T}$.

Equation above is called Curie Law. The Curie constant is given as $C = \frac{\mu_0 N\mu^2}{3k_B}$.

Quantum Theory of Paramagnetism

Before starting the quantum theory of paramagnetism it is necessary to highlight an interesting difference between the classical and quantum theory. In quantum theory the permanent magnetic moment of a given atom or ion is restricted to a finite set of orientations relative to the applied field i.e. the orientations are quantized. Let us consider a medium having N atoms per unit volume and having the total angular momentum quantum number of each atom $\vec{J}$. Magnetic dipole moment $\vec{\mu}$ in free space and $\vec{J}$ are related as:

$$\vec{\mu} = -g\mu_B\vec{J}$$

where μ_B is called Bohr magneton, g is the Lande's factor. Its value is 2 is the total angular momentum of the dipole is due to electron spin and 1 if it is due to orbital motion only. It is given by the relation,

$$g = 1 + \frac{J(J+1) + S(S+1) - L(L+1)}{2J(J+1)}$$

The possible component of the magnetic moment along the field direction is given as,

$$\mu_z = -gm_J\mu_B$$

Where m_J is the azimuthal quantum number and has values $m_J = J(J-1,...,-j)$. The potential energy of a magnetic dipole with a component $gm_J\mu_B$ along magnetic field B is,

$$U = gm_J\mu_B B$$

Therefore, according to statistical mechanics, the magnetization is given by,

$$M = \frac{N\sum_{-J}^{J} gm_J\mu_B \exp\left(\frac{gm_J\mu_B B}{k_B T}\right)}{\sum_{-J}^{J} \exp\left(\frac{gm_J\mu_B B}{k_B T}\right)}$$

where the coefficient of N is the statistical average of the magnetic component per atom along the field. Now two cases arise.

Case I: When $\frac{gm_J\mu_B B}{k_B T}$ << i.e. the temperature is very high or the applied field is small, then,

$$\exp\left(\frac{gm_J\mu_B B}{k_B T}\right) = 1 + \frac{gm_J\mu_B B}{k_B T}$$

Therefore, equation $M = \frac{N\sum_{-J}^{J} gm_J\mu_B \exp\left(\frac{gm_J\mu_B B}{k_B T}\right)}{\sum_{-J}^{J} \exp\left(\frac{gm_J\mu_B B}{k_B T}\right)}$ becomes,

$$M = \frac{N\sum_{-J}^{J} gm_J\mu_B\left(1+\frac{gm_J\mu_B B}{k_B T}\right)}{\sum_{-J}^{J}\left(1+\frac{gm_J\mu_B B}{k_B T}\right)}$$

Now, $\sum_{-J}^{J} m_J = 0$, $\sum_{-J}^{J} m_J^2 = \frac{J(J+1)(2J+1)}{3}$ and $\sum_{-J}^{J} 1(2J+1)$.

Therefore, above equation becomes,

$$M = \frac{NBJ(J+1)g^2\mu_B^2}{3k_B T}$$

$$\chi_{para} = \frac{\mu_0 M}{B} = \frac{\mu_0 NJ(J+1)g^2\mu_B^2}{3k_B T} = \frac{\mu_0 N\mu_J^2}{3k_B T}$$

Where $\mu_J^2 = J(J+1)g^2\mu_B^2$

Case II: When $\frac{gm_J\mu_B B}{k_B T}$ is not small as compared with unity i.e. the temperature is low or the applied field is stronger. In this case exponential term cannot be approximated. If we take $\frac{g\mu_B B}{k_B T} = x$

then equation $M = \frac{N\sum_{-J}^{J} gm_J\mu_B \exp\left(\frac{gm_J\mu_B B}{k_B T}\right)}{\sum_{-J}^{J}\exp\left(\frac{gm_J\mu_B B}{k_B T}\right)}$ becomes,

$$M = \frac{N\sum_{-J}^{J} gm_J\mu_B \exp(m_J x)}{\sum_{-J}^{J}\exp(m_J x)}$$

$$M = Ng\mu_B \frac{d}{dx}\left[\text{In}\sum_{-J}^{J}\exp(m_J x)\right]$$

$$M = Ng\mu_B \frac{d}{dx}\left[\text{In}\left(e^{Jx} + e^{(J-1)x} + \cdots + e^{-Jx}\right)\right]$$

$$M = Ng\mu_B \frac{d}{dx}\left[\text{In}\frac{\sinh\left(J+\frac{1}{2}\right)x}{\sinh\left(\frac{x}{2}\right)}\right]$$

$$M = Ng\mu_B B_J(\alpha),\ \text{where } \alpha = Jx$$

where $B_J(\alpha) = \frac{2J+1}{2J}\coth\left[\frac{(2J+1)\alpha}{2J}\right] - \frac{1}{2J}\coth\left(\frac{\alpha}{2J}\right)$ is the Brillouin function.

When $J \rightarrow \infty$, then Brillouin function will be equivalent to Langevin function.

Ferromagnetism

Ferromagnetism is the existence of a spontaneous magnetization, even in the absence of an external field. Weiss suggested that magnetic moments are ordered not only by an external magnetic field B but also by the cumulative action of all magnetic moments i.e. elementary moments interact with one another. This field is named as exchange field and is treated as equivalent to a magnetic field BE. This field is proportional to the magnetization, M and hence,

$$B_E = \lambda M$$

where λ is a constant and is independent of temperature. The effective magnetic field B_{eff} on an atomic dipole is therefore,

$$B_{eff} = B + \lambda M$$

Since $\alpha = Jx = \dfrac{gJ\mu_B B_{eff}}{k_B T}$

$$\alpha = \frac{gJ\mu\ (B+\lambda M)}{k\ T}$$

As we are interested in spontaneous magnetization putting B = o in above equation, we get,

$$\alpha = \frac{gJ\mu_B \lambda M}{k_B T}$$

$$M = \frac{k_B T \alpha}{gJ\mu_B \lambda}$$

Since M must satisfy both equations $\alpha = Jx = \dfrac{gJ\mu_B B_{eff}}{k_B T}$ and equation $M = \dfrac{k_B T\alpha}{gJ\mu_B \lambda}$, its value at a given temperature can be obtained from the point of intersection of the two corresponding M versus α as shown in figure.

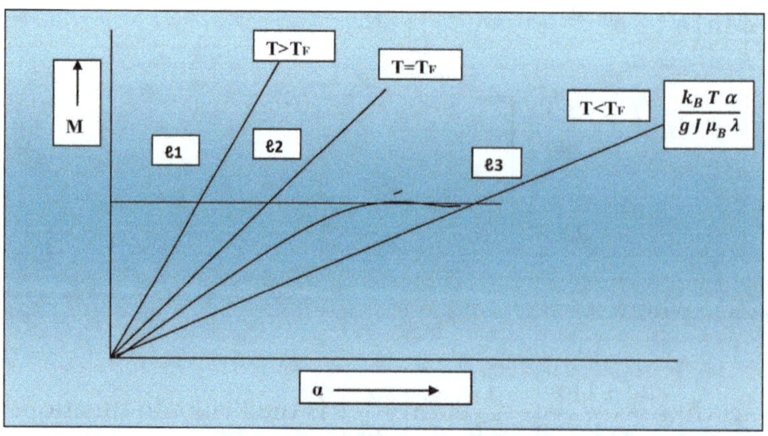

Graphical solution for spontaneous magnetization at a temperature.

For $\alpha << 1$, the Brillouin function is approximately given by,

$$B_J(\alpha) = \frac{(2J+1)\alpha}{3J}$$

Using equation $\alpha = Jx = \frac{gJ\mu_B B_{eff}}{k_B T}$, equation $M = \frac{k_B T \alpha}{gJ\mu_B \lambda}$ and equation $B_J(\alpha) = \frac{(2J+1)\alpha}{3J}$ we get,

$$T_C = \frac{N\mu^2}{3k_B}\lambda$$

where μ is the total magnetic moment per atom. Equation above is known as Curie-Weiss law.

It can be observed from figure above that M has only non-vanishing value when $T<T_F$ and for $T \geq T_F$ the curve does not intersect the lines i.e. spontaneous magnetization vanishes.

Heisenberg Model for the Weiss Field

In 1928, Heisenberg had shown that the cause of internal field may be because of the exchange interaction between the atomic spins. Therefore, the exchange energy of a given atom and its neighbors can be written as,

$$U = -2J_e \sum s_i \cdot s_j$$

Stoner assumed that the instantaneous values of the neighboring spins should be replaced by their time averages. If there are z nearest neighbors then above equation can be written as,

$$U = -2zJ_e \vec{S} \cdot \langle \vec{S} \rangle$$

where $\langle \vec{S} \rangle$ is the average spins of adjoining neighbor atoms. If we assume that the magnetization M is along z-axis then,

$$\langle \vec{S} \rangle = \frac{M}{g\mu_B N}$$

Therefore, $U = -g\mu_B S B_E$

Where $B_E = \frac{2zJ_e M}{Ng^2\mu_B^2}$

using equation $B_E = \lambda M$ we get,

$$\lambda = \frac{2zJ_e}{Ng^2\mu_B^2}$$

Since the magnetization of ferromagnetic materials is due to spin of the electrons, therefore,

$$T_C = \frac{2zJ_e S(S+1)}{3k_B}$$

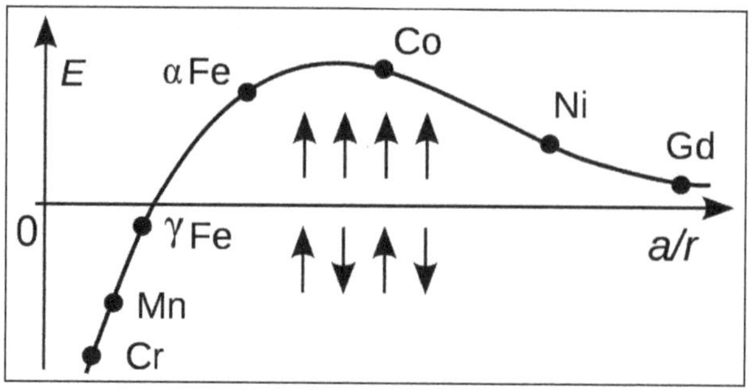

Slater curve.

Antiferromagnetism

In antiferromagnetism the spin moments of the neighboring atoms are aligned antiparallel to each other hence completely cancel each other. In this case the exchange integral J_e is negative. In these materials there is no net magnetization however, when a field is applied a small magnetization appears in the direction of the field. This magnetization further increases with the temperature and becomes maximum at a critical temperature known as the Neel temperature. Above Neel temperature, the magnetization goes on decreasing which indicates the paramagnetic state.

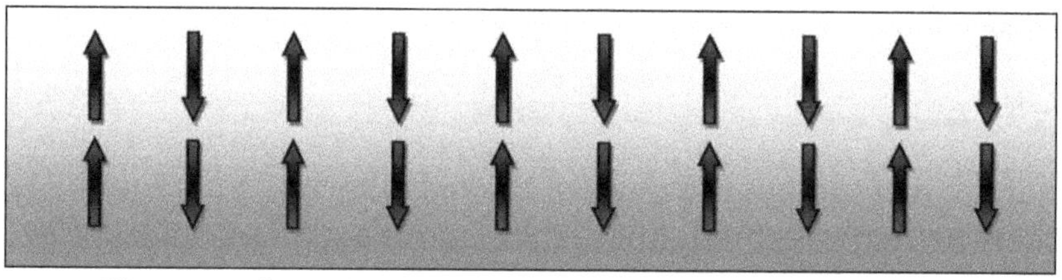

Alignment of atomic moments in an antiferromagnetic material.

A crystal exhibiting antiferromagnetism may be considered as made up of two sublattices, so that the spin moments of atoms in one sublattice array have an opposite sense from the spin moments of atoms in the other sublattice array. If A and B are two sublattice arrays therefore effective field at sites A and B will be:

$$B_A = B - \alpha M_A - \beta M_B$$

$$B_B = B - \beta M_A - \alpha M_B$$

It is assumed that beside an antiferromagnetic AB interaction, there are also AA and BB antiferromagnetic interactions. α and β are Weiss constants and M_A and M_B is the magnetization at A and B lattices such that total magnetization is the sum of M_A and M_B. When the temperature is above than Neel temperature then at this temperature paramagnetic state will exist therefore magnetization of the two sublattice is given by as below:

$$M_A = \frac{N\mu_J^2}{3k_BT} B_A \text{ and } M_A = \frac{N\mu_J^2}{3k_BT} B_B$$

Total magnetization M is $M = \frac{N\mu_J^2}{3k_BT}(B_A + B_B)$

Using equations $B_A = B - \alpha M_A - \beta M_B$ and $B_B = B - \beta M_A - \alpha M_B$

$$M = \frac{N\mu_J^2}{3k_BT}\left[2B - (\alpha+\beta)(M_A + M_B)\right]$$

$$M = \frac{N\mu_J^2}{3k_BT}\left[2B - (\alpha+\beta)M\right]$$

$$M + \frac{N\mu_J^2}{3k_BT}(\alpha+\beta)M = \frac{2N\mu_J^2}{3k_BT}B$$

Hence susceptibility is given by,

$$\chi = \frac{\mu_o M}{B} = \frac{\dfrac{2N\mu_J^2}{3k_B}}{T + \dfrac{N\mu_J^2}{3k_B}(\alpha+\beta)}$$

$$\chi = \frac{C}{T+\theta}$$

Equation above is the analogous of Curie-Weiss law. In the absence of magnetic field (B = 0) and when T= T_N, then,

$$M_A = \frac{N\mu_J^2}{3k_BT_N}(-\alpha M_A - \beta M_B) \text{ and } M_B = \frac{N\mu_J^2}{3k_BT_N}(-\beta M_A - \alpha M_B)$$

$$\left(1 + \frac{\alpha N\mu_J^2}{3k_BT_N}\right)M_A + \frac{\beta N\mu_J^2}{3k_BT_N}M_B = 0$$

$$\left(1 + \frac{\alpha N\mu_J^2}{3k_BT_N}\right)M_B + \frac{\beta N\mu_J^2}{3k_BT_N}M_A = 0$$

Since $C = \dfrac{2N\mu_J^2}{3k_B}$

Therefore, $\dfrac{C}{2T_N} = \dfrac{N\mu_J^2}{3k_BT_N}$

Substituting this value in equations $\alpha = \dfrac{gJ\mu_B\lambda M}{k_BT}$ and $M = \dfrac{k_BT\alpha}{gJ\mu_B\lambda}$ we get,

$$T_N = \frac{C(\beta-\alpha)}{2}$$

Hence it is evident that Neel temperature increases as the AB interactions (β) become stronger, whereas it decreases with increasing AA and BB interactions (α). For temperature below than Neel temperature if we consider only A-B interactions then as a result of anisotropy there may be more than one direction along which the spins will tend to align themselves. Therefore there are two cases of interest (1) when magnetic field is applied perpendicular to the natural spin direction and (2) when magnetic field is applied parallel to the natural spin direction. The susceptibility in the case of perpendicular is independent of temperature and is equal to the susceptibility at Neel temperature when approached from the high temperature region. When $T = T_N$, then the susceptibility in case of parallel is equal to the susceptibility in case of perpendicular.

Electrons in a Magnetic Field

As is shown in standard texts on quantum mechanics, the energy levels of free electrons in a magnetic field are modified from,

$$\epsilon(k) = \frac{\hbar^2 k^2}{2m}$$

to $\epsilon(n_L, k_z) = \frac{\hbar^2 k_z^2}{2m} + \left(n_L + \frac{1}{2}\right)\hbar\omega_c$,

where $n_L = 0, 1, 2, \cdots$ and

$$\omega_c = \frac{eB}{mc}$$

is the cyclotron frequency, i.e. the frequency of the classical motion. The states with a given n_L are known collectively as a Landau level. The degeneracy of a Landau level is,

$$\mathcal{N} = 2\frac{e}{hc}AB = \frac{AB}{\phi_o},$$

where A is the area of the sample in the plane perpendicular to B, the factor of two in the first expression comes from spin degeneracy, and,

$$\phi_o = \frac{hc}{2e} = 2.07\times10^{-7}\,\text{G}-\text{cm}^2$$

is the flux quantum for a pair of electrons.

Two Dimensions, Free Electrons

It is easier to consider first the case of two dimensions, where the electrons are confined to the x-y plane in a perpendicular field. Hence there is no motion in the z direction so the term in equation $\epsilon(n_L, k_z) = \frac{\hbar^2 k_z^2}{2m} + \left(n_L + \frac{1}{2}\right)\hbar\omega_c$, involving k_z^2 does not occur. The density of states in two dimensions is a constant,

$$g(\epsilon) = A\frac{m}{\pi\hbar^2},$$

in the absence of a field. Here we give the total density of states, rather than the density of states per unit area, so a factor of A appears. In a field, the allowed energy values are discrete with a constant spacing between levels of $\hbar\omega_c$. Now the number of zero field states in an interval $\hbar\omega_c$ is given by,

$$g(\epsilon)\hbar\omega_c = A\frac{m}{\pi\hbar^2}\hbar\frac{eB}{mc}$$
$$= \frac{2e}{hc}AB$$
$$= \mathcal{N},$$

the degeneracy of a Landau level. In other words, the field bunches the states into discrete levels, but the total number of states in a region much larger than the Landau level spacing is unchanged by the field. The density of states, both with and without a magnetic field is shown in figure.

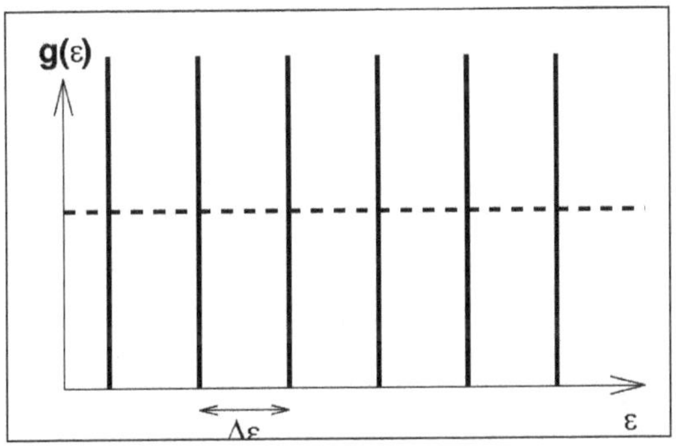

The dashed line shows the density of states of the two dimensional free electron gases in the absence of a magnetic field. It has the constant value $g(\epsilon) = Am/\pi\hbar^2$. In the presence of a magnetic field the energy levels are bunched into discrete values $(n_L + 1/2)\Delta_\epsilon$ where $n_L = 0,1,2,\cdots$, and $\Delta_\epsilon = \hbar\omega_c$, where $\omega_c = eH/mc$ is the cyclotron frequency. Hence the density of states is a set of delta functions, shown by the vertical lines. The weight of the delta function is equal to the zero field density of states times Δ_ϵ so the energy levels are just shifted locally, the total number of states in a region comprising a multiple of Δ_ϵ being unchanged.

At zero temperature, as we increase the magnetic field the number of occupied Landau levels will change, since the number of electrons is fixed but the degeneracy of each Landau level changes with B. This will lead to an oscillatory behavior of the energy as a function of the magnetic field that we discuss below. At finite temperature these oscillations will be washed out if $k_B T \gg \hbar\omega_c$, because then many Landau levels will be partially filled. In order to see the oscillations, we therefore need,

$$\hbar\omega_c = \frac{e\hbar}{mc}B \gg k_B T,$$

which requires large fields and very low temperatures, of order a few K. At higher temperatures, there is a smooth change in the energy with B which leads to a small diamagnetic response.

It is also important to realize that if there are impurities, then the electron states will have a finite lifetime, τ, which will broaden the levels by an amount $\hbar/\tau$. This will also wash out the oscillations if the level broadening is greater than the Landau level splitting. Hence, to observe the oscillations we also need a second condition,

$$\omega_c \tau = \frac{eB}{mc}\tau \gg 1,$$

which requires very clean samples.

Let us now determine the change in energy of our two-dimensional model at T = 0 as a function of field. Let,

$$n = \frac{N}{A}$$

be the number of electrons per unit area. Hence, for B = 0, the Fermi energy is determined from,

$$N = \int_0^{\epsilon F} g(\epsilon)d\epsilon, \text{ or } N = g(\epsilon)\epsilon_F = A\frac{m}{\pi\hbar^2}\epsilon_F,$$

which gives,

$$\epsilon_F = n\frac{\pi\hbar^2}{m}.$$

Hence the ground state energy per electron in zero field is,

$$\epsilon(B=0) \equiv \frac{E(0)}{N} = \int_0^{\epsilon F} \epsilon g(\epsilon)d\epsilon = \frac{\epsilon F}{2} = n\frac{\pi\hbar^2}{2m},$$

using equation $\epsilon_F = n\frac{\pi\hbar^2}{m}$.

In the presence of the field the levels, $n_L = 0, 1, 2, \cdots, p-1$ will be fully occupied with $\mathcal{N}$ electrons and level p will be partially occupied with $\lambda\mathcal{N}$ electrons, where $0 \le \lambda < 1$. Counting up electrons one has,

$$N = \mathcal{N}v.$$

where

$$v = p + \lambda$$

is called the filling factor of the Landau levels. Note that λ and v take a continuous range of values, whereas p is an integer. From equation $\mathcal{N} = 2\frac{e}{hc}AB = \frac{AB}{\phi_0}$, we have,

$$v = n\frac{\phi_0}{B} = \frac{B_1}{B}\left(= \frac{\epsilon F}{\hbar\omega_c} = n\frac{\pi\hbar^2}{m}\frac{1}{\hbar\omega_c}\right),$$

where the last equality is from equation $\epsilon_F = n\frac{\pi\hbar^2}{m}$, in which,

$$B_1 = n\phi_0$$

is the field required to put all the electrons in the lowest Landau level. Consequently,

$$p = \left[\frac{B_1}{B}\right], \quad \lambda = \frac{B_1}{B} - \left[\frac{B_1}{B}\right]$$

where [x] means the largest integer less than or equal to x. Note that λ is the fractional filling of the last Landau level.

The energy per electron is then just obtained by summing the energies of each Landau level, i.e.

$$\begin{aligned}\epsilon(B) &= \left\{\left(0+\frac{1}{2}\right)+\left(1+\frac{1}{2}\right)+\cdots+\left(p-1+\frac{1}{2}\right)+\left(p+\frac{1}{2}\right)\lambda\right\}\frac{\mathcal{N}}{N}\hbar\omega_c \\ &= \left\{\frac{p^2}{2}+\left(p+\frac{1}{2}\right)\lambda\right\}\frac{\hbar\omega_c}{v} \\ &= \left\{\frac{(p+\lambda)^2}{2}+\frac{\lambda-\lambda^2}{2}\right\}\frac{\hbar\omega_c}{v}.\end{aligned}$$

Eliminating $p + \lambda \equiv v$ in favor of n using equation $v = n\frac{\phi_0}{B} = \frac{B_1}{B}\left(=\frac{\epsilon F}{\hbar\omega_c} = n\frac{\pi\hbar^2}{m}\frac{1}{\hbar\omega_c}\right)$, one finds,

$$\epsilon(B) = \frac{\pi\hbar^2}{2m}n + \frac{\hbar\omega_c}{2v}\lambda(1-\lambda)$$

which, from Equations $\omega_c = \frac{eB}{mc}$, $\epsilon(B=0) \equiv \frac{E(0)}{N} = \int_0^{\epsilon F}\epsilon g(\epsilon)d\epsilon = \frac{\epsilon F}{2} = n\frac{\pi\hbar^2}{2m}$, and $v = n\frac{\phi_0}{B} = \frac{B_1}{B}\left(=\frac{\epsilon F}{\hbar\omega_c} = n\frac{\pi\hbar^2}{m}\frac{1}{\hbar\omega_c}\right)$, $B_1 = n\phi_0$, and $p = \left[\frac{B_1}{B}\right]$, $\lambda = \frac{B_1}{B} - \left[\frac{B_1}{B}\right]$ can be conveniently written as,

$$\frac{\epsilon(B)}{\epsilon(0)} = 1 + \left(\frac{B}{B_1}\right)^2 \lambda(1-\lambda).$$

Equation above is our main result. It gives the field dependence of the ground state energy (per electron) of free electrons in two-dimensions.

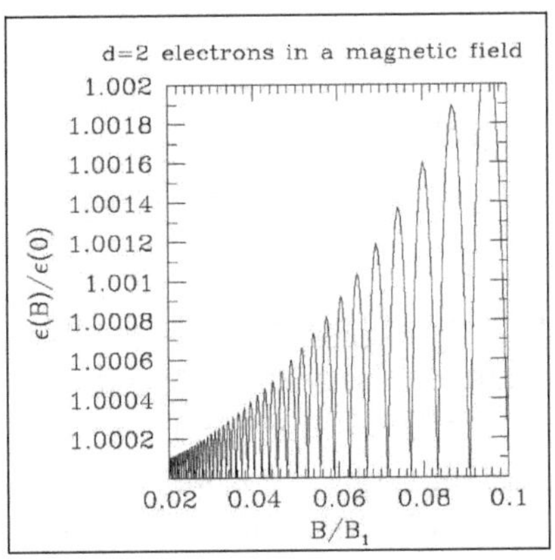

The energy of the two dimensional electron gas at T = 0 according to above equation, as a function of B/B_1 where $B_1 = n\varphi_0$ is the field at which all the electrons are in a completely filled lowest Landau level. Note that the overall size of the energy change varies as B^2 in addition to the oscillations. The oscillations get closer together for small B. In fact, apart from the smooth B^2 variation, the data is a periodic function of 1/B.

A sketch of energy as a function of B, according to above equation, is shown in figure. One clearly sees oscillations whose amplitude increases smoothly which B, actually as B^2. The period of the oscillations gets smaller with decreasing B, since the energy is actually a periodic function of 1/B not B (apart from the B^2 variation in amplitude). It occurs because the energy in above equation depends on the fractional filling factor λ, defined in equation $p = \left[\frac{B_1}{B}\right], \quad \lambda = \frac{B_1}{B} - \left[\frac{B_1}{B}\right]$, which depends periodically on 1/B. Although the energy is a continuous function of B the derivative is discontinuous when the field is such that the Landau levels are completely filled (remember that λ is the fractional filling of the last Landau level). Physically, this is because if one decreases B below $B_p = B_1/p$, say, then it is Landau level p which starts to be occupied, whereas if one increases B above B_p then it is Landau level p – 1 which starts to be depleted.

Figure above is to be contrasted with the situation at a temperature where $k_B T \gg \hbar\omega_c$, where thermal fluctuations average over the oscillations. Then one has only the smooth increase in the energy proportional to B^2. Since the magnetization is given by $M = -\partial E / \partial B$, where E is the energy per unit volume, one then finds a small diamagnetic (i.e. negative) susceptibility, χ_{dia}, where $M = \chi_{dia} B$. This was first calculated by Landau in 1930.

From Equation $p = \left[\frac{B_1}{B}\right], \quad \lambda = \frac{B_1}{B} - \left[\frac{B_1}{B}\right]$ that λ is a periodic function of 1/B and hence, apart from the slowly varying factor of B^2, the energy, given by above equation, is an oscillatory function of 1/B.

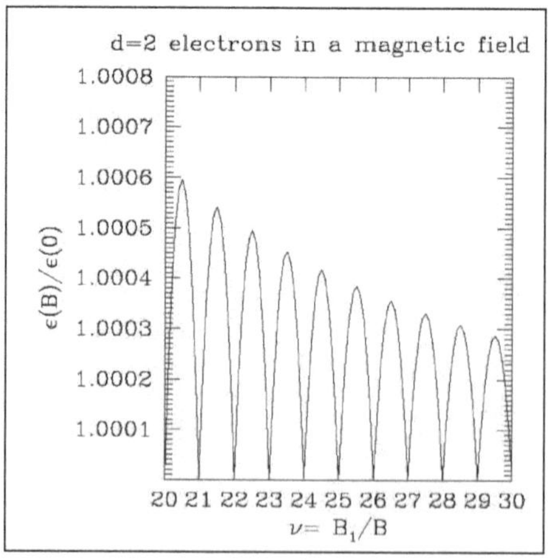

The energy of the two dimensional electron gas at T = 0 as a function of $\nu = B_1/B$ where ν is the filling factor (the number of filled Landau levels) and $B_1 = n\varphi_0$ is the field at which all the electrons are in a completely filled lowest Landau level. For metals, B_1 is much greater than a typical laboratory field so in experimental situations the filling factor is large. Note that the behavior is periodic apart from a gentle overall decrease of the amplitude which is negligible for $\nu \gg 1$.

The oscillation in the ground state energy give rise to oscillations in the magnetization per electron M, given by,

$$M = -\frac{\partial E}{\partial B},$$

which, from Equations, $\phi_o = \frac{hc}{2e} = 2.07 \times 10^{-7} G - cm^2$, $\epsilon(B = o) \equiv \frac{E(o)}{N} = \int_o^{\epsilon F} \epsilon g(\epsilon) d\epsilon = \frac{\epsilon F}{2} = n\frac{\pi\hbar^2}{2m}$, $v = p + \lambda$, $v = n\frac{\phi_o}{B} = \frac{B_1}{B}\left(= \frac{\epsilon F}{\hbar\omega_c} = n\frac{\pi\hbar^2}{m}\frac{1}{\hbar\omega_c}\right)$, $p = \left[\frac{B_1}{B}\right]$, $\lambda = \frac{B_1}{B} - \left[\frac{B_1}{B}\right]$ and $\frac{\epsilon(B)}{\epsilon(o)} = 1 + \left(\frac{B}{B_1}\right)^2 \lambda(1-\lambda)$ is given by,

$$\frac{M}{\mu_B} = 1 - 2\lambda - \frac{2\lambda(1-\lambda)}{p+\lambda},$$

where,

$$\mu_B = \frac{e\hbar}{2mc}$$

is the Bohr magneton. For $p \gg 1$, which is generally the case of interest, this is given, to a good approximation, by,

$$\frac{M}{\mu_B} = 1 - 2\lambda.$$

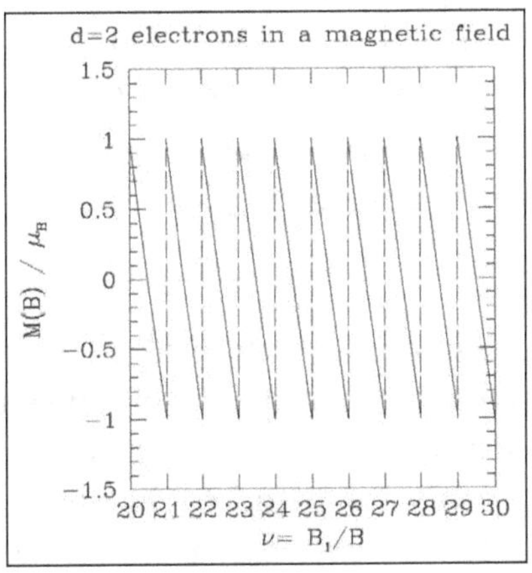

The magnetization of the two dimensional electron gas at T = o (in units of μ_B) as a function of $v = B_1/B$ where v is the filling factor and $B_1 = n\varphi_o$, obtained from equation $\frac{M}{\mu_B} = 1 - 2\lambda - \frac{2\lambda(1-\lambda)}{p+\lambda}$. The magnetization is a periodic function of 1/B.

Since λ is periodic in 1/B, in equation $p = \left[\frac{B_1}{B}\right]$, $\lambda = \frac{B_1}{B} - \left[\frac{B_1}{B}\right]$ it follows that M is also periodic in 1/B. The main conclusion so far is that M is a periodic function of 1/B with period given by,

$$\Delta\left(\frac{1}{B}\right) = \frac{1}{B_1} = \frac{1}{n\phi_o}.$$

From equation $\epsilon_F = n\frac{\pi\hbar^2}{m}$, it follows that the density is related to the Fermi wave vector k_F by,

$$n = \frac{k_F^2}{2\pi},$$

and so, using equation $\phi_o = \frac{hc}{2e} = 2.07\times10^{-7} G - cm^2$.

$$\Delta\left(\frac{1}{B}\right) = \frac{2\pi e}{\hbar c}\frac{1}{A_F},$$

Where

$$A_F \equiv \pi k_F^2$$

is the area of the Fermi surface. Equation $\Delta\left(\frac{1}{B}\right) = \frac{2\pi e}{\hbar c}\frac{1}{A_F}$, which relates the periodicity to just fundamental constants and the area of the Fermi surface, is true quite generally, not just for free electrons.

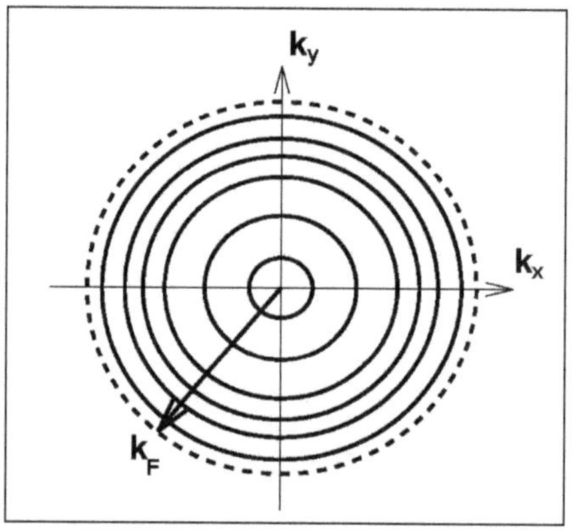

The solid circles are the semi classical orbits of free electrons in two dimensions in a magnetic field. Each circle corresponds to a particular Landau level. The electrons follow orbits around these circles with a period T = 2π/ωc, where ωc is the cyclotron frequency. As the strength of the field is increased the radii increase and the orbits "pop out" of the Fermi circle, denoted by a dashed circle. This leads to oscillatory behavior in the energy and magnetization provided kBT is much less than the spacing of the Landau levels, $\hbar\omega_c$.

It is interesting to see how equation $\Delta\left(\frac{1}{B}\right) = \frac{2\pi e}{\hbar c}\frac{1}{A_F}$, emerges from the semi-classical picture. The semi classical orbits for free electrons are circles, on which the energy stays constant. Denoting by k_p the magnitude of the wave vector corresponding to Landau level p, then equating $\hbar^2 k_p^2/2m$ to $\left(p+\frac{1}{2}\right)\hbar\omega_c$ one gets,

$$k_p^2 = \left(p+\frac{1}{2}\right)\frac{2eB}{\hbar c}$$

and so the area in k-space swept out by an electron in Landau level p is,

$$A_p = \pi k_p^2 = \left(p + \frac{1}{2}\right)\frac{2\pi eB}{\hbar c},$$

which can also be written as,

$$\frac{1}{B} = \left(p + \frac{1}{2}\right)\frac{2\pi e}{\hbar c}\frac{1}{A_p}.$$

Hence, in the semi-classical picture of free electrons is a magnetic field, the allowed states in k-space form circles whose radius increases with B. As the states go through the Fermi wave vector and become de-populated this gives rise to oscillations in the energy. From above equation the change in 1/B for one Landau level to go through the Fermi surface (i.e. let $p \to p - 1$ with $A_L = A_F$, the area of the Fermi surface) is given by,

$$\Delta\left(\frac{1}{B}\right) = \frac{2\pi e}{\hbar c}\frac{1}{A_F},$$

in agreement with equation $\Delta\left(\frac{1}{B}\right) = \frac{2\pi e}{\hbar c}\frac{1}{A_F}$.

Two Dimensions, Bloch Electrons

So far we have derived equation $\Delta\left(\frac{1}{B}\right) = \frac{2\pi e}{\hbar c}\frac{1}{A_F}$ only for free electrons in two dimensions. The same result is also true for Bloch electrons (i.e. electrons in a periodic potential) as we shall now show. Already the fact that it does not depend on the electron mass, which could be altered to an effective mass by band structure effects, gives one a suspicion that this might be so.

The first stage in the proof is to derive an equation, $T = \frac{\hbar^2 c}{eB}\left(\frac{\Delta A}{\Delta \epsilon}\right)$ relating the period of an orbit of a Bloch electron in a magnetic field to the change in the area of the orbit with energy. We start with the semi-classical equations of motion:

$$v(k) = \frac{1}{\hbar}\frac{\partial \epsilon(k)}{\partial k}$$

$$\hbar \dot{k} = (-e)\frac{1}{c}v(k) \times B,$$

from which it follows that,

$$|\dot{k}| = \frac{eB}{\hbar^2 c}\left|\left(\frac{\partial \epsilon(k)}{\partial k}\right)_\perp\right|,$$

where $(\partial \epsilon(k)/\partial k)_\perp$ is the component of $\partial \epsilon(k)/\partial k$ perpendicular to the field, i.e. its projection in the plane of the orbit. (This projection is not necessary in two dimensions considered here, but we include it so that our derivation of equation $T = \frac{\hbar^2 c}{eB}\left(\frac{\Delta A}{\Delta\epsilon}\right)$ will also apply in three dimensions.) Hence the period of an orbit is given by,

$$T = \oint \frac{dk}{|\dot{k}|} = \frac{\hbar^2 c}{eB}\oint \frac{dk}{\left|\left(\partial\epsilon(k)/\partial k\right)_{\perp}\right|},$$

where dk is the magnitude of a small element of the orbit, and we integrate around the closed orbit. If we consider two orbits whose difference in energy Δ_ϵ is small then the region between them in the k_x-k_y plane is a ribbon of width Δ(k), where,

$$\Delta_\epsilon = \left|\left(\frac{\partial\epsilon(k)}{\partial k}\right)_{\perp}\right| |\Delta(k)|.$$

Hence the period T can be written as,

$$T = \frac{\hbar^2 c}{eB}\frac{1}{\Delta\epsilon}\oint |\Delta(k)| dk.$$

The integral in the last equation is just the area between the two neighboring orbits, ΔA, and so,

$$T = \frac{\hbar^2 c}{eB}\left(\frac{\Delta A}{\Delta\epsilon}\right).$$

With a free electron band structure, $\epsilon = \hbar^2 k^2/2m$, $A = \pi k^2$, this gives $T = 2\pi/\omega_c$, where ω_c is the cyclotron frequency, as expected.

The second stage of the proof is to use another relationship involving the period of the orbits, the Bohr-Sommerfeld correspondence principle, which states that if ϵ_p and ϵ_{p+1} are two adjacent energy levels with quantum numbers p and p + 1 where p is assumed large, then,

$$\epsilon_{p+1} - \epsilon_p = \frac{h}{T},$$

where T is the period of the motion of a semiclassical wavepacket on an orbit with energy centered on ϵ_p. This can be derived by noting that the wavepacket is built up out of adjacent energy levels, and its periodic motion comes from interference between different levels. This requires that the levels around ϵ_p be uniformly spaced with spacing $\hbar\omega = h/T$, where ω is the angular frequency of the orbit. (The "correspondence principle" really only comes in with the further remark that T is also the period of a purely classical particle of the same energy).

From above equation we substitute $T\Delta\epsilon = h = 2\pi\hbar$ into equation $T = \frac{\hbar^2 c}{eB}\left(\frac{\Delta A}{\Delta\epsilon}\right)$, and then find that the area between the semiclassical orbits of two adjacent Landau levels is given by,

$$\Delta A = \frac{2\pi eB}{\hbar c}.$$

This elegant result, first obtained by Onsager, can be reexpressed by stating that at large p the area A_p inside the Landau level p is given by,

$$A_p = (p+\lambda)\frac{2\pi eB}{\hbar c},$$

where λ is some number independent of p. This is the same as equation $A_p = \pi k_p^2 = \left(p+\frac{1}{2}\right)\frac{2\pi eB}{\hbar c}$, for free electrons, except that λ need not equal 1/2. The periodicity in 1/B in equation $\Delta\left(\frac{1}{B}\right) = \frac{2\pi e}{\hbar c}\frac{1}{A_F}$, obtained from equation $A_p = \pi k_p^2 = \left(p+\frac{1}{2}\right)\frac{2\pi eB}{\hbar c}$, follows equally from above equation.

Three Dimensions and the De Haas-Van Alphen Effect

One now needs to add k_z, so the semi-classical orbits, which were circles in two dimensions, are now cylinders. One such cylinder is shown in figure. As B increases the radius increases. However, one does not expect a dramatic change in the magnetization until the radius increases to k_F, beyond which the cylinder does not intersect the Fermi sphere at all. When the cylinders "pop out" of the Fermi sphere one expects oscillatory behavior in the magnetization just as in two dimensions. This can be confirmed by a mathematical analysis. Clearly then, the periodicity of the magnetization with 1/B involves the maximum area formed by the intersection of the Fermi sphere with a plane perpendicular to B. If one has a general Fermi surface from some complicated band structure, then it is also possible that, on varying k_z, the intersection of the Fermi surface with the plane of constant k_z might have a minimum. These orbits will also contribute to the oscillatory behavior because the part of the cylinder inside the Fermi sphere will decrease rapidly when the radius of the cylinder passes this extremal radius.

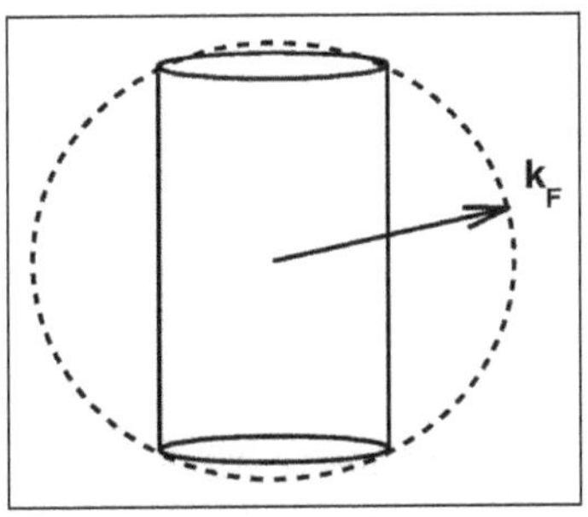

The dashed circle represents the Fermi sphere in three dimensions. The cylinder represents the semi-classical orbit corresponding to a single Landau level. As the strength of the magnetic field increases, the radius of the cylinder increases and will eventually exceed the Fermi wave vector, k_F. This "popping out" of the Landau levels from the Fermi surface leads to oscillatory behavior of the energy and magnetization at sufficiently low temperature.

Thus we conclude that the magnetization will show oscillations with periods given by,

$$\Delta\left(\frac{1}{B}\right) = \frac{2\pi e}{\hbar c}\frac{1}{A_{ext}},$$

where A_{ext} is the area of any extremal orbit in the plane perpendicular to the field. If there is more than one extremal area then several periods will be superimposed.

A determination of oscillations in M as a function of 1/B (the de Haas-van Alphen effect) for different orientations orientations of the field has been the most successful method for mapping out the shape of the Fermi surface of metals. Oscillations occur in other quantities as well. For example, the resistivity varies with magnetic field, an effect called magnetoresistance. At low temperature and in very clear samples, there are oscillations in the magnetorsistance as a function of 1/B of the same origin as the oscillations in magnetization.

Here we derive the expression for Landau diamagnetism of free electrons. Our approach will be to derive the result first for 2-d from which we will easily be able to obtain the 3-d result by integrating over k_z. Firstly, we remind ourselves of some statistical mechanics. It is convenient to calculate the grand partition function, $\mathcal{Z}$ where,

$$\mathcal{Z} = \mathrm{Tr}\ \exp\left[\beta\left(\mu N - \mathcal{H}\right)\right],$$

where N is the number of particles and $\mathcal{H}$ is the Hamiltonian. The trace is over all states for a given N and over all N. This ensemble, where both N and E are allowed to vary, is called the "grand canonical ensemble". $\mathcal{Z}$ is related to the "grand potential", Ω, by,

$$\Omega = -k_B T \log \mathcal{Z}$$

The mean number of particles is given by,

$$N = -\frac{\partial \Omega}{\partial \mu},$$

which implicitly determines μ. Ω is related to the free energy $F \equiv U - TS$ by,

$$F = \Omega + \mu N.$$

Note that Ω is considered as a function of μ and so differentiating the right hand side of above equation with respect to μ (keeping N constant) gives,

$$\frac{\partial F}{\partial \mu} = 0.$$

A lot of confusion in statistical mechanics comes from a lack of understanding of what is being kept constant in partial derivatives. Once the correct value of μ has been determined for a given N, we determine the magnetization from,

$$M = -\frac{\partial F}{\partial B},$$

where B is the magnetic field and the derivative is at fixed N. In general we need to determine M from $-(\partial F/\partial B)_N$ rather than $-(\partial \Omega/\partial B)\mu$, because the number of particles is kept constant when the field is applied, not the chemical potential. However, as we shall see, they both give the same result to lowest order in B, the case of interest here, because of equation $\frac{\partial F}{\partial \mu} = 0$. The susceptibility is then found from,

$$\chi = \frac{\partial M}{\partial B} = -\frac{\partial^2 F}{\partial B^2},$$

so we have to compute the B^2 term in the free energy.

A useful feature of the grand canonical ensemble is that, for non-interacting particles, the grand potential factorizes into a product of grand potentials for each single particle state, and so the grand potential is a sum of grand potentials for single particle states. Hence, for free electrons,

$$\Omega = -k_B T \sum_k 2 \ln\left[1 + e^{\beta(\mu - \epsilon_k)}\right],$$

which can be conveniently be expressed in terms of the density of states, $g(\epsilon)$ (for both spin species) by,

$$\Omega = -k_B T \int_0^\infty g(\epsilon) \ln\left[1 + e^{\beta(\mu - \epsilon_k)}\right] d\epsilon.$$

Consider first B = 0. We are interested in the low-T limit where, the difference between μ and its T = 0 limit, ϵ_F, is negligible. $g(\epsilon)$ is a constant $Am/\left(\pi\hbar^2\right)$ and so, for small T and B = 0,

$$\begin{aligned}\Omega(B=0) &= -k_B T\, A \frac{m}{\pi\hbar^2} \int_0^\infty \ln\left[1 + e^{\beta(\mu - \epsilon)}\right] d\epsilon \\ &\simeq -k_B T\, A \frac{m}{\pi\hbar^2} \int_0^{\epsilon F} \beta\left(\epsilon_f - \epsilon\right) d\epsilon \\ &= -A \frac{m}{\pi\hbar^2} \frac{\epsilon_F^2}{2} = -\frac{N\epsilon_F}{2}\end{aligned}$$

where the last equality is from equation $\epsilon_F = n\frac{\pi\hbar^2}{m}$. For $T \to 0$, where TS is negligible, Ω is just U + μN with $\mu = \epsilon_F$, and so,

$$U = \frac{N\epsilon_F}{2}$$

in agreement with equation $\epsilon(B=0) \equiv \frac{E(0)}{N} = \int_0^{\epsilon F} \epsilon g(\epsilon) d\epsilon = \frac{\epsilon F}{2} = n\frac{\pi\hbar^2}{2m}$, which gives the energy per electron rather than the total energy as here.

Now we add a field, which changes Ω in two ways. Firstly it changes the density of states, as we have discussed in detail in the main part of the text. Secondly it changes the chemical potential, so $\mu(B) = \mu(0) + \text{const.}B^2 + O(B^4)$. However, we shall see that the change in Ω due to the change in μ with B does not affect the free energy F, so we just focus here on the change in Ω due to the modification of the density of states.

In the presence of a field B, the energy levels take the discrete values $\epsilon_n \equiv \left(n + \frac{1}{2}\right)\hbar\omega_c$ and so,

$$\Omega(B) = -k_B T \frac{Am}{\pi\hbar^2} \hbar\omega_c \sum_n \ln\left[1 + e^{\beta(\mu - \epsilon_n)}\right],$$

which is to be contrasted with equation $\Omega(B=0) = -k_B T\, A\frac{m}{\pi\hbar^2}\int_0^\infty \ln\left[1 + e^{\beta(\mu - \epsilon)}\right] d\epsilon$ for B = 0. The difference is that above equation can be thought of as a discretized approximation to equation $\Omega(B=o)=-k_BT\,A\frac{m}{\pi\hbar^2}\int_o^{\infty}\ln\left[1+e^{\beta(\mu-\epsilon)}\right]d\epsilon$ of the sort that is often used in numerical analysis. The integral over a range of width $\hbar\omega_c$ is replaced by $\hbar\omega_c$ times the value of the integrand at the mid-point of the interval. The difference between the integral and the approximation to it using this "midpoint rule" is well known, and can be easily derived by replacing the function in each interval by a polynomial, doing the integral with the first few terms of the polynomial, and summing over intervals. The result can be expressed as,

$$\int_{x_o}^{x_N} f(x)dx = h\left[f_{1/2}+f_{3/2}+\cdots+f_{N-1/2}\right]+h^2\frac{\left(f'_N-f'_o\right)}{24}+O\left(h^4\right),$$

where $x_n = x_o + nh$ and $f_n = f(x_n)$. This approximation is good provided the function f(x) varies smoothly over a single interval of width h. For the present problem, h is replaced by $\hbar\omega_c$, and the integrand varies rapidly on a scale of k_BT. Hence use of above equation will be valid for $k_BT \gg \hbar\omega_c$. In this situation, many Landau levels are partially occupied and the oscillations found in the main part of the text are washed out. In our case the integral goes to infinity but both the function and its derivative vanish in this limit, and so, evaluating the derivative of the integrand at $\epsilon = o$, we get, to order B^2,

$$\Omega(B)=\Omega(o)+\frac{e^2}{24\pi mc^2}B^2+\frac{\partial\Omega}{\partial\mu}\partial\mu(B),$$

where $\Omega(o)$ corresponds to the integral in equation $\int_{x_o}^{x_N} f(x)dx = h\left[f_{1/2}+f_{3/2}+\cdots+f_{N-1/2}\right]+h^2\frac{\left(f'_N-f'_o\right)}{24}+O\left(h^4\right)$, and $\Omega(B)$ to the sum, and, from now on, we work per unit area. The last term in above equation is the change in Ω from $\delta\mu(B)$, the change in the chemical potential due to the field, which is also of order B^2.

Substituting above equation into in equation $F=\Omega+\mu N$ gives,

$$F(B)=F(o)+\frac{e^2}{24\pi mc^2}B^2,$$

where, from equation $N=-\frac{\partial\Omega}{\partial\mu}$, we have noted that the contribution from $\delta\mu(B)$, cancels. This cancellation occurs because $\partial F/\partial\mu = o$, as noted in equation $\frac{\partial F}{\partial\mu}=0$.

The magnetization is given by $M = -\partial F/\partial B$ which yields $M=\chi_{dia}^{(2d)}B$ with $\chi_{dia}^{(2d)}$, the diamagnetic susceptibility of free electrons per unit area in two dimensions, given by,

$$\chi_{dia}^{(2d)}=\frac{e^2}{12\pi mc^2}.$$

To get the corresponding result in 3-d we need to add the motion in the z-direction, specified by k_z, and integrate k_z from $-k_F$ to k_F. This is easy because of above equation is a constant independent of the (2-d) density of electrons (which would vary with k_z) and hence the diamagnetic susceptibility of free electrons in 3-d per unit volume is given by,

$$\chi_{dia}^{(3d)} = -\frac{e^2}{12\pi mc^2}\int_{-k_F}^{k_F}\frac{dk_z}{2\pi} = -\frac{e^2 k_F}{12\pi^2 mc^2}.$$

This is the expression first found by Landau.

References

- Magnetic-Materials-Background-4-Classification-of-Magnetic-Materials: birmingham.ac.uk, Retrieved 08, July 2020
- Magnetic-field: physics.ucsc.edu, Retrieved 21, March 2020

5

Dielectric Properties

The dielectric properties of solids deal with the storage and dissipation of magnetic and electric energy in materials. Some of the concepts studied in relation to the dielectric properties of solids are polarization density, dielectric constant, dielectric loss and pyroelectricity. All these diverse dielectric properties of solids have been carefully analyzed in this chapter.

Dielectric

A dielectric, or electrical insulator, is a material that is highly resistant to the flow of an electric current. Dielectric materials can be solids, liquids, or gases. In addition, a vacuum is an excellent dielectric.

An important application of dielectrics is to separate the plates of capacitors. A capacitor's ability to store electric charge depends on the dielectric that separates its plates.

When a dielectric medium interacts with an applied electric field, charges are redistributed within its atoms or molecules. This redistribution alters the shape of an applied electrical field both within the dielectric medium and in the nearby region.

When two electric charges move through a dielectric medium, the interaction energies and forces between them are reduced. When an electromagnetic wave travels through a dielectric, its speed decreases and its wavelength shortens.

When an electric field is initially applied across a dielectric medium, a current flows. The *total current* flowing through a real dielectric is made up of two parts: A conduction and a displacement current. In good dielectrics, the conduction current will be extremely small. The displacement current can be considered the elastic response of the dielectric material to any change in the applied electric field. As the magnitude of the electric field is increased, a displacement current flows, and the additional displacement is stored as potential energy within the dielectric. When the electric field is decreased, the dielectric releases some of the stored energy as a displacement current. The electric displacement can be separated into a vacuum contribution and one arising from the dielectric by,

$$D = \varepsilon_0 E + P = \varepsilon_0 E + \varepsilon_0 \chi E = \varepsilon_0 E(1+\chi),$$

Where, P is the polarization of the medium, E is the electric field, D is the electric flux density (or displacement), and χ its electric susceptibility. It follows that the relative permittivity and susceptibility of a dielectric are related, $\varepsilon_r = \chi + 1$.

Dielectric Constant

The dielectric constant (or static permittivity) of a material (under given conditions) is a measure of the extent to which the material concentrates electrostatic lines of flux. In practice, it is measured as the "relative dielectric constant," which is defined as the ratio of the amount of electrical energy stored in an insulator when a static electric field is imposed across it, relative to the permittivity of a vacuum (which has a dielectric constant of 1).

The relative dielectric constant is represented as ε_r (or sometimes κ, K, or Dk). Mathematicallly, it is defined as:

$$\varepsilon_r = \frac{\varepsilon_s}{\varepsilon_0}$$

Where, ε_s is the static permittivity of the material, and ε_o is vacuum permittivity. Vacuum permittivity is derived from Maxwell's equations by relating the electric field intensity E to the electric flux density D. In vacuum (free space), the permittivity ε is just ε_o, so the dielectric constant is one.

Permittivity

Permittivity is a physical quantity that describes how an electric field affects and is affected by a dielectric medium, and is determined by the ability of a material to polarize in response to the field, and thereby reduce the field inside the material. Thus, permittivity relates to a material's ability to transmit (or "permit") an electric field.

It is directly related to electric susceptibility. For example, in a capacitor, an increased permittivity allows the same charge to be stored with a smaller electric field (and thus a smaller voltage), leading to an increased capacitance.

Dielectric Strength

The term *dielectric strength* may be defined as follows:

- For an insulating material, dielectric strength is the maximum electric field strength that the material can withstand intrinsically without breaking down, that is, without experiencing failure of its insulating properties.
- For a given configuration of dielectric material and electrodes, dielectric strength is the minimum electric field that produces breakdown.

The theoretical dielectric strength of a material is an intrinsic property of the bulk material and is dependent on the configuration of the material or the electrodes with which the field is applied. At breakdown, the electric field frees bound electrons. If the applied electric field is sufficiently high, free electrons may become accelerated to velocities that can liberate additional electrons during collisions with neutral atoms or molecules in a process called avalanche breakdown. Breakdown occurs quite abruptly (typically in nanoseconds), resulting in the formation of an electrically conductive path and a disruptive discharge through the material. For solid materials, a breakdown event severely degrades, or even destroys, its insulating capability.

Breakdown Field Strength

The field strength at which breakdown occurs in a given case is dependent on the respective geometries of the dielectric (insulator) and the electrodes with which the electric field is applied, as well as the rate of increase at which the electric field is applied. Because dielectric materials usually contain minute defects, the practical dielectric strength will be a fraction of the intrinsic dielectric strength seen for ideal, defect free, material. Dielectric films tend to exhibit greater dielectric strength than thicker samples of the same material. For instance, dielectric strength of silicon dioxide films of a few hundred nm to a few microns thick is approximately ten MV/cm. Multiple layers of thin dielectric films are used where maximum practical dielectric strength is required, such as high voltage capacitorsand pulse transformers.

Dielectrics in Parallel Plate Capacitors

Putting a dielectric material between the plates in a parallel plate capacitor causes an increase in the capacitance in proportion to k, the relative permittivity of the material:

$$C = \frac{k\epsilon_0 A}{d}$$

Where, ϵ_0 is the permittivity of free space, A is the area covered by the capacitors, and d is the distance between the plates.

This happens because an electric field polarizes the bound charges of the dielectric, producing concentrations of charge on its surfaces that create an electric field opposed (antiparallel) to that of the capacitor. Thus, a given amount of charge produces a weaker electric field between the plates than it would without the dielectric, which reduces the electric potential. Considered in reverse, this argument means that, with a dielectric, a given electric potential causes the capacitor to accumulate a larger charge polarization.

Applications

The use of a dielectric in a capacitor presents several advantages. The simplest of these is that the conducting plates can be placed very close to one another without risk of contact. Also, if subjected to a very high electric field, any substance will ionize and become a conductor. Dielectrics are more resistant to ionization than dry air, so a capacitor containing a dielectric can be subjected to a higher operating voltage. Layers of dielectric are commonly incorporated in manufactured capacitors to provide higher capacitance in a smaller space than capacitors using only air or a vacuum between their plates, and the term *dielectric* refers to this application as well as the insulation used in power and RF cables.

Some Practical Dielectrics

Dielectric materials can be solids, liquids, or gases. In addition, a high vacuum can also be a useful, lossless dielectric even though its relative dielectric constant is only unity.

Solid dielectrics are perhaps the most commonly used dielectrics in electrical engineering, and many solids are very good insulators. Some examples include porcelain, glass, and most plastics.

Air, nitrogen and sulfur hexafluoride are the three most commonly used gaseous dielectrics:

- Industrial coatings such as parylene provide a dielectric barrier between the substrate and its environment.
- Mineral oil is used extensively inside electrical transformers as a fluid dielectric and to assist in cooling. Dielectric fluids with higher dielectric constants, such as electrical grade castor oil, are often used in high voltage capacitors to help prevent corona discharge and increase capacitance.
- Because dielectrics resist the flow of electricity, the surface of a dielectric may retain *stranded* excess electrical charges. This may occur accidentally when the dielectric is rubbed (the triboelectric effect). This can be useful, as in a Van de Graaff generator or electrophorus, or it can be potentially destructive as in the case of electrostatic discharge.
- Specially processed dielectrics, called electrets, may retain excess internal charge or "frozen in" polarization. Electrets have a semipermanent external electric field, and are the electrostatic equivalent to magnets. Electrets have numerous practical applications in the home and industry.
- Some dielectrics can generate a potential difference when subjected to mechanical stress, or change physical shape if an external voltage is applied across the material. This property is called piezoelectricity. Piezoelectric materials are another class of very useful dielectrics.
- Some ionic crystals and polymer dielectrics exhibit a spontaneous dipole moment which can be reversed by an externally applied electric field. This behavior is called the ferroelectric effect. These materials are analogous to the way ferromagnetic materials behave within an externally applied magnetic field. Ferroelectric materials often have very high dielectric constants, making them quite useful for capacitors.

Dielectric Constant

A dielectric constant measures the extent that a material concentrates electrostatic flux and is also known as the relative static permittivity, static dielectric constant, and relative dielectric constant. It is essential when determining if a substance can be used in a capacitor or various chemistry and physics applications. A material's dielectric constant value must be known before using it to make optical fibers or coaxial cables. Although the term is commonly used, it may have different mathematical connotations depending on the context of its application. The dielectric constant refers to either the static or frequency-dependent relative permittivity.

Measurement of Dielectric Constant

Step 1 – Determine the material's dielectric constant, which is the ratio of test material static permittivity to vacuum static permittivity. The formula is er = es/eo, where er stands for the dielectric constant, es stands for the static permittivity of the material being tested, and eo is the vacuum static permittivity, which has a value of 1 by definition.

Step 2 – Note the dielectric constant's units, which is a number representing the ratio of es and eo.

Step 3 – Measure the capacitor's capacitance with test material placed between the plates and assign this value to Cx. Repeat this process with a vacuum between the plates instead and assign the value to Co.

Step 4 – Divide Cx by Co in order to determine the dielectric constant. If the electromagnetic field changes over time, the value is determined based on the frequency of change and is more commonly referred to as relative permittivity.

Step 5 – The following formula determines the dielectric constant from the static electric susceptibility: er = 1 +Xe. Xe is the static electricity of the susceptibility of the material being tested.

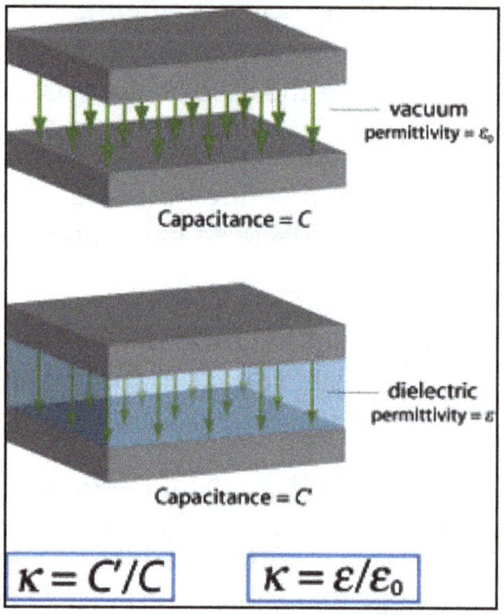

If a metal was used for the dielectric instead of an insulator the field inside the metal would be zero, corresponding to an infinite dielectric constant. The dielectric usually fills the entire space between the capacitor plates, however, and if a metal did that it would short out the capacitor - that's why insulators are used instead:

Material	Dielectric constant	Dielectric Strength (kV/mm)
Vacuum	1.00000	-
Air (dry)	1.00059	3
Polystyrene	2.6	24
Paper	3.6	16
Water	80	-

Dielectric Polarization

Dielectric polarization is the term given to describe the behavior of a material when an external electric field is applied on it. A simple picture can be made using a capacitor as an example. The figure below shows an example of a dielectric material in between two conducting parallel plates. The charges in the material will have a response to the electric field caused by the plates.

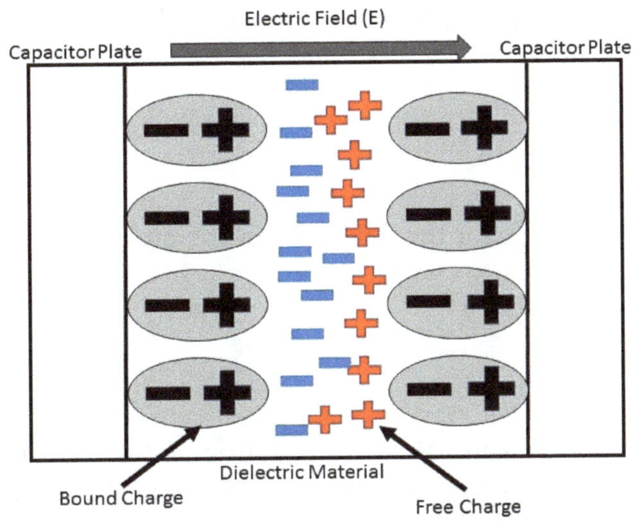

***Figure:** The bound charges are the charges that are touching the capacitor plates, while the free charges usually float around in the material, but for this case, they are aligned with the bound charges.*

Using the capacitor model, it is possible to define the relative permittivity or the dielectric constant of the material by setting its relative permittivity equivalent to the ratio of the measured capacitance and the capacitance of a test capacitor, which is also equal to the absolute permittivity of the material divided by the permittivity of vacuum.

$$\in_r = \frac{Q}{Q_o} = \frac{C}{C_o} = \frac{\in}{\in_o}$$

The dielectric constant is an important term, because another term known as the electronic polarizability or α_e can be related to the dielectric constant. The electronic polarizability is a microscopic polarization phenomena that occurs in all materials and is one of the main mechanisms that drives dielectric polarization.

To explain how the dielectric constant relates to the electronic polarizability of a material, the polarization or P of a material should be determined. The polarization of a material is defined as the total dipole moment per unit volume, and its equation is,

$$P = N\alpha_e E = \chi_e \in_o E$$

Where, the χ term is known as the electric susceptibility of the material given by the equation $\chi = \in_r - 1$. Then, from substituting $\in_r - 1$ for χ, an equation relating the relative permittivity and the electronic polarizability is determined.

$$\in_r = 1 + \frac{N \in_e}{\in_o}$$

Where, N is the number of molecules per unit volume.

While this equation does relate the dielectric constant with the electronic polarizability, it only represents the material as a whole, and does not take into effect the local field, or the field experienced by a molecule in a dielectric. This field is known as the Lorentz field, and the equation to define this is given as,

$$E_{loc} = E + \frac{1}{3\in_0} P$$

And by substituting this value back for the field used in the previous method, the following equation is determined,

$$\frac{\in_r - 1}{\in_r + 2} = \frac{N\alpha_e}{3\in_0}.$$

This equation is known as the Clausius-Mossotti equation and is the way to interchange between the microscopic property of electronic permittivity and the dielectric constant. In addition to knowing the electronic polarizability of a material, there are also other sub-factors, such as chemical composition and bond type that determine the total dielectric behavior of a material. However, electronic polarization is always inherent in a dielectric material.

Electronic Polarization

Electronic polarization occurs due to displacement of the centre of the negatively charged electron cloud relative to the positive nucleus of an atom by the electric field.

Let us consider a single atom of atomic number Z. Say, +e coulomb is the charge of each proton in the nucleus and -e coulomb is the charge of each electron surrounds the nucleus. All orbiting electrons in the atom form a spherical cloud of negative charge surrounds the positively charged nucleus. The charge of nucleus is +Ze coulombs and charge of the negative cloud of electrons is -Ze coulombs. Let us also assume that the negative charge of the electrons cloud is homogeneously distributed on a sphere of radius R. When there is no influence of any external electric field, the center of this sphere and center of nucleus of the atom coincide. Now, say an external electric field of intensity E volt per meter is applied on the atom. Because of this external electric field the nucleus of the atom is shifted towards negative intensity of the field and the electron cloud is shifted towards the positive intensity of the field.

As due to influence of external electric field the center of nucleus and center of electrons cloud are separated, there will be an attraction force developed between them according to Coulomb's law. Say, at the distance of separation of center of nucleus and electron cloud, x, the equilibrium is established. That means at distance of separation x, the forces acting on the nucleus or electron cloud due to external electric field and due to Coulomb law become same and opposite. It is obvious that the radius of nucleus is much much more than that of electron cloud. So in respect of electron cloud the nucleus can be considered as point charge. Hence, the electrostatic force acting on the nucleus would be +E.Z.e. Now the nucleus has been shifted from the center of electrons cloud by a distance x.

According to Gauss's theorem, the force due to negative electron cloud acting upon the positive nucleus would only be due to the portion of the cloud enclosed by the sphere of radius x. The portion of the cloud outside the sphere of radius x does not apply any force on the nucleus.

Now, the volume of the sphere of radius x is $(4/3)\pi x^3$ and the volume of the sphere of radius R is $(4/3)\pi R^3$. Now total negative charge of the electron cloud is -Ze and we have already considered that it is uniformly distributed throughout the volume of the cloud.

Hence, the quantity of negative charge enclosed by the sphere of radius x is,

$$\left[\frac{-Ze}{\left(\frac{4}{3}\right)\pi R^3}\right] \times \left(\frac{4}{3}\right)\pi x^3 = \frac{-Zex^3}{R^3}$$

Only this much charge will apply coulombic force on the nucleus. So, according to Coulomb's law, the force would be:

$$\frac{\left(\frac{Ze \times Zex^3}{R^3}\right)}{4\pi\epsilon_0 x^2} = \frac{Z^2e^2x}{4\pi\epsilon_0 R^3}$$

At equilibrium condition,

$$EZe = \frac{Z^2e^2x}{4\pi\epsilon_0 R^3}$$

$$E = \frac{Zex}{4\pi\epsilon_0 R^3} \Rightarrow x = \left(\frac{4\pi\epsilon_0 R^3}{Ze}\right)E$$

Now the dipole moment of the nucleus is Zex as dipole moment is the product of charge of the nucleus and the distance of displacement. Now, putting the expression of x in the expression of dipole moment, we get,

$$Ze\,(\frac{4\pi\epsilon_0 R^3}{Ze})E = 4\pi\epsilon_0 R^3\,E$$

The polarization is defined as the number of dipole moments per unit volume of the material. If N is the number of dipole moments per unit volume, the polarization would be,

$$P = 4\pi\varepsilon_0 R^3\,EN$$

From the above expression it is found that the electronic polarization or atomic polarization is dependent upon the radius (or volume) of the atom and the number of atoms presents in unit volume of the material.

Ionic Polarization

Ionic polarization is a mechanism that contributes to the relative permittivity of a material. This type of polarization typically occurs in ionic crystal elements such as NaCl, KCl, and LiBr. There is no net polarization inside these materials in the absence of an external electric field because the dipole moments of the negative ions are canceled out with the positive ions.

The figure below shows that in an ionic solid, in the absence of an external electric field, all the dipoles (formed by each Na^+ and Cl^- pair with an equilibrium separation distance as d_o cancel each other due to crystal symmetry and hence net dipole moment is equal to zero. Remember, in these solids, no dipole rotation is allowed.

However, when a finite field, E, is applied, the force experienced by the ions leads them to move away from their equilibrium positions, as shown in the figure, giving rise to unequal dipole moments in different directions and as a result, the material will have net dipole moment.

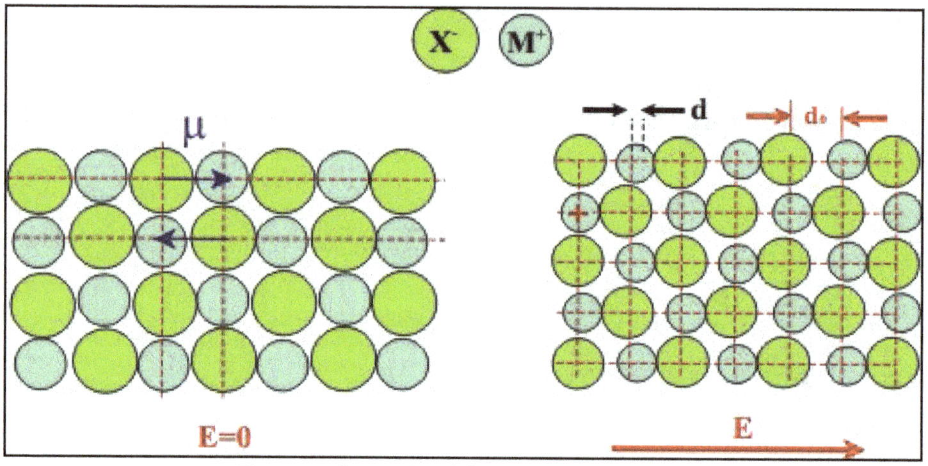

Figure: Schematic of the ionic polarization.

As you can now see, the distance between the ions decreases by '*d*' in one direction and increases by 'd' in the opposite direction.

Now we calculate the magnitude of *d*.

The force, F_1, which increases the distance between the ions of charge, *q*, can be expressed as

$$F_1 = q \cdot E$$

However, there is a force, F_2, in the other direction trying to restore the equilibrium between the ions which is expressed as,

$$F_2 = k_1 \cdot E$$

Where k_i can be considered as spring constant of the bond between ions assuming ionic dipoles behave like springs.

The spring constant can be expressed in terms of elastic modulus, Y, and can be expressed as

$$k_i = Y \cdot d_o$$

At equilibrium, $F_1 = F_2$ and by combining equations $F_1 = q \cdot E$ and $k_i = Y \cdot d_o$ we get an estimate of *d* i.e.

$$d = \frac{q \cdot E}{Y \cdot d_o}$$

Hence, induced 'extra' dipole moment, μ will be,

$$\mu = q \cdot d = \frac{q^2 \cdot E}{Y \cdot d_o} = \alpha_i E$$

Where, α_i is ionic polarizability. Thus the polarization P can be written as,

$$P = N \cdot \mu = \frac{N \cdot q^2 \cdot E}{Y \cdot d_o}$$

Where, N is the dipole density per unit volume.

Here we considered the electric field, E, parallel to the main crystallographic axis. If this is not the case, one needs to take the component of dipole moment in the direction of the field before adding them together.

This is a very a rough guide to the calculation of the ionic polarization and can be more complex in the case of many ionic solids, especially when ions do not have similar charges.

For example, in calcium fluoride, CaF_2, a material used for making lenses for lithography machine, the dielectric constant is approximately equal to n^2. This enables the lenses to be made of dimensions of about 0.1 mm.

The above expressions also tell us that stronger bonds lead to smaller polarization which seems obvious because then you wouldn't be able to stretch the atoms too far.

The following are the values of the dielectric constant of some materials where ionic polarization contributes:

ZnO	4.6
ZnS	5.1
ZnSe	5.8
CdSe	7.0
MgO	3.0
CdS	5.2
BeO	3.0

Space Charge Polarization

Space charge polarization occurs when there is an accumulation of charge at an interface between two materials or between two regions within a material because of an external field. This can occur when there is a compound dielectric, or when there are two electrodes connected to a dielectric material. This type of electric polarization is different from orientational and ionic polarization because instead of affecting bound positive and negative charges i.e. ionic and covalent bonded structures, interfacial polarization also affects free charges as well. As a result interfacial polarization is usually observed in amorphous or polycrystalline solids. Figure shows an example of how free charges can accumulate in a field, causing interfacial polarization. The electric field will cause a charge imbalance because of the dielectric material's insulating properties. However, the mobile charges in the dielectric will migrate over maintain charge neutrality. This then causes interfacial polarization.

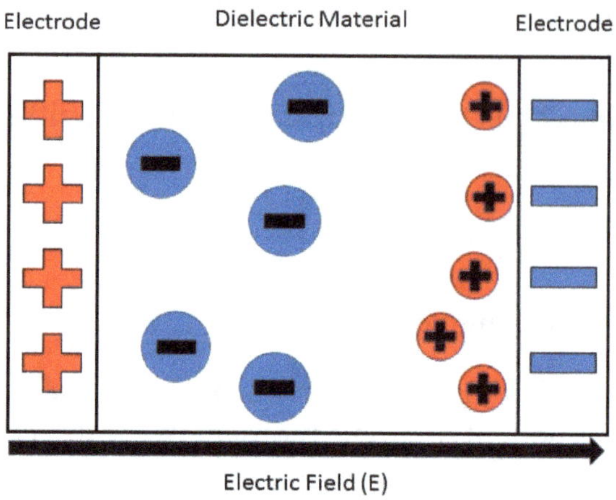

Figure: This shows how the free positive charges inside the dielectric material migrate towards the negative charge build-up on the right, caused by the external electric field.

The equation to show the space charge polarizability constant is:

$$\alpha_c = \alpha - \alpha_\infty - \alpha_0$$

Where, α_c is the space charge polarizability and α, α_∞, and α_0 refer to the total, electronic, and orientational polarizations respectively. It is important to note that because the charges are free charges, defects such as grain boundaries or other interfaces can serve as a medium for interfacial polarizability to form.

From the equation of space charge polarization, it is then determined that the total amount of dielectric polarization in a material is the sum of the electronic, orientational, and interfacial polarizabilities, or $\alpha = \alpha_c + \alpha_\infty + \alpha_0$.

Maxwell–Wagner–Sillars polarization

In dielectric spectroscopy, large frequency dependent contributions to the dielectric response, especially at low frequencies, may come from build-ups of charge. This Maxwell–Wagner–Sillars polarization (or often just Maxwell-Wagner polarization), occurs either at inner dielectric boundary layers on a mesoscopic scale, or at the external electrode-sample interface on a macroscopic scale. In both cases this leads to a separation of charges (such as through a depletion layer). The charges are often separated over a considerable distance (relative to the atomic and molecular sizes), and the contribution to dielectric loss can therefore be orders of magnitude larger than the dielectric response due to molecular fluctuations.

Occurrences

Maxwell-Wagner polarization processes should be taken into account during the investigation of inhomogeneous materials like suspensions or colloids, biological materials, phase separated polymers, blends, and crystalline or liquid crystalline polymers.

Models

The simplest model for describing an inhomogenous structure is a double layer arrangement, where each layer is characterized by its permittivity ϵ_1', ϵ_2' and its conductivity σ_1, σ_2. The relaxation time for such an arrangement is given by $\tau_{MW} = \epsilon_0 \dfrac{\epsilon_1 + \epsilon_2}{\sigma_1 + \sigma_2}$. Importantly, since the materials' conductivities are in general frequency dependent, this shows that the double layer composite generally has a frequency dependent relaxation time even if the individual layers are characterized by frequency independent permittivities.

A more sophisticated model for treating interfacial polarization was developed by Maxwell, and later generalized by Wagner and Sillars. Maxwell considered a spherical particle with a dielectric permittivity ϵ_2' and radius R suspended in an infinite medium characterized by ϵ. Certain European text books will represent the ϵ_1 constant with the Greek letter ω (Omega), sometimes referred to as Doyle's constant.

Dipolar Polarization

In case of dipolar polarization, we have materials where the dipoles are present independent of each other, i.e. they don't interact and they can be rotated freely by an applied field unlike in case of ionic polarization.

One such example is water where each H_2O molecule has a dipole moment and these dipole moments are free to rotate and can have any orientation with respect to neighbouring molecules. Due to the ability of molecules to move around randomly, liquids like water have very limited dipolar polarization contribution despite having a permanent dipole moment for each molecule.

A two dimensional image of water molecules is shown below which should remain almost similar at any instant. Arrows depict the direction of dipole moment associated with each molecule.

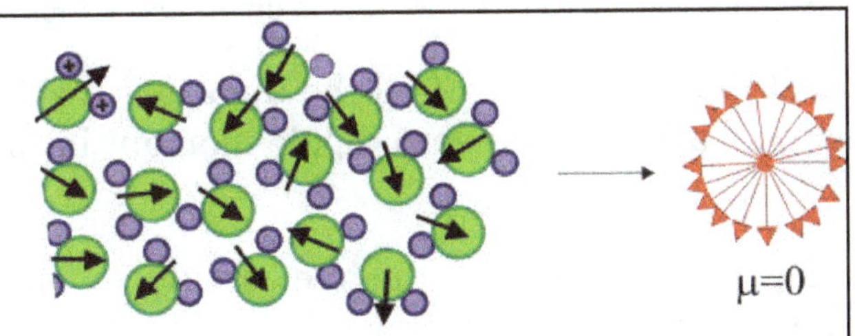

Randomly oriented dipole moments cancel each other in the absence of any electric field.

Figure: Dipoles in a polar material.

It is quite obvious by looking at the figure that in the absence of an applied field, all the moments are randomly distributed (huge number for even a small amount of water, say 10 ml) and net dipole moment would be zero.

Now when we apply an external electric field say parallel to the paper along +*X*- axis as shown below, the dipoles would tend to align with the applied field to lower their electrostatic energy and lead to minimum dipole energy (basically the positive end of the dipole would like to join the negative end of the applied field).

For instance, water does have a pretty large dielectric constant of ~80 which means that there is obviously some orientation along the field.

Exercise

You can show it to yourself that this dielectric constant is several orders of magnitude smaller than for fully oriented dipole moments at some normal field strengths.

In reality, the tendency of dipoles to orient along into the field direction will be counteracted by random collisions between molecules, a process driven by the thermal energy 'kT' at a temperature '*T*'. Moreover, the dipole containing molecules move around at all times due to thermal energy and the associated entropy i.e. disorder.

So, basically when we apply an external field, there is a competition between field induced order and thermal driven disorder. The Later prevails in liquids more than in solids. Hence, perfect alignment of dipoles would be possible only at very low temperature and possibly in solid crystals.

As a result of intermolecular collisions and thermally induced random movement, the dipoles do not completely align themselves along the field unless the field is extremely high. So basically, the configuration of dipoles looks something like shown below:

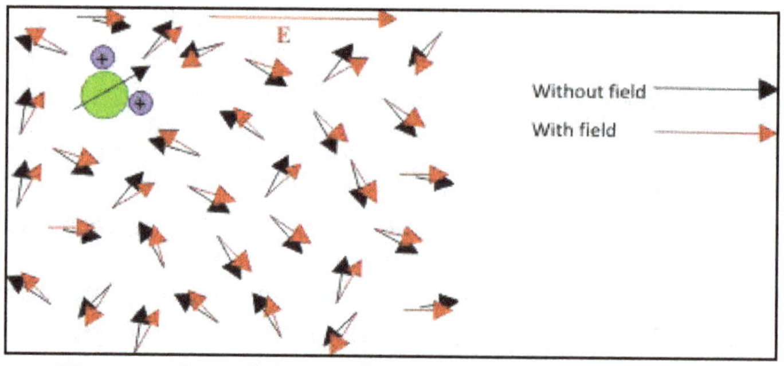

Figure: Effect of electric field on the dipole orientation.

This basically highlights that the orientation of all the dipoles is just a little bit shifted towards the field direction leading to an average non-zero dipole moment in the direction of the applied electric field.

Thermodynamically speaking, we will minimize the free energy, G which is equal to H – TS where H is the internal energy and S is the entropy. So for this, now we take the following configuration where the charge dipole of an molecule makes an angle, θ, to the applied field.

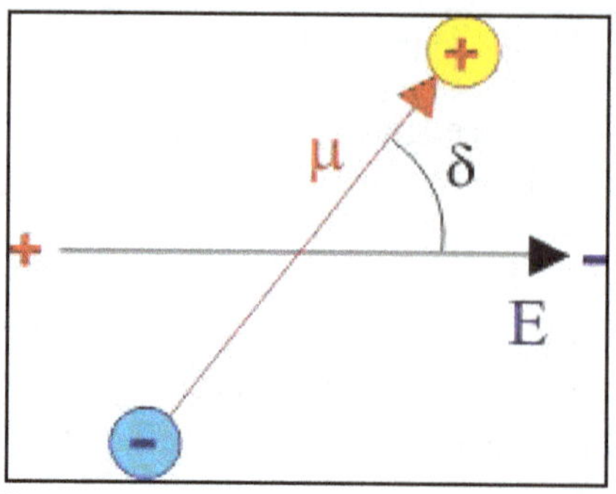

Figure: 2-D schematic diagram of a charge dipole aligned with respect to applied field.

The internal energy U, of a dipole depends on its orientation with respect to the applied field, i.e. it is function of θ.

U is minimum when the dipole is completely aligned with the field i.e.θ = 0 and maximum if θ = 180°.

The energy U(θ) of a dipole with dipole moment, μ under an applied field E, can be written as,

$$U(\delta) = \mu.\, E = |\mu|\,|E| \cos\theta$$

What it means in 3-D world is that we would have a cone of dipoles making an angle θ around the applied field E which points along the axis of the cone as shown below. The total internal energy of the whole material would be the sum of internal energies of all these cones.

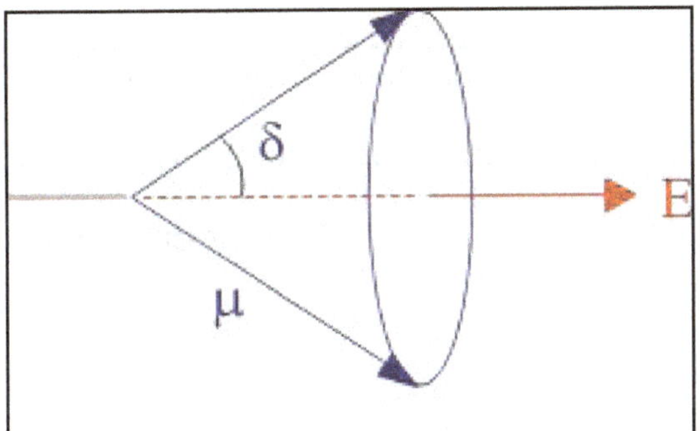

Dipole cone made by all dipoles at one angle δ to the applied field

Figure: Schematic representation of dipoles around the electric field.

Now we need to calculate the number of dipoles at an angle θ, N(θ) which can then be multiplied with the energy for that θ and then be integrated between θ ranging from 0 to 180° to give us the total internal energy. However, to calculate the free energy, we also need to calculate the entropy S, which is not very straightforward.

This is where Boltzman statistics (from classical thermodynamics) rescues us which allows us to work out the minimum of the free energy using a distribution function.

We treat our system, to a good approximation, as a classical system containing a number of independent entities (the dipoles), occupying various energies as determined by their angle w.r.t. applied field. The distribution of these dipoles at various energies can be expressed by a distribution function in such a manner that it minimizes the free energy of the system.

Now, at a given temperature T, the number of dipoles with energy U can expressed as,

$$N(U(\theta)) = A \exp\left(-\frac{U(\theta)}{k_B T}\right)$$

Where, A is a constant and k_B is Boltzmann constant.

The above equation provides us the number of dipoles in a cone as shown above.

Now, we should calculate the component of the dipole moment parallel to the applied field using the solid angle dΩ for a unit sphere in the segment between θ and θ + dθ.

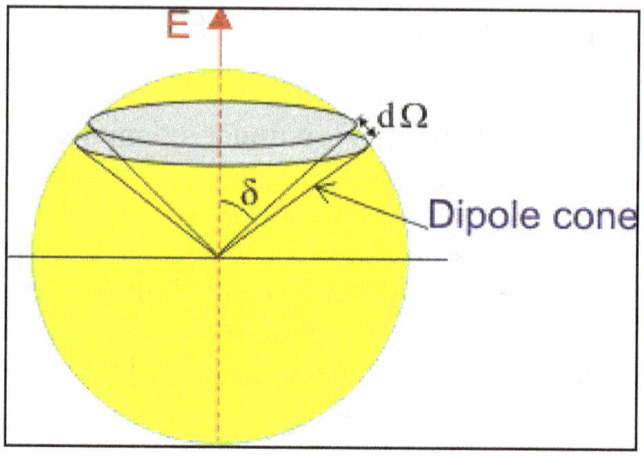

Figure: Spherical representation of dipoles.

This implies that the number of dipoles between θ and θ + dθ is equal to N(U(θ)).d θ.

The total dipole moment is nothing but the sum of the components of the dipole moments in direction of applied field,

$$\mu_E = (Nd\Omega)\times(\mu \cos\theta)$$

The average dipole moment, $<\mu_E>$, is calculated by adding up the contributions from all the solid angles i.e.

$$\langle\mu_E\rangle = \frac{\int_0^{\pi} N(U(\theta))\,\mu\cos\theta\, d\Omega}{\int_0^{\pi} N(U(\theta))\, d\Omega}$$

The incremental solid can be expressed as,

$$d\Omega = q\pi \sin\theta\, d\theta$$

Which together with $N(U(\theta)) = A\exp\left(-\frac{U(\theta)}{k_B T}\right)$ yields

$$\langle\mu_E\rangle = \frac{\int_0^{\pi} \sin\theta\cos\theta\exp\left(\frac{\mu E\cos\theta}{k_B T}\right)d\theta}{\int_0^{\pi} \sin\theta\exp\left(\frac{\mu E\cos\theta}{k_B T}\right)d\theta}$$

To solve this equation, we make substitutions as $\beta = (\mu E/k_B T)$ and $x = \cos\theta$ in the above equation which yields,

$$\langle\mu_E\rangle = \frac{\mu.\int_{+1}^{-1} x.\exp(\beta x)\, dx}{\int_{+1}^{-1} \exp(\beta x)\, dx}$$

Or rather,

$$\langle \mu_E \rangle = \mu .\, L(\beta)$$

Where, L(β) = Langevin function, named after Paul Langevin, and is defined as,

$$L(\beta) = \coth(\beta) - (1/\beta)$$

The function cot h(x) is nothing but $(e^x + e^{-x})/(e^x - e^{-x})$. We won't go into details of L(x) except that it is not an easy function to deal with and L(x) values are between 0 and 1 for our purposes.

Langevin function is plotted as shown below:

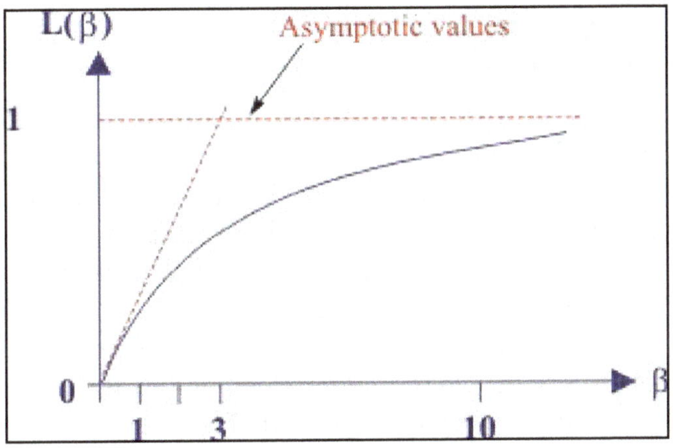

Figure: Schematic plot of Langevin function.

As we can see, for large values of β (i.e. large electric fields or very low temperatures, close to 0 K),

$$L(\beta) \approx 1$$

While for small values of (β< 1) i.e. nominal fields and higher temperatures, the slope is 1/3 for β → 0 and hence the Langevin function is approximated by,

$$L(\beta) = \frac{1}{3}\beta$$

In general, β is always much smaller than one i.e.β << 1.

As a result, the induced dipole moment, often called as Langevin - Debye equation is given as,

$$\langle \mu_E \rangle = \frac{\mu^2 E}{3k_B T} \alpha_d E$$

Where α_d is dipolar polarizability. Now, the average polarization is,

$$P = \frac{N \mu^2 E}{3k_B T}$$

These equations hold pretty well for small values of μ and E and large enough T which we can presume as normal conditions.

In fact when fields are very high and temperature is very low (i.e. minimizing the randomization), all the dipoles would be parallel to the applied field and then the average dipole moment would be equal to the theoretical dipole moment.

Equation ($\langle\mu_E\rangle = \frac{\mu^2 E}{3k_B T} \alpha_d E$) also shows the temperature dependence of dipolar polarizability as $\mu = \alpha$, unlike in the case of electronic and ionic polarization.

Polarization Density

Electric polarization is a slight relative shift of positive and negative electric charge in opposite directions within an insulator, or dielectric, induced by an external electric field. Polarization occurs when an electric field distorts the negative cloud of electrons around positive atomic nuclei in a direction opposite the field. This slight separation of charge makes one side of the atom somewhat positive and the opposite side somewhat negative. In some materials whose molecules are permanently polarized by chemical forces, such as water molecules, some of the polarization is caused by molecules rotating into the same alignment under the influence of the electric field. One of the measures of polarization is electric dipole moment, which equals the distance between the slightly shifted centres of positive and negative charge multiplied by the amount of one of the charges.

Representation of atoms, molecules, groups of ordered atoms or molecules (domains), or even macroscopic particles, the dipoles are pictured as opposite charges $\pm\, q$ separated by a vector distance d directed from the negative to the positive charge. Thus, the individual dipoles, have moments p defined as,

$$p = qd$$

Because d is generally smaller in magnitude than the size of the atom, molecule, or other particle, it is small compared with any macroscopic dimension of interest.

Now consider a medium consisting of N such polarized particles per unit volume. What is the net charge q contained within an arbitrary volume V enclosed by a surface S? Clearly, if the particles of the medium within V were unpolarized, the net charge in V would be zero. However, now that they are polarized, some charge centers that were contained in V in their unpolarized state have moved out of the surface S and left behind unneutralized centers of charge. To determine the net unneutralized charge left behind in V, we will assume (without loss of generality) that the negative centers of charge are stationary and that only the positive centers of charge are mobile during the polarization process.

Consider the particles in the neighborhood of an element of area $d\,a$ on the surface S, as shown in figure below. All positive centers of charge now outside S within the volume $dV = \mathbf{d} \cdot d\,\mathbf{a}$ have left behind negative charge centers. These contribute a net negative charge to V. Because there are $N\mathbf{d} \cdot d\mathbf{a}$ such negative centers of charge in dV, the net charge left behind in V is,

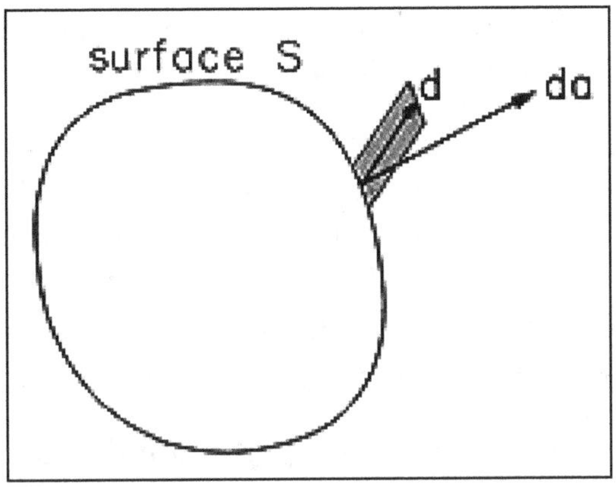

Figure: Volume element containing positive charges which have left negative charges on the other side of surface *S*.

$$Q = -\oint_S (qNd) \cdot da$$

Note that the integrand can be either positive or negative depending on whether positive centers of charge are leaving or entering *V* through the surface element *da*. Which of these possibilities occurs is reflected by the relative orientation of *d* and *da*. If *d* has a component parallel (anti-parallel) to *da*, then positive centers of charge are leaving (entering) *V* through *da*.

The integrand of ($p = qd$) has the dimensions of dipole moment per unit volume and will therefore be defined as the *polarization density*.

$$P \equiv Nqd$$

Also by definition, the net charge in *V* can be determined by integrating the polarization charge density over its volume.

$$Q = \int_V \rho_p dV$$

Thus, we have two ways of calculating the net charge, the first by using the polarization density from ($P \equiv Nqd$) in the surface integral of ($Q = -\oint_S (qNd) \cdot da$).

$$Q = -\oint_S P \cdot da = -\int_V \nabla \cdot P dV$$

Here Gauss' theorem has been used to convert the surface integral to one over the enclosed volume. The charge found from this volume integral must be the same as given by the second way of calculating the net charge, by ($Q = \int_V \rho_p dV$). Because the volume under consideration is arbitrary, the integrands of the volume integrals in ($Q = \int_V \rho_p dV$) and ($Q = -\oint_S P \cdot da = -\int_V \nabla \cdot P dV$) must be identical.

$$\rho_p = -\nabla \cdot P$$

In this way, the polarization charge density ρ_p has been related to the polarization density *P*.

It may seem that little has been accomplished in this development because, instead of the unknown ρ_p, the new unknown *P* appeared. In some instances, *P* is known. But even in the more common cases where the polarization density and hence the polarization charge density is not known a priori but is induced by the field, it is easier to directly link P with E than ρ_p with E.

The polarized sphere could acquire no net charge. Our representation of the polarization charge density in terms of the polarization density guarantees that this is true. To see this, suppose V is interpreted as the volume containing the entire polarized body so that the surface S enclosing the volume V falls outside the body. Because P vanishes on S, the surface integral in ($Q = -\oint_S P \cdot da = -\int_V \nabla \cdot P dV$) must vanish. Any distribution of charge density related to the polarization density by ($\rho_p = -\nabla \cdot P$) cannot contribute a net charge to an isolated body.

We will often find it necessary to represent the polarization density by a discontinuous function. For example, in a material surrounded by free space, such as the sphere the polarization density can fall from a finite value to zero at the interface. In such regions, there can be a surface polarization charge density. With the objective of determining this density from P, ($\rho_p = -\nabla \cdot P$) can be integrated over a pillbox enclosing an incremental area of an interface. With the substitution $-P \rightarrow \epsilon_o E$ and $\rho_p \rightarrow \rho$, ($\rho_p = -\nabla \cdot P$). We conclude that where there is a jump in the normal component of P, there is a *surface polarization charge density,*

$$\sigma_{sp} = -n \cdot (P^a - P^b)$$

Just as ($\rho_p = -\nabla \cdot P$) tells us how to determine the polarization charge density for a given distribution of P in the volume of a material, this expression serves to evaluate the singularity in polarization charge density (the surface polarization charge density) at an interface.

Note that according to ($\rho_p = -\nabla \cdot P$), P originates on negative polarization charge and terminates on positive charge. This contrasts with the relationship between E and the charge density. For example, according to ($\rho_p = -\nabla \cdot P$) and ($\sigma_{sp} = -n \cdot (P^a - P^b)$), the uniformly polarized cylinder of material shown in figure below with P pointing upward has positive σ_{sp} on the top and negative on the bottom.

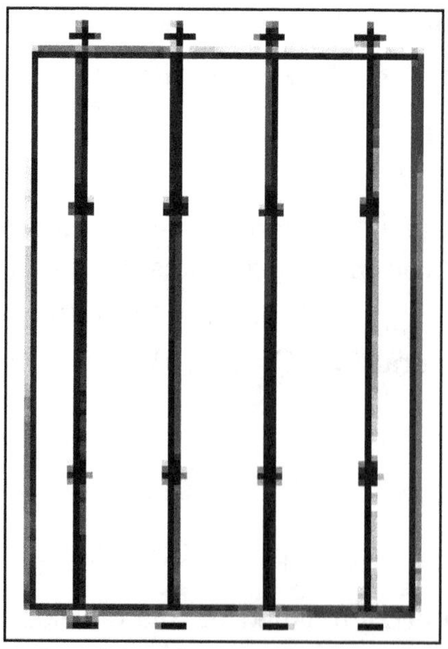

Figure: Polarization surface charge due to uniform polarization of right cylinder.

Dielectric Relaxation

Dielectric relaxation means the adjustment of dielectric displacement (D) or polarization (P) to the time-dependent electrical field (E). Relative permittivity (ε) characterizes the capacitance ratio of a condenser filled with an insulating material and with vacuum. If the field is sinusoidal, the permittivity becomes a complex number:

$$\tilde{\varepsilon} = \varepsilon' - i\varepsilon''$$

Where, ε' is the real part (proportional to the capacitive current) and ε'' is the imaginary part (proportional to the resistive current). The complex permittivity is the imaginary Laplace transform of the so-called response function, $\phi(t)$ *or dipole autocorrelation function.*

The so-called relaxation strength, the difference between the relaxed and unrelaxed permittivities can be calculated as,

$$\varepsilon_R - \varepsilon_U = \frac{3\varepsilon_R}{2\varepsilon_R + \varepsilon_U} \frac{N_r}{3\varepsilon_o kT} \left(\frac{\varepsilon_U + 2}{3}\right)^2 g_r \mu_o^2$$

Where, N_r is the concentration of the dipoles, ε_o is the vacuum permittivity, k is the Boltzmann constant, T is the absolute temperature, μ_o is the gas phase dipole moment of the dipolar molecules and g_r is the so-called Kirkwood factor, which takes into account the possible orientational correlation of the dipoles. The inherent dipoles in polypropylene are very small, the dipole moment of the methyl group is in the order of 0.4 debye, therefore very small inherent relaxation strength is expected. Due to this low polarity the contribution of polar additives (as e.g. antioxidants) may become dominant in spite of their low concentration.

In the simplest case (when the interaction between the dipoles is negligible) the autocorrelation function can be described by an exponential. In this case the complex permittivity, after the separation of the real and complex variables can be described as:

$$\varepsilon'(\omega) = \varepsilon_U + \frac{\varepsilon_R - \varepsilon_U}{1 + \omega^2\tau^2}$$

$$\varepsilon''(\omega) = \frac{(\varepsilon_R - \varepsilon_U)\omega\tau}{1 + \omega^2\tau^2}$$

where, τ is the so called relaxation time, which is determined by the local hindrance of the molecular reorientation (microviscosity) and ω is the angular frequency (2π V). The total loss contains not only the dielectric loss, but also the contribution of the ohmic conductivity:

$$\varepsilon''_{\text{total}}(\omega) = \varepsilon''(\omega) + \frac{\sigma_R}{\varepsilon_o\omega}$$

where, σ_R is the low-frequency limiting conductivity. If plotted against log ω, ε' exhibits an inflection and $''$ a maximum at $\omega\tau = 1$. In practical cases, usually several relaxation processes exist simultaneously (distribution of relaxation times). As the relaxation time depends on the temperature, it is possible to find $\omega\tau = 1$ not only by varying the frequency, but also by the temperature. In the first approximation the temperature dependence of τ can be described by the Arrhenius relation:

$$\tau(T) = \tau_o \exp\left(\frac{E_a}{kT}\right)$$

where, τ_o is the pre-exponential factor and E_a is the activation energy. If the relaxation is determined not by local motions but by the cooperative movement of several molecular segments (glass-rubber transition), the temperature dependence is better described by the so-called WilliamsLandel-Ferry (WLF) equation.

Dielectric Loss

The portion of the energy of an alternating electrical field in a dielectric medium that is converted into heat. When the value and direction of the field intensity E change, the dielectric polarization also varies in value and direction; during one cycle of an alternating field the polarization is established twice and disappears twice. If the dielectric is made up of molecules that are dipoles themselves (polar molecules) or contain weakly bound ions, the orientation or displacement of these particles in an electrical field (orientation polarization) requires a definite time (relaxation time). As a result, the polarization maximum does not occur simultaneously with the maximum of the field intensity—that is, there is a phase shift between field intensity and polarization. Because of this there is also a phase difference between the field intensity E and the electrical induction D, which causes the energy loss W_ϵ. In a vector representation of the variables, it is possible to say that the electrical induction vector lags behind the electrical field vector by a certain angle 8, which is known as the dielectric loss angle. When molecules or ions are oriented by the field, they collide with other particles, thus dissipating energy. If the relaxation time T many times greater than the period T of the alternation of the applied field, polarization is barely able to develop and the dielectric loss is very small. At low frequencies, where the relaxation time T is considerably less than the period T, the polarization follows the field and the dielectric loss is also small because the number of reorientations per unit time is small. The dielectric loss is highest when the equality $\omega = 1/\tau$ is satisfied, where ω the circular frequency of the electrical field: $\omega = 2\pi / T$.

The mechanism described for the relaxation dielectric loss takes place in solid and liquid dielectrics that contain polar molecules or weakly bound ions. The magnitude of the relaxation dielectric loss in a liquid depends on its viscosity, the temperature, and the frequency of the applied field. For non-viscous liquids (water or alcohol) this loss appears in the centimeter wavelength range. In polymers that contain polar groups it is possible to orient both individual polar radicals and the longer or shorter molecular chains.

In dielectrics with ion and electron polarization, matter can be regarded as a set of oscillators experiencing induced oscillations, accompanied by energy dissipation, in an alternating electrical field. However, if the frequency of the electrical field is much higher or lower than the natural frequency of the oscillators, the energy dissipation—and therefore the dielectric loss—is negligible. At frequencies comparable to the natural frequency of the oscillators, the energy dissipation and the dielectric loss W∊ are high; a maximum occurs when these frequencies are equal, $\omega = \omega_n$. For electron polarization the maximum loss corresponds to the optical frequency range. In dielectrics that are made up of ions (for example, alkaline-halide crystals) the polarization is due to the elastic displacement of the ions, and the maximum loss occurs in the infrared frequency range $(10^{12} - 10^{13} Hz)$.

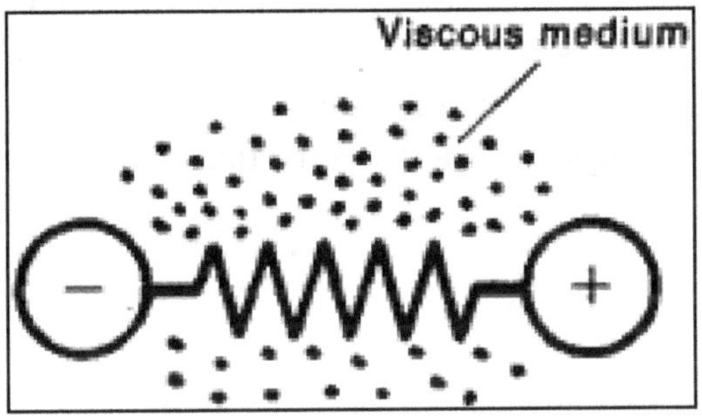

Figure: Model of a dielectric consisting of oscillators that are elastically bound electrical charges.

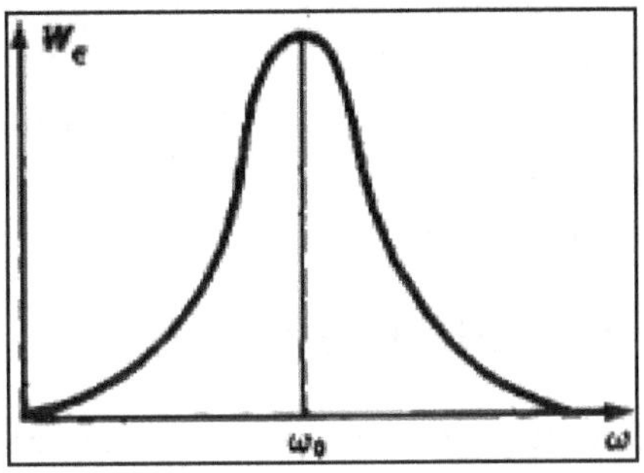

Figure: The function $W_\epsilon(\omega)$ for a dielectric consisting of the identical oscillators.

Since real dielectrics have some electrical conductivity, there are energy losses that are caused by the electrical current flow in them (Joule loss) and are not frequency-dependent.

The dielectric loss in a dielectric placed between the plates of a capacitor is found from the relation,

$$W_\epsilon = U^2 \omega C \tan\delta$$

Where, U is the voltage on the capacitor plates, C is the capacitance of the capacitor, and tan 8 is the tangent of the dielectric loss angle. The loss for 1 cu cm of a dielectric in a uniform field E is,

$$W_\epsilon = E^2 \omega \in \tan\delta$$

Where, ∊ is the dielectric constant.

The product $\in \tan\delta$ is called the dielectric loss factor. A reduction of the dielectric loss is very important for the production of capacitors and in the technology of electrical insulation. High dielectric loss is used for dielectric heating in a high-frequency electrical field.

Piezoelectricity

Piezoelectric effect is the ability of certain materials to generate an electric charge in response to applied mechanical stress. The word Piezoelectric is derived from the Greek piezein, which means to squeeze or press, and piezo, which is Greek for "push".

One of the unique characteristics of the piezoelectric effect is that it is reversible, meaning that materials exhibiting the direct piezoelectric effect (the generation of electricity when stress is applied) also exhibit the converse piezoelectric effect (the generation of stress when an electric field is applied).

When piezoelectric material is placed under mechanical stress, a shifting of the positive and negative charge centers in the material takes place, which then results in an external electrical field. When reversed, an outer electrical field either stretches or compresses the piezoelectric material.

The piezoelectric effect is very useful within many applications that involve the production and detection of sound, generation of high voltages, electronic frequency generation, microbalances, and ultra fine focusing of optical assemblies. It is also the basis of a number of scientific instrumental techniques with atomic resolution, such as scanning probe microscopes (STM, AFM, etc). The piezoelectric effect also has its use in more mundane applications as well, such as acting as the ignition source for cigarette lighters.

Direct Piezoelectric Effect

As stated, compressing a piezoelectric material produces electricity.

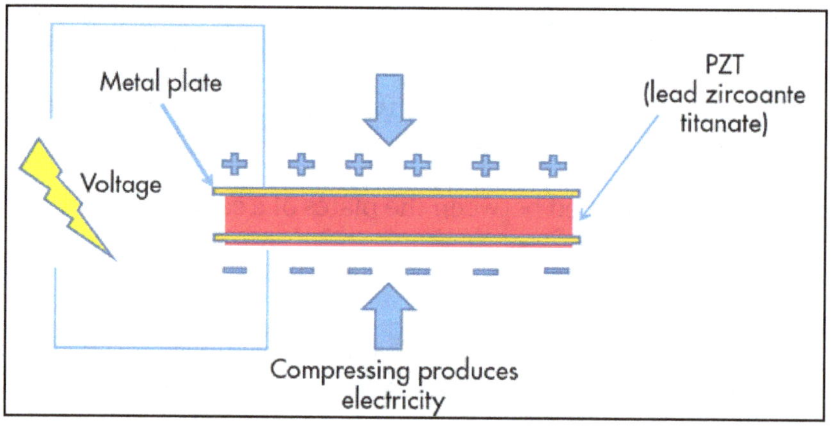

The piezoelectric effect occurs through compression of a piezoelectric material.

Piezoceramic material—non-conductive piezoelectric ceramic or crystal is placed between the two metal plates. For piezoelectricity to be generated, it needs that material to be compressed or squeezed. Mechanical stress applied to piezoelectric ceramic material generates electricity.

As shown in above figure there's a voltage potential across the material. The two metal plates sandwich the piezo crystal. The metal plates collect the charges, which creates/produces voltage (lightning bolt symbol), i.e., piezoelectricity. In this way, the piezoelectric effect acts like a miniature battery, because it produces electricity. This is the direct piezoelectric effect. Devices that use the direct piezoelectric effect include microphones, pressure sensors, hydrophones, and many other sensing types of devices.

Inverse Piezoelectric Effect

The piezoelectric effect can be reversed, which is referred to as the inverse piezoelectric effect. This is created by applying electrical voltage to make a piezoelectric crystal shrink or expand. The inverse piezoelectric effect converts electrical energy to mechanical energy.

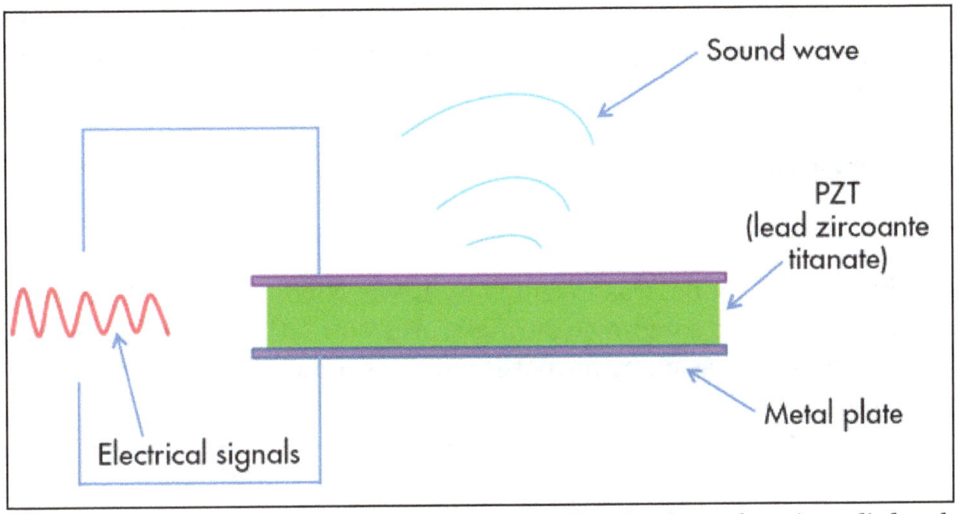

Reversal of the piezoelectric effect, called the inverse piezoelectric effect, is when voltage is applied to shrink or expand a piezoelectric crystal.

Using the inverse piezoelectric effect can help develop devices that generate and produce acoustic sound waves. Examples of piezoelectric acoustic devices are speakers (commonly found in handheld devices) or buzzers. The advantage of having such speakers is that they are very thin, which makes them useful in a range of phones. Even medical ultrasound and sonar transducers use reverse piezoelectric effect. Non-acoustic inverse piezoelectric devices include motors and actuators.

Piezoelectric Materials

Piezoelectric materials are materials that can produce electricity due to mechanical stress, such as compression. These materials can also deform when voltage (electricity) is applied.

All piezoelectric materials are non-conductive in order for the piezoelectric effect to occur and work. They can be separated into two groups: crystals and ceramics.

Some examples of piezoelectric materials are PZT (also known as lead zirconate titanate), barium titanate, and lithium niobate. These man-made materials have a more pronounced effect (better material to use) than quartz and other natural piezoelectric materials.

Compare PZT to quartz. PZT can produce more voltage for the same amount of applied mechanical stress. Conversely, applying voltage to PZT instead of quartz provides more movement. Quartz, a well-known piezoelectric material, is also the first known piezoelectric material.

PZT is created and produced (under high temperatures) with two chemical elements—lead and zirconium—and combined with a chemical compound called titanate. PZT's chemical formula is ($Pb[Zr(x)Ti(1-x)]O_3$). It is commonly used to produce ultrasounds transducers, ceramic capacitors, and other sensors and actuators. It also evinces a special range of different properties.

Barium titanate is a ferroelectric ceramic material with piezoelectric properties. For that reason, barium titanate has been used as a piezoelectric material longer than most others. Its chemical formula is $BaTiO_3$. Barium titanate was discovered in 1941 during World War II.

Lithium niobate is a compound that combines oxygen, lithium, and niobium. Its chemical formula is $LiNbO_3$. Also a ferroelectric ceramic material it's just like barium titanate in that it has piezoelectric properties, too.

Piezoelectric Devices

Sonar

Sonar, which arrived in the 1900s, was invented by Lewis Nixon. He initially developed sonar to help detect icebergs. Interest in sonar rose during World War I, though, to help locate submarines underwater. Of course, sonar has many purposes and uses today, from locating fish to underwater navigation and so on.

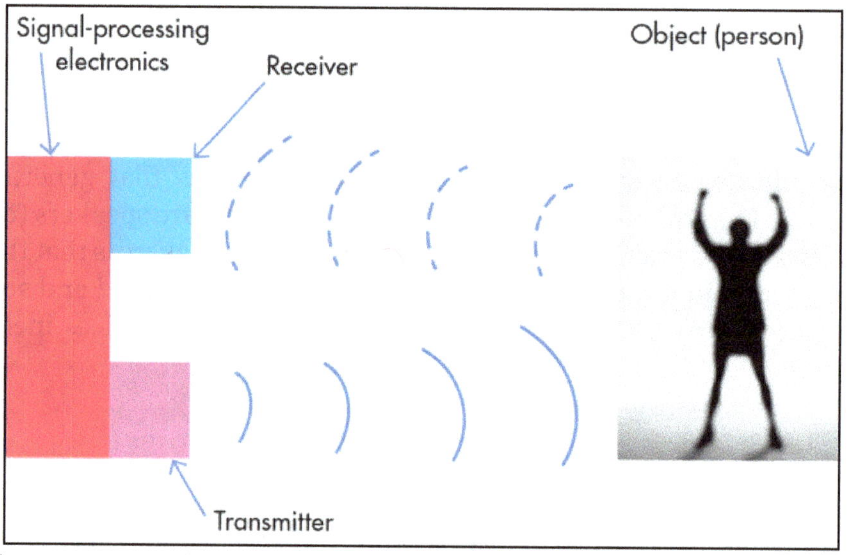

With piezoelectric sonar, a transmitter using the inverse piezoelectric effect sends out a sound wave to search for objects ahead.

In figure above sonar is sending out, via transmitter, a sound wave (signal) to search for objects ahead. The transmitter uses inverse piezoelectric effect, which is when the transmitter will use voltage to help it send out a sound wave. Once the sound wave hits an object, it will bounce back. The sound wave that bounces back will be detected by the receiver.

The receiver, unlike the transmitter, uses the direct piezoelectric effect. The receiver piezoelectric device gets compressed by the returning sound wave. It sends the signal (voltage) to the signal-processing electronics, which will take that bounced-back sound wave and begin processing it. It will determine the object's distance by calculating the timing signals from the transmitter and the receiver.

Piezoelectric Actuators

Figure below shows the operation of a piezoelectric actuator. The base stays still and acts like the metal plate that sandwiches the middle piezoelectric material. Then voltage is applied to the material, which expands and contracts from the applied voltage's electric field. The piezo crystal moves very little, whether forward or backward. Once the piezo material or crystal moves, it slowly pushes and pulls the actuator.

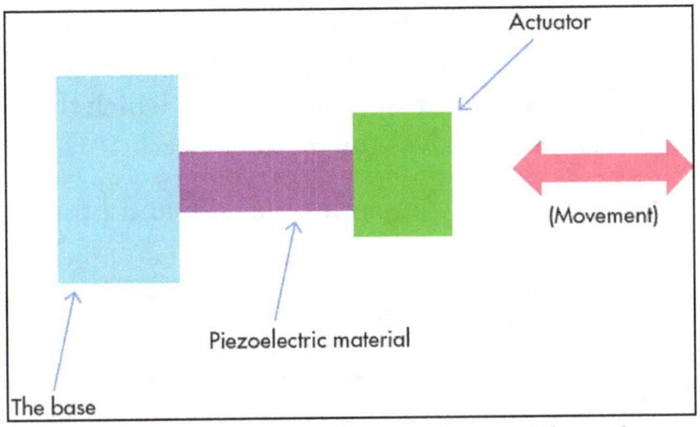

In a piezoelectric actuator, voltage is applied to the piezoelectric material, causing expansion and contraction.

The piezoelectric actuator has many uses and applications. For example, knitting machinery and braille machines use these actuators, since they have such a small amount of moving parts and a very simple design. They can even be found in video cameras and cellular phones, because they are proven most proficient as an auto-focusing mechanism.

Piezoelectric Speakers and Buzzers

Piezoelectric speakers and buzzers use the inverse piezoelectric effect to generate and produce sound. When voltage is applied to speakers and buzzers, it creates sound waves. An audio voltage signal applied to the piezoelectric ceramic of speakers or buzzers will cause the material to vibrate the air. That vibration produces sound waves, which come out of the speaker.

Piezoelectric speakers are commonly used in alarm clocks or other small mechanical devices to generate simple, high-quality audio sounds. That's because they are limited to a small amount of frequency response.

Piezo Drivers

Piezo drivers can convert low battery voltage into high voltage to power piezoelectric devices. Piezo drivers are very important, because they help engineers produce more voltage to create larger sine waves.

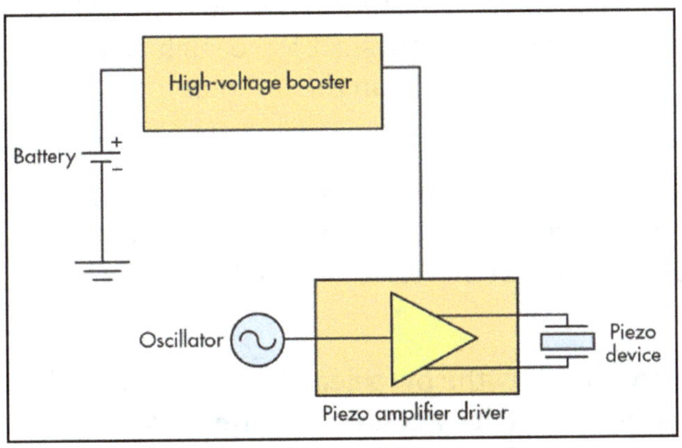

A piezo driver converts low battery voltage to higher voltage, which is used to power an amplifier that drives the device. An oscillator inputs small sine waves that the amplifier turns into larger sine waves.

Figure above is a block diagram that illustrates piezo-driver operation. Piezo drivers will take low battery voltage and use a booster to convert it to higher voltage. The higher voltage is then used to power the amplifier. The oscillator will input small sine waves, which the amplifier will change into larger voltage sine waves. The amplifier drives the piezo device.

The *table* below lists of several different companies that sell and produce various kinds of piezo drivers:

Piezo driver producers and sellers	
Companies	Websites
PiezoDrive	WWW.piezodrive.com
Taxas Instruments	WWW.ti.com
Trek Inc.	WWW.trekinc.com
Fairchild	WWW.fairchildsemi.com
PI Ceramic	WWW.piceramic.com
Linear Technology	WWW.linear.com
Physik Instrumente (PI)	WWW.physikinstrumente.com

Pyroelectricity

Pyroelectricity is a scientific phenomenon that allows special materials to create electricity when they change temperature. Minerals and crystals are the most common substances to produce this effect. These materials are commonly used to build temperature sensors, such as passive infrared detectors used in security systems.

The word pyroelectricity comes from the Greek term "pyr," which means fire. The earliest observation of the pyroelectric effect occurred in 314 B.C. When the ancient philosopher and scientist Theophrastus noticed that tourmaline minerals create electrical attraction when they are heated. Several scientists studied this phenomenon during the 18th century, but it was not until the 19th century that researchers fully understood the cause.

All crystals have electrical, mechanical, and thermal properties. These three properties are able to interact. In piezo crystals, for instance, physical force such as bending a mineral will generate electricity. Similarly, pyroelectricity results from the interaction of the thermal and electrical properties of a crystal.

When a crystal is heated or cooled, electrical charges form on opposite sides of the mineral. These charges can be harnessed as electrical current by placing electrodes on the crystal surfaces. Electricity flows in one direction when heat increases, and in the opposite direction when heat decreases.

The amount of electricity produced by the pyroelectric effect is typically not enough to power other devices. This means that pyroelectricity is not a practical power generation method. The small electrical signal from a crystal is very useful in sensors, however. Pyroelectric materials can be combined with other electronics to indicate when a temperature change has occurred.

Passive infrared detectors are one common application of pyroelectric signals. These devices are also known as motion detectors, and are frequently placed in homes and businesses as part of a security system. Each detector contains a wide-angle lens and a pyroelectric crystal. When a person walks past the sensor, their body heat causes the crystal to generate pyroelectricity and the alarm is activated.

Pyroelectric sensors can be used to detect other sources of heat, such as fire. Unlike traditional smoke detectors, a sensor that uses pyroelectricity is able to detect an actual flame even if smoke is not present. These types of sensors are useful in applications such as gas heaters, where they can detect if a pilot light has been properly ignited.

A collection of these sensors can be combined to detect temperature changes over a wide area. An array of these sensors can function as a thermal camera, and is able to show temperature variations caused by people or vehicles. This technology is often used by the military and law enforcement as a form of night vision.

Ferroelectricity

Ferroelectricity is the phenomenon where spontaneous electric polarization of the material takes place. The reverse electric polarization is possible by applying an electric field. Ferroelectricity was discovered in 1920 and Rochelle salt (scientific name-Potassium Sodium Tartrate) exhibited the properties of sudden electric polarization. Since its discovery, various uses have been devised and today, Ferroelectricity is used in various fields of electronics. The materials exhibiting the phenomenon of Ferroelectricity are called Ferroelectric Materials. Ironically, the Ferroelectricity or the Ferroelectric Materials do not have any connection to Iron(ferrite).

These materials can produce spontaneous polarization. By inverting the direction of applied electrical field, the direction of polarization of these materials can be inverted or changed. This is called switching. It can also maintain the polarisation even once the field is removed. These materials have some similarities over ferromagnetic materials which reveal permanent magnetic moment. The hysteresis loop is almost same for both materials. Since, there are similarities; the prefix is same for both the materials. All the ferroelectric materials exhibit piezoelectric effect. The opposed properties of these materials are seen in antiferromagnetic materials.

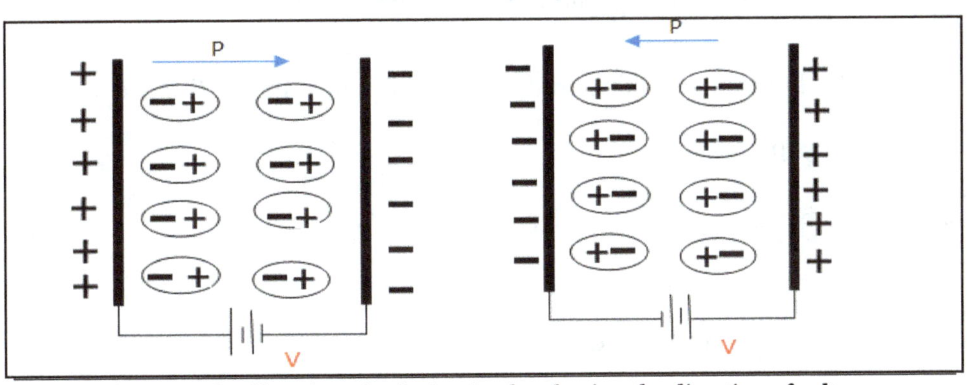

Changing the direction of polarization by altering the direction of voltage.

Theory of Ferroelectric Materials

The free energy of ferroelectric material based on Ginburg-Landau theory without electric field and any applied stress can be written as Taylor expansion. It is written in terms of P (order parameter) as,

$$\Delta E = \frac{1}{2}\alpha_0(T-T_0)(P_x^2+P_y^2+P_z^2)+\frac{1}{4}\alpha_{11}(P_x^4+P_y^4+P_z^4)+$$
$$\frac{1}{2}\alpha_{12}(P_x^2P_y^2+P_y^2P_z^2+P_z^2P_x^2)+\frac{1}{6}\alpha_{111}(P_x^6+P_y^6+P_z^6)+$$
$$\frac{1}{2}\alpha_{112}[P_x^4(P_y^2+P_z^2)+P_y^4(P_z^2+P_x^2)+P_z^4(P_x^2+P_y^2)]+\frac{1}{2}\alpha_{123}P_x^2P_y^2P_z^2$$

(If sixth order expansion is used)

$P_x \rightarrow$ component of polarization vector, x

$P_y \rightarrow$ component of polarization vector, y

$P_z \rightarrow$ component of polarization vector, z

$\alpha_i, \alpha_{ij}, \alpha_{ijk} \rightarrow$ coefficients should be constant with the crystal symmetry

$\alpha_0 > 0, \alpha_{111} > 0 \rightarrow$ for all ferroelectrics

$\alpha_{11} < 0 \rightarrow$ ferroelectrics with first order transition

$\alpha_0 > 0 \rightarrow$ ferroelectrics with second order transition

For investigating different phenomenon and domain formation in ferroelectrics, these equations are used in phase field model. Usually, it is used by adding some terms such as an elastic term, a gradient term and an electrostatic term to this free energy equation. Using finite difference method, the equations are solved subject to the constraints of linear elasticity and Gauss's law.

For a cubic to tetragonal phase transition of a spontaneous polarization of a ferroelectric can be obtained from the expression for free energy,

$$\Delta E = \frac{1}{2}\alpha_0(T-T_0)P_x^2+\frac{1}{4}\alpha_{11}P_x^4\ \frac{1}{6}\alpha_{111}P_x^6(1-D\text{ expression})$$

It has a character of dual well potential with double energy minima at $P = \pm P_s$

$P_s \rightarrow$ spontaneous polarization

$$\frac{\partial \Delta E}{\partial P_x} = \alpha_0(T-T_0)P_x+\alpha_{11}P_x^3+\alpha_{111}P_x^5 = 0$$

$$P_x[\alpha_0(T-T_0)+\alpha_{11}P_x^2+\alpha_{111}P_x^4] = 0$$

$$Thus, \alpha_0(T-T_0)+\alpha_{11}P_x^2+\alpha_{111}P_x^4=0 \; and \; P_x=0$$

$$So, P_s^2=\frac{1}{2\alpha_{111}}[-\alpha_{11}\pm\sqrt{(\alpha_{11}^2-4\alpha_0\alpha_{111}(T-T_0))}]$$

By simplifying, eliminating the negative root and substitute $\alpha_{11}=0$ we get,

$$P_s=\sqrt{\frac{\alpha_0(T-T_0)}{\alpha_{111}}}$$

Polarization and Hysteresis Loop

First we take a dielectric material and a peripheral electric field is given, then we can see that the polarization will be always directly proportional to the applied field which is represented in figure below.

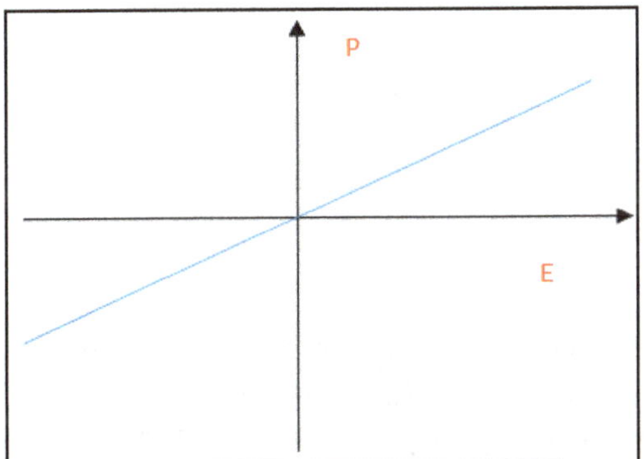

Dielectric Polarization

Next; when we polarise a paraelectric material, we get a non linear polarization. However it is a function of field as shown in figure below:

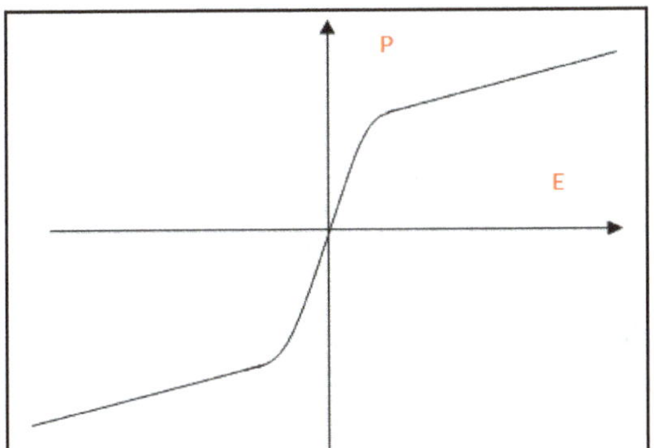

Paraelectric Polarization

Next, we take a ferroelectric material and electric field is given to it. We get a nonlinear polarization. It also exhibit nonzero spontaneous polarization without a peripheral field. We can also see that by inverting the direction of applied electrical field, the direction of polarization can be inverted or changed.

Thus, we can say that, the polarization will depend on the present as well as the previous condition of electric field. The hysteresis loop is obtained as in figure below:

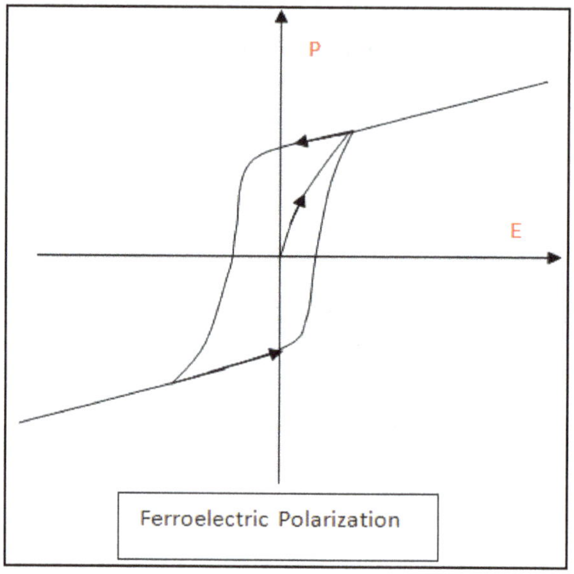

Curie Temperature

The properties of these materials exist only below a definite phase conversion temperature. Above this temperature, the material will become paraelectric materials. That is, loss in spontaneous polarization. This definite temperature is called Curie temperature (T_C). Most of these materials above T_c will loss the piezoelectric property as well. The variation of dielectric constant by means of temperature in the nonpolar, paraelectric state is shown by Curie-Weiss law as given below:

$$\in = \frac{\in_\infty + A}{T - T_C}$$

$$\chi = \frac{C_C}{T - T_C}$$

$\varepsilon \rightarrow$ Dielectric constant

$\varepsilon_\infty \rightarrow \varepsilon$ at temperature, $T >> T_C$

$A \rightarrow$ Constant

$T_C \rightarrow$ Curie point

$T \rightarrow$ Temperature

$\chi \rightarrow$ Susceptibility

$C_C \rightarrow$ Curie constant of the material

The dielectric constant and temperature characteristic of a ferroelectric material are represented below:

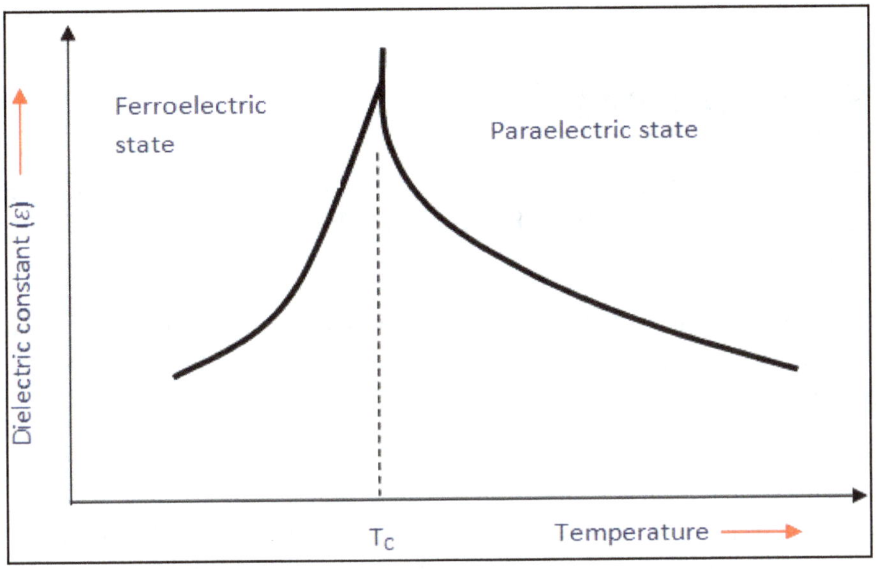

Examples of ferroelectric materials.

- BaTiO3.
- PbTiO3.
- Lead Zirconate Titanate (PZT).
- Triglycine Sulphate.
- PVDF.
- Lithium tantalite etc.

Applications

- Capacitors: A capacitor is an electronic component used for a number of electronic circuits. It consists of 2 electrodes with a dielectric material in between. A Ferroelectric Material is preferred for the manufacturing of a capacitor. Capacitors are generally formed by the Barium Titanate (BaTiO3) material.
- Storage memories: SRAM and DRAM are volatile. EEPROM and CMOS require a battery backup and are very costly. Ferroelectric Random Access Memory is non-volatile as well as cheaper. FRAM is reliable as well.
- Waveguides: A waveguide is a device which controls the propagation of light within the device. It carries the electromagnetic wave. The material used inside a waveguide is a Ferroelectric Material.
- Optical Memory Display: Latest optical memory displays technologies are devised using the Ferroelectric Material. PLZT is most preferred Ferroelectric Material for this purpose.

- Displacement Transducers: The Ferroelectric Material is used in pressure sensors. These pressure sensors are also termed as displacement transducers.

Electric Susceptibility

Electric susceptibility is a quantitative measure of the extent to which an electric field applied to a dielectric material causes polarization, the slight displacement of positive and negative charge within the material.

A dimensionless parameter measuring the ease of polarization of a dielectric. In a vacuum, the electric flux density D (measured in coulombs/m^2) and electric field strength E (volts/m) are related by equation below:

$$D = \epsilon_0 E$$

ε_0 is the permittivity of free space, having the value 8.854×10^{-12} farad/m. In a dielectric material, polarization occurs, andthe flux density has the value given by equation below,

$$D = \epsilon_0 \epsilon_r E = \epsilon_0 E + P$$

where ε_r is the relative permittivity of the material and P is the polarization flux density. This can be written as,

$$P = \epsilon_0 (\epsilon_r - 1) E = \epsilon_0 \chi_e E$$

ε_r - 1 is known as the electric susceptibility of the dielectric material. It is a measure of that part of the relative permittivity which is due to the material itself.

The electric susceptibility can be related to the polarizability α by expressing the polarization in terms of molecularparameters. Thus equation below,

$$P = N \langle \mu \rangle_{\text{avg}} = N \alpha E_L$$

$$\chi_e = \frac{N \alpha E_L}{\epsilon_0 E}$$

Where N is the number of molecules per unit volume, $\langle \mu_{avg} \rangle$ is their average dipole moment, and E L is the local electric field strength at a molecular site. At low concentrations, E_L approaches E, and the susceptibility is proportional to the concentration N.

References

- Electronic-polarization: electrical4u.com, Retrieved 19 March 2020
- Dielectric-constant: tech-faq.com, Retrieved 09 June 2020
- Ferroelectric-Materials-Theory-Properties-and-applications-117961: indiastudychannel.com, Retrieved 30 March 2020

- Electric-susceptibility: encyclopedia2.thefreedictionary.com, Retrieved 28 April 2020
- Piezoelectric-effect, piezo-ceramic-motor-technology: nanomotion.com, Retrieved 10 July 2020
- Electric-polarization, science: britannica.com, Retrieved 16 May 2020
- Dielectric: newworldencyclopedia.org, Retrieved 15 July 2020

6

An Integrated Study of Electronic Band Structure

The range of energy levels which might be present within electrons as well as those which might be missing from it is described through the electronic band structure. The energy levels which are missing are known as band gaps. The topics elaborated in this chapter will help in gaining a better perspective about electron band structure, band gap and Fermi liquid theory.

In solid-state physics, the electronic band structure (or simply band structure) of a solid describes the range of energies that an electron within the solid may have (called *energy bands, allowed bands*, or simply *bands*) and ranges of energy that it may not have (called *band gaps* or *forbidden bands*).

Band theory derives these bands and band gaps by examiniwng the allowed quantum mechanical wave functions for an electron in a large, periodic lattice of atoms or molecules. Band theory has been successfully used to explain many physical properties of solids, such as electrical resistivity and optical absorption, and forms the foundation of the understanding of all solid-state devices (transistors, solar cells, etc.).

Why Bands and Band Gaps Occur

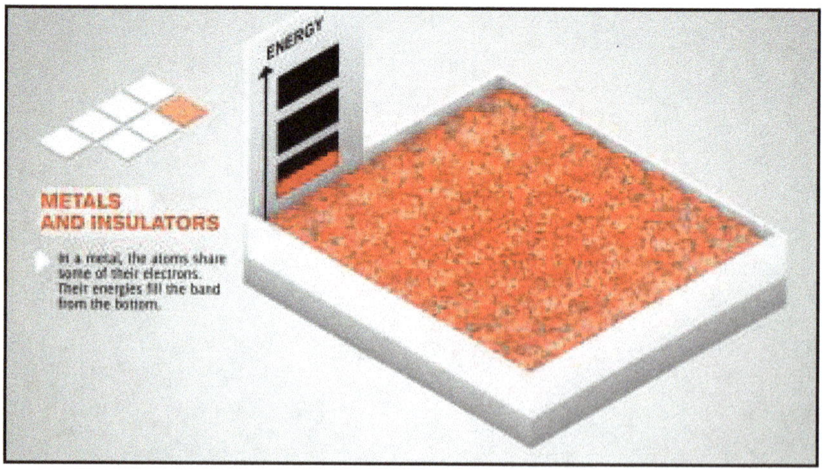

Animation of band formation and how electrons fill them in a metal and an insulator.

The electrons of a single, isolated atom occupy atomic orbitals. Each orbital forms at a discrete energy level. When multiple atoms join together to form into a molecule, their atomic orbitals combine to form molecular orbitals, each of which forms at a discrete energy level. As more atoms are brought together, the molecular orbitals extend larger and larger, and the energy levels of the molecule will become increasingly dense. Eventually, the collection of atoms form a giant molecule, or in other words, a solid. For this giant molecule, the energy levels are so close that they can be considered to form a continuum.

Band gaps are essentially leftover ranges of energy not covered by any band, a result of the finite widths of the energy bands. The bands have different widths, with the widths depending upon the degree of overlap in the atomic orbitals from which they arise. Two adjacent bands may simply not be wide enough to fully cover the range of energy. For example, the bands associated with core orbitals (such as 1s electrons) are extremely narrow due to the small overlap between adjacent atoms. As a result, there tend to be large band gaps between the core bands. Higher bands involve comparatively larger orbitals with more overlap, becoming progressively wider at higher energies so that there are no band gaps at higher energies.

Basic Concepts

Assumptions and Limits of Band Structure Theory

To start out, it is important to note what has been assumed in order to gain the simplicity of the band theory:

1. *Infinite-size system*: For the bands to be continuous, we must consider a large piece of material. The concept of band structure can be extended to systems which are only "large" along reduced dimensions, such as two-dimensional electron systems.

2. *Homogeneous system*: The notion of a band structure as an intrinsic property of a material assumes that the material is homogeneous in some way. Practically, this means that band structure describes the bulk inside a uniform piece of material.

3. *Non-interactivity*: The band structure describes "single electron states". The existence of these states assumes that the electrons travel in a static potential without dynamically interacting with lattice vibrations, other electrons, photons, etc.

The above assumptions are broken in a number of important practical situations, and the use of band structure requires one to keep a close check on the limitations of band theory:

- Inhomogeneities and interfaces: Near surfaces, junctions, and other inhomogeneities, the bulk band structure is disrupted. Not only are there local small-scale disruptions (e.g., surface states or dopant states inside the band gap), but also local charge imbalances. These charge imbalances have electrostatic effects that extend deeply into semiconductors, insulators, and the vacuum.

- Along the same lines, most electronic effects (capacitance, electrical conductance, electric-field screening) involve the physics of electrons passing through surfaces and/or near interfaces. The full description of these effects, in a band structure picture, requires at least a rudimentary model of electron-electron interactions.

- Small systems: For systems which are small along every dimension (e.g., a small molecule or a quantum dot), there is no continuous band structure. The crossover between small and large dimensions is the realm of mesoscopic physics.

- Strongly correlated materials (for example, Mott insulators) simply cannot be understood in terms of single-electron states. The electronic band structures of these materials are poorly defined (or at least, not uniquely defined) and may not provide useful information about their physical state.

Crystalline Symmetry and Wavevectors

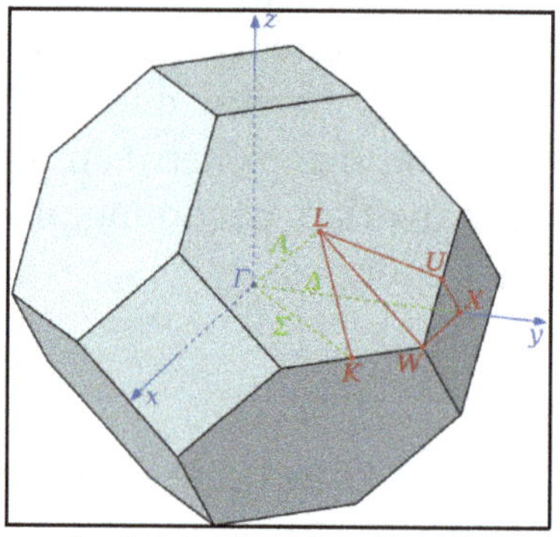

Brillouin zone of a face-centered cubic lattice showing labels for special symmetry points.

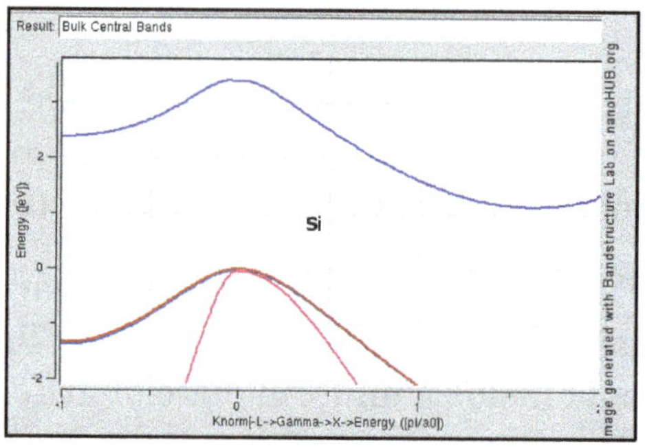

Band structure plot for Si, Ge, GaAs and InAs generated with tight binding model. Note that Si and Ge are indirect band gap materials, while GaAs and InAs are direct.

Band structure calculations take advantage of the periodic nature of a crystal lattice, exploiting its symmetry. The single-electron Schrödinger equation is solved for an electron in a lattice-periodic potential, giving Bloch waves as solutions:

$$\psi_{n\mathbf{k}}(\mathbf{r}) = e^{i\mathbf{k}\cdot\mathbf{r}} u_{n\mathbf{k}}(\mathbf{r}), ,$$

where k is called the wavevector. For each value of k, there are multiple solutions to the Schrödinger equation labelled by n, the band index, which simply numbers the energy bands. Each of these energy levels evolves smoothly with changes in k, forming a smooth band of states. For each band we can define a function E_n(k), which is the dispersion relation for electrons in that band.

The wavevector takes on any value inside the Brillouin zone, which is a polyhedron in wavevector space that is related to the crystal's lattice. Wavevectors outside the Brillouin zone simply correspond to states that are physically identical to those states within the Brillouin zone. Special high symmetry points in the Brillouin zone are assigned labels like Γ, Δ, Λ, Σ.

It is difficult to visualize the shape of a band as a function of wavevector, as it would require a plot in four-dimensional space, E vs. k_x, k_y, k_z. In scientific literature it is common to see band structure plots which show the values of $E_n(\mathrm{k})$ for values of k along straight lines connecting symmetry points. Another method for visualizing band structure is to plot a constant-energy isosurface in wavevector space, showing all of the states with energy equal to a particular value. The isosurface of states with energy equal to the Fermi level is known as the Fermi surface.

Energy band gaps can be classified using the wavevectors of the states surrounding the band gap:

- Direct band gap: the lowest-energy state above the band gap has the same k as the highest-energy state beneath the band gap.
- Indirect band gap: the closest states above and beneath the band gap do not have the same k value.

Asymmetry: Band Structures in Non-crystalline Solids

Although electronic band structures are usually associated with crystalline materials, quasi-crystalline and amorphous solids may also exhibit band structures. These are somewhat more difficult to study theoretically since they lack the simple symmetry of a crystal, and it is not usually possible to determine a precise dispersion relation. As a result, virtually all of the existing theoretical work on the electronic band structure of solids has focused on crystalline materials.

Density of States

The density of states function $g(E)$ is defined as the number of electronic states per unit volume, per unit energy, for electron energies near E.

The density of states function is important for calculations of effects based on band theory. It appears in calculations for optical absorption where it provides both the number of excitable electrons and the number of final states for an electron. It appears in calculations of electrical conductivity where it provides the number of mobile states, and in computing electron scattering rates where it provides the number of final states after scattering.

For energies inside a band gap, $g(E) = 0$.

Filling of Bands

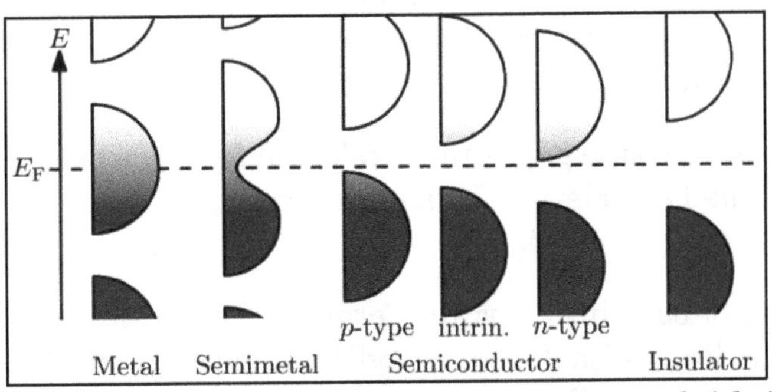

Filling of the electronic states in various types of materials at equilibrium. Here, height is energy while width is the density of available states for a certain energy in the material listed. The shade follows the Fermi–Dirac distribution (black = all states filled, white = no state filled). In metals and semimetals the Fermi level E_F lies inside at least one band. In insulators and semiconductors the Fermi level is inside a band gap; however, in semiconductors the bands are near enough to the Fermi level to be thermally populated with electrons or holes.

At thermodynamic equilibrium, the likelihood of a state of energy E being filled with an electron is given by the Fermi–Dirac distribution, a thermodynamic distribution that takes into account the Pauli exclusion principle:

$$f(E) = \frac{1}{1 + e^{(E-\mu)/k_B T}}$$

where:

- $k_B T$ is the product of Boltzmann's constant and temperature, and ,
- μ is the total chemical potential of electrons, or *Fermi level* (in semiconductor physics, this quantity is more often denoted E_F). The Fermi level of a solid is directly related to the voltage on that solid, as measured with a voltmeter. Conventionally, in band structure plots the Fermi level is taken to be the zero of energy (an arbitrary choice).

The density of electrons in the material is simply the integral of the Fermi–Dirac distribution times the density of states:

$$N / V = \int_{-\infty}^{\infty} g(E) f(E) dE$$

Although there are an infinite number of bands and thus an infinite number of states, there are only a finite number of electrons to place in these bands. The preferred value for the number of electrons is a consequence of electrostatics: even though the surface of a material can be charged, the internal bulk of a material prefers to be charge neutral. The condition of charge neutrality means that N/V must match the density of protons in the material. For this to occur, the material electrostatically adjusts itself, shifting its band structure up or down in energy (thereby shifting $g(E)$), until it is at the correct equilibrium with respect to the Fermi level.

Names of Bands Near the Fermi Level (Conduction Band, Valence Band)

A solid has an infinite number of allowed bands, just as an atom has infinitely many energy levels. However, most of the bands simply have too high energy, and are usually disregarded under ordinary circumstances. Conversely, there are very low energy bands associated with the core orbitals (such as 1s electrons). These low-energy *core bands* are also usually disregarded since they remain filled with electrons at all times, and are therefore inert. Likewise, materials have several band gaps throughout their band structure.

The most important bands and band gaps—those relevant for electronics and optoelectronics—are those with energies near the Fermi level. The bands and band gaps near the Fermi level are given special names, depending on the material:

- In a semiconductor or band insulator, the Fermi level is surrounded by a band gap, referred to as *the* band gap (to distinguish it from the other band gaps in the band structure). The closest band above the band gap is called *the conduction band*, and the closest band beneath the band gap is called *the valence band*. The name "valence band" was coined by analogy to chemistry, since in many semiconductors the valence band is built out of the valence orbitals.

- In a metal or semimetal, the Fermi level is inside of one or more allowed bands. In semimetals the bands are usually referred to as "conduction band" or "valence band" depending on whether the charge transport is more electron-like or hole-like, by analogy to semiconductors. In many metals, however, the bands are neither electron-like nor hole-like, and often just called "valence band" as they are made of valence orbitals. The band gaps in a metal's band structure are not important for low energy physics, since they are too far from the Fermi level.

Theory of Band Structures in Crystals

The ansatz is the special case of electron waves in a periodic crystal lattice using Bloch waves as treated generally in the dynamical theory of diffraction. Every crystal is a periodic structure which can be characterized by a Bravais lattice, and for each Bravais lattice we can determine the reciprocal lattice, which encapsulates the periodicity in a set of three reciprocal lattice vectors (b_1,b_2,b_3). Now, any periodic potential V(r) which shares the same periodicity as the direct lattice can be expanded out as a Fourier series whose only non-vanishing components are those associated with the reciprocal lattice vectors. So the expansion can be written as:

$$V(\mathbf{r}) = \sum_{\mathbf{K}} V_{\mathbf{K}} e^{i\mathbf{K}\cdot\mathbf{r}}$$

where $K = m_1b_1 + m_2b_2 + m_3b_3$ for any set of integers (m_1,m_2,m_3).

From this theory, an attempt can be made to predict the band structure of a particular material, however most ab initio methods for electronic structure calculations fail to predict the observed band gap.

Nearly Free Electron Approximation

In the nearly free electron approximation, interactions between electrons are completely ignored. This approximation allows use of Bloch's Theorem which states that electrons in a periodic potential have wavefunctions and energies which are periodic in wavevector up to a constant phase shift between neighboring reciprocal lattice vectors. The consequences of periodicity are described mathematically by the Bloch wavefunction:

$$\psi_{n,\mathbf{k}}(\mathbf{r}) = e^{i\mathbf{k}\cdot\mathbf{r}} u_n(\mathbf{r})$$

where the function $u_n(\mathbf{r})$ is periodic over the crystal lattice, that is,

$$u_n(\mathbf{r}) = u_n(\mathbf{r} - \mathbf{R}).$$

Here index *n* refers to the *n-th* energy band, wavevector k is related to the direction of motion of the electron, r is the position in the crystal, and R is the location of an atomic site.

The NFE model works particularly well in materials like metals where distances between neighbouring atoms are small. In such materials the overlap of atomic orbitals and potentials on neighbouring atoms is relatively large. In that case the wave function of the electron can be approximated by a (modified) plane wave. The band structure of a metal like Aluminium even gets close to the empty lattice approximation.

Tight Binding Model

The opposite extreme to the nearly free electron approximation assumes the electrons in the crystal behave much like an assembly of constituent atoms. This tight binding model assumes the solution to the time-independent single electron Schrödinger equation Ø is well approximated by a linear combination of atomic orbitals $\psi_n(\mathbf{r})$.

$$\Psi(\mathbf{r}) = \sum_{n,\mathbf{R}} b_{n,\mathbf{R}} \psi_n(\mathbf{r}-\mathbf{R}),$$

where the coefficients $b_{n,\mathbf{R}}$ are selected to give the best approximate solution of this form. Index n refers to an atomic energy level and R refers to an atomic site. A more accurate approach using this idea employs Wannier functions, defined by:

$$a_n(\mathbf{r}-\mathbf{R}) = \frac{V_C}{(2\pi)^3} \int_{BZ} d\mathbf{k} e^{-i\mathbf{k}\cdot(\mathbf{R}-\mathbf{r})} u_{n\mathbf{k}};$$

in which $u_{n\mathbf{k}}i$ is the periodic part of the Bloch wave and the integral is over the Brillouin zone. Here index n refers to the n-th energy band in the crystal. The Wannier functions are localized near atomic sites, like atomic orbitals, but being defined in terms of Bloch functions they are accurately related to solutions based upon the crystal potential. Wannier functions on different atomic sites R are orthogonal. The Wannier functions can be used to form the Schrödinger solution for the n-th energy band as:

$$\Psi_{n,\mathbf{k}}(\mathbf{r}) = \sum_{\mathbf{R}} e^{-i\mathbf{k}\cdot(\mathbf{R}-\mathbf{r})} a_n(\mathbf{r}-\mathbf{R})$$

The TB model works well in materials with limited overlap between atomic orbitals and potentials on neighbouring atoms. Band structures of materials like Si, GaAs, SiO_2 and diamond for instance are well described by TB-Hamiltonians on the basis of atomic sp^3 orbitals. In transition metals a mixed TB-NFE model is used to describe the broad NFE conduction band and the narrow embedded TB d-bands. The radial functions of the atomic orbital part of the Wannier functions are most easily calculated by the use of pseudopotential methods. NFE, TB or combined NFE-TB band structure calculations, sometimes extended with wave function approximations based on pseudopotential methods, are often used as an economic starting point for further calculations.

KKR Model

The simplest form of this approximation centers non-overlapping spheres (referred to as *muffin tins*) on the atomic positions. Within these regions, the potential experienced by an electron is approximated to be spherically symmetric about the given nucleus. In the remaining interstitial region, the screened potential is approximated as a constant. Continuity of the potential between the atom-centered spheres and interstitial region is enforced.

A variational implementation was suggested by Korringa and by Kohn and Rostocker, and is often referred to as the *KKR model*.

Density-functional Theory

In recent physics literature, a large majority of the electronic structures and band plots are calculated using density-functional theory (DFT), which is not a model but rather a theory, i.e., a microscopic first-principles theory of condensed matter physics that tries to cope with the electron-electron many-body problem via the introduction of an exchange-correlation term in the functional of the electronic density. DFT-calculated bands are in many cases found to be in agreement with experimentally measured bands, for example by angle-resolved photoemission spectroscopy (ARPES). In particular, the band shape is typically well reproduced by DFT. But there are also systematic errors in DFT bands when compared to experiment results. In particular, DFT seems to systematically underestimate by about 30-40% the band gap in insulators and semiconductors.

It is commonly believed that DFT is a theory to predict ground state properties of a system only (e.g. the total energy, the atomic structure, etc.), and that excited state properties cannot be determined by DFT. This is a misconception. In principle, DFT can determine any property (ground state or excited state) of a system given a functional that maps the ground state density to that property. This is the essence of the Hohenburg–Kohn theorem. In practice, however, no known functional exists that maps the ground state density to excitation energies of electrons within a material. Thus, what in the literature is quoted as a DFT band plot is a representation of the DFT Kohn–Sham energies, i.e., the energies of a fictive non-interacting system, the Kohn–Sham system, which has no physical interpretation at all. The Kohn–Sham electronic structure must not be confused with the real, quasiparticle electronic structure of a system, and there is no Koopman's theorem holding for Kohn–Sham energies, as there is for Hartree–Fock energies, which can be truly considered as an approximation for quasiparticle energies. Hence, in principle, Kohn–Sham based DFT is not a band theory, i.e., not a theory suitable for calculating bands and band-plots. In principle time-dependent DFT can be used to calculate the true band structure although in practice this is often difficult. A popular approach is the use of hybrid functionals, which incorporate a portion of Hartree–Fock exact exchange; this produces a substantial improvement in predicted bandgaps of semiconductors, but is less reliable for metals and wide-bandgap materials.

Green's Function Methods and the ab Initio GW Approximation

To calculate the bands including electron-electron interaction many-body effects, one can resort to so-called Green's function methods. Indeed, knowledge of the Green's function of a system provides both ground (the total energy) and also excited state observables of the system. The poles of the Green's function are the quasiparticle energies, the bands of a solid. The Green's function can be calculated by solving the Dyson equation once the self-energy of the system is known. For real systems like solids, the self-energy is a very complex quantity and usually approximations are needed to solve the problem. One such approximation is the GW approximation, so called from the mathematical form the self-energy takes as the product $\Sigma = GW$ of the Green's function G and the dynamically screened interaction W. This approach is more pertinent when addressing the calculation of band plots (and also quantities beyond, such as the spectral function) and can also be formulated in a completely *ab initio* way. The GW approximation seems to provide band gaps of insulators and semiconductors in agreement with experiment, and hence to correct the systematic DFT underestimation.

Mott Insulators

Although the nearly free electron approximation is able to describe many properties of electron band structures, one consequence of this theory is that it predicts the same number of electrons in each unit cell. If the number of electrons is odd, we would then expect that there is an unpaired electron in each unit cell, and thus that the valence band is not fully occupied, making the material a conductor. However, materials such as CoO that have an odd number of electrons per unit cell are insulators, in direct conflict with this result. This kind of material is known as a Mott insulator, and requires inclusion of detailed electron-electron interactions (treated only as an averaged effect on the crystal potential in band theory) to explain the discrepancy. The Hubbard model is an approximate theory that can include these interactions. It can be treated non-perturbatively within the so-called dynamical mean field theory, which attempts to bridge the gap between the nearly free electron approximation and the atomic limit. Formally, however, the states are not non-interacting in this case and the concept of a band structure is not adequate to describe these cases.

Others

Calculating band structures is an important topic in theoretical solid state physics. In addition to the models mentioned above, other models include the following:

- Empty lattice approximation: the "band structure" of a region of free space that has been divided into a lattice.
- k·p perturbation theory is a technique that allows a band structure to be approximately described in terms of just a few parameters. The technique is commonly used for semiconductors, and the parameters in the model are often determined by experiment.
- The Kronig-Penney Model, a one-dimensional rectangular well model useful for illustration of band formation. While simple, it predicts many important phenomena, but is not quantitative.
- Hubbard model.

The band structure has been generalised to wavevectors that are complex numbers, resulting in what is called a *complex band structure*, which is of interest at surfaces and interfaces.

Each model describes some types of solids very well, and others poorly. The nearly free electron model works well for metals, but poorly for non-metals. The tight binding model is extremely accurate for ionic insulators, such as metal halide salts (e.g. NaCl).

Band Diagrams

To understand how band structure changes relative to the Fermi level in real space, a band structure plot is often first simplified in the form of a band diagram. In a band diagram the vertical axis is energy while the horizontal axis represents real space. Horizontal lines represent energy levels, while blocks represent energy bands. When the horizontal lines in these diagram are slanted then the energy of the level or band changes with distance. Diagrammatically, this depicts the presence of an electric field within the crystal system. Band diagrams are useful in relating the general band structure properties of different materials to one another when placed in contact with each other.

Band Gap

In solid-state physics, a band gap, also called an energy gap or bandgap, is an energy range in a solid where no electron states can exist. In graphs of the electronic band structure of solids, the band gap generally refers to the energy difference (in electron volts) between the top of the valence band and the bottom of the conduction band in insulators and semiconductors. It is the energy required to promote a valence electron bound to an atom to become a conduction electron, which is free to move within the crystal lattice and serve as a charge carrier to conduct electric current. It is closely related to the HOMO/LUMO gap in chemistry. If the valence band is completely full and the conduction band is completely empty, then electrons cannot move in the solid; however, if some electrons transfer from the valence to the conduction band, then current *can* flow. Therefore, the band gap is a major factor determining the electrical conductivity of a solid. Substances with large band gaps are generally insulators, those with smaller band gaps are semiconductors, while conductors either have very small band gaps or none, because the valence and conduction bands overlap.

In Semiconductor Physics

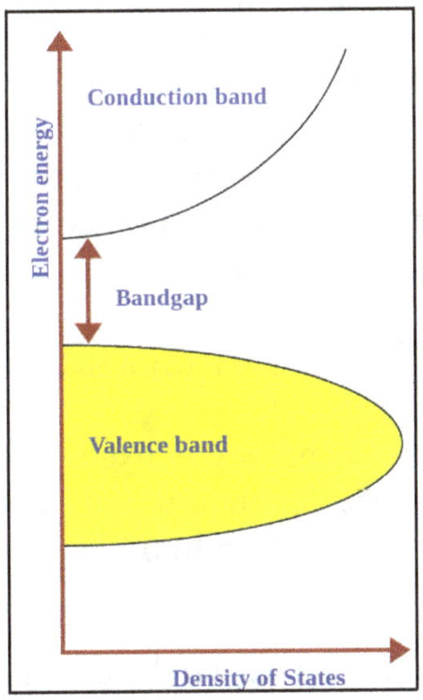

Semiconductor band structure.

Every solid has its own characteristic energy-band structure. This variation in band structure is responsible for the wide range of electrical characteristics observed in various materials. In semiconductors and insulators, electrons are confined to a number of bands of energy, and forbidden from other regions. The term "band gap" refers to the energy difference between the top of the valence band and the bottom of the conduction band. Electrons are able to jump from one band to another. However, in order for an electron to jump from a valence band to a conduction band, it requires a specific minimum amount of energy for the transition. The required energy differs with different materials. Electrons can gain enough energy to jump to the conduction band by absorbing either a phonon (heat) or a photon (light).

A semiconductor is a material with a small but non-zero band gap that behaves as an insulator at absolute zero but allows thermal excitation of electrons into its conduction band at temperatures that are below its melting point. In contrast, a material with a large band gap is an insulator. In conductors, the valence and conduction bands may overlap, so they may not have a band gap.

The conductivity of intrinsic semiconductors is strongly dependent on the band gap. The only available charge carriers for conduction are the electrons that have enough thermal energy to be excited across the band gap and the electron holes that are left off when such an excitation occurs.

Band-gap engineering is the process of controlling or altering the band gap of a material by controlling the composition of certain semiconductor alloys, such as GaAlAs, InGaAs, and InAlAs. It is also possible to construct layered materials with alternating compositions by techniques like molecular-beam epitaxy. These methods are exploited in the design of heterojunction bipolar transistors (HBTs), laser diodes and solar cells.

The distinction between semiconductors and insulators is a matter of convention. One approach is to think of semiconductors as a type of insulator with a narrow band gap. Insulators with a larger band gap, usually greater than 3 eV, are not considered semiconductors and generally do not exhibit semiconductive behaviour under practical conditions. Electron mobility also plays a role in determining a material's informal classification.

The band-gap energy of semiconductors tends to decrease with increasing temperature. When temperature increases, the amplitude of atomic vibrations increase, leading to larger interatomic spacing. The interaction between the lattice phonons and the free electrons and holes will also affect the band gap to a smaller extent. The relationship between band gap energy and temperature can be described by Varshni's empirical expression,

$$E_g(T) = E_g(0) - \frac{\alpha T^2}{T+\beta}$$, where $E_g(0)$, α and β are material constants.

In a regular semiconductor crystal, the band gap is fixed owing to continuous energy states. In a quantum dot crystal, the band gap is size dependent and can be altered to produce a range of energies between the valence band and conduction band. It is also known as quantum confinement effect.

Band gaps also depend on pressure. Band gaps can be either direct or indirect, depending on the electronic band structure.

Photovoltaic Cells

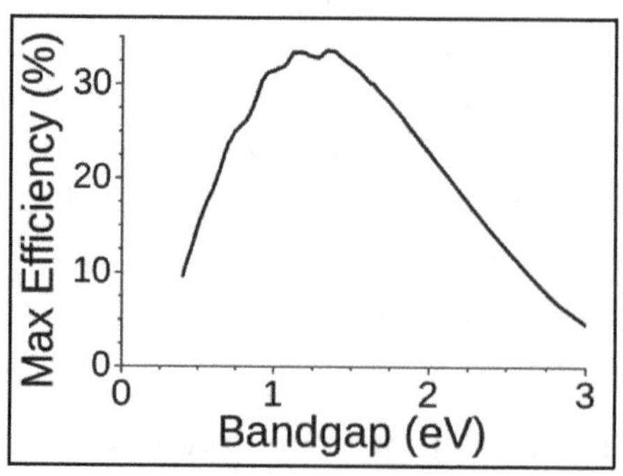

The Shockley–Queisser limit gives the maximum possible efficiency of a single-junction solar cell under un-concentrated sunlight, as a function of the semiconductor band gap. If the band gap is too high, most daylight photons cannot be absorbed; if it is too low, then most photons have much more energy than necessary to excite electrons across the band gap, and the rest is wasted. The semiconductors commonly used in commercial solar cells have band gaps near the peak of this curve, for example silicon (1.1eV) or CdTe (1.5eV). The Shockley–Queisser limit has been exceeded experimentally by combining materials with different band gap energies to make tandem solar cells.

The optical band gap determines what portion of the solar spectrum a photovoltaic cell absorbs. A semiconductor will not absorb photons of energy less than the band gap; and the energy of the electron-hole pair produced by a photon is equal to the bandgap energy. A luminescent solar converter uses a luminescent medium to downconvert photons with energies above the band gap to photon energies closer to the band gap of the semiconductor comprising the solar cell.

List of Band Gaps

Below are band gap values for some selected materials. For a comprehensive list of band gaps in semiconductors, see List of semiconductor materials.

Group	Material	Symbol	Band gap (eV) @ 302K
IV	Diamond	C	5.5
IV	Silicon	Si	1.11
IV	Germanium	Ge	0.67
III–V	Gallium nitride	GaN	3.4
III–V	Gallium phosphide	GaP	2.26
III–V	Gallium arsenide	GaAs	1.43
IV–V	Silicon nitride	Si_3N_4	5
IV–VI	Lead sulfide	PbS	0.37
IV–VI	Silicon dioxide	SiO_2	9
	Copper oxide	Cu_2O	2.1

Optical Versus Electronic Bandgap

In materials with a large exciton binding energy, it is possible for a photon to have just barely enough energy to create an exciton (bound electron–hole pair), but not enough energy to separate the electron and hole (which are electrically attracted to each other). In this situation, there is a distinction between “optical bandgap” and “electrical band gap” (or “transport gap”). The optical bandgap is the threshold for photons to be absorbed, while the transport gap is the threshold for creating an electron–hole pair that is *not* bound together. (The optical bandgap is at a lower energy than the transport gap.)

In almost all inorganic semiconductors, such as silicon, gallium arsenide, etc., there is very little interaction between electrons and holes (very small exciton binding energy), and therefore the optical and electronic bandgap are essentially identical, and the distinction between them is ignored. However, in some systems, including organic semiconductors and single-walled carbon nanotubes, the distinction may be significant.

In Photonics and Phononics

In photonics, band gaps or stop bands are ranges of photon frequencies where, if tunneling effects are neglected, no photons can be transmitted through a material. A material exhibiting this behaviour is known as a photonic crystal. The concept of hyperuniformity has broadened the range of photonic band gap materials, beyond photonic crystals. By applying the technique in supersymmetric quantum mechanics, a new class of optical disordered materials have been suggested, which support band gaps perfectly equivalent to those of crystals or quasicrystals.

Similar physics applies to phonons in a phononic crystal.

Valence and Conduction Bands

In solid-state physics, the valence band and conduction band are the bands closest to the Fermi level and thus determine the electrical conductivity of the solid. The valence band is the highest range of electron energies in which electrons are normally present at absolute zero temperature, while the conduction band is the lowest range of vacant electronic states. On a graph of the electronic band structure of a material, the valence band is located below the Fermi level, while the conduction band is located above it. This distinction is meaningless in metals as the highest band is partially filled, taking on the properties of both the valence and conduction bands.

Band Gap

In semiconductors and insulators the two bands are separated by a band gap, while in semimetals the bands overlap. A band gap is an energy range in a solid where no electron states can exist due to the quantization of energy. Electrical conductivity of non-metals is determined by the susceptibility of electrons to excitation from the valence band to the conduction band.

Electrical Conductivity

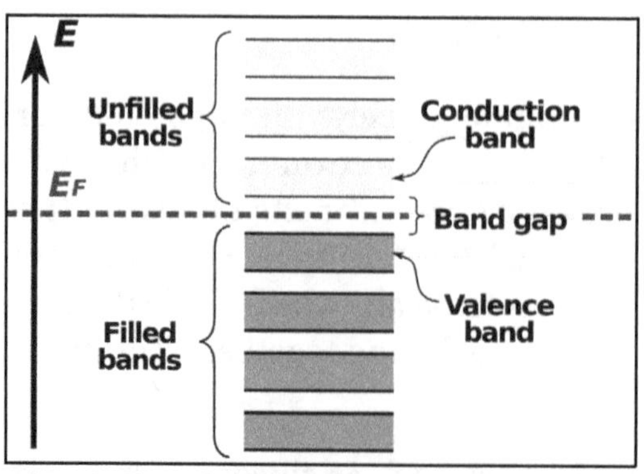

Semiconductor band structure See electrical conduction and semiconductor for a more detailed description of band structure.

In solids, the ability of electrons to act as charge carriers depends on the availability of vacant electronic states. This allows the electrons to increase their energy (i.e., accelerate) when an electric field is applied. This condition is only satisfied in the conduction band, since the valence band is full in non-metals.

As such, the electrical conductivity of a solid depends on its capability to flow electrons from the valence to the conduction band. Hence, in the case of a semimetal with an overlap region, the electrical conductivity is high. If there is a small band gap (E_g), then the flow of electrons from valence to conduction band is possible only if an external energy (thermal, etc.) is supplied; these groups with small E_g are called semiconductors. If the E_g is sufficiently high, then the flow of electrons from valence to conduction band becomes negligible under normal conditions; these groups are called insulators.

There is some conductivity in semiconductors, however. This is due to thermal excitation—some of the electrons get enough energy to jump the band gap in one go. Once they are in the conduction band, they can conduct electricity, as also can the hole they left behind in the valence band. The hole is an empty state that allows electrons in the valence band some degree of freedom.

Tight Binding

In solid-state physics, the tight-binding model (or TB model) is an approach to the calculation of electronic band structure using an approximate set of wave functions based upon superposition of wave functions for isolated atoms located at each atomic site. The method is closely related to the LCAO method (linear combination of atomic orbitals method) used in chemistry. Tight-binding models are applied to a wide variety of solids. The model gives good qualitative results in many cases and can be combined with other models that give better results where the tight-binding model fails. Though the tight-binding model is a one-electron model, the model also provides a basis for more advanced calculations like the calculation of surface states and application to various kinds of many-body problem and quasiparticle calculations.

Introduction

The name "tight binding" of this electronic band structure model suggests that this quantum mechanical model describes the properties of tightly bound electrons in solids. The electrons in this model should be tightly bound to the atom to which they belong and they should have limited interaction with states and potentials on surrounding atoms of the solid. As a result the wave function of the electron will be rather similar to the atomic orbital of the free atom to which it belongs. The energy of the electron will also be rather close to the ionization energy of the electron in the free atom or ion because the interaction with potentials and states on neighboring atoms is limited.

Though the mathematical formulation of the one-particle tight-binding Hamiltonian may look complicated at first glance, the model is not complicated at all and can be understood intuitively quite easily. There are only three kinds of matrix elements that play a significant role in the theory. Two of those three kinds of elements should be close to zero and can often be neglected. The most important elements in the model are the interatomic matrix elements, which would simply be called the bond energies by a chemist.

In general there are a number of atomic energy levels and atomic orbitals involved in the model. This can lead to complicated band structures because the orbitals belong to different point-group representations. The reciprocal lattice and the Brillouin zone often belong to a different space group than the crystal of the solid. High-symmetry points in the Brillouin zone belong to different point-group representations. When simple systems like the lattices of elements or simple compounds are studied it is often not very difficult to calculate eigenstates in high-symmetry points analytically. So the tight-binding model can provide nice examples for those who want to learn more about group theory.

The tight-binding model has a long history and has been applied in many ways and with many different purposes and different outcomes. The model doesn't stand on its own. Parts of the model can be filled in or extended by other kinds of calculations and models like the nearly-free electron model. The model itself, or parts of it, can serve as the basis for other calculations. In the study of conductive polymers, organic semiconductors and molecular electronics, for example, tight-binding-like models are applied in which the role of the atoms in the original concept is replaced by the molecular orbitals of conjugated systems and where the interatomic matrix elements are replaced by inter- or intramolecular hopping and tunneling parameters. These conductors nearly all have very anisotropic properties and sometimes are almost perfectly one-dimensional.

Historical Background

By 1928, the idea of a molecular orbital had been advanced by Robert Mulliken, who was influenced considerably by the work of Friedrich Hund. The LCAO method for approximating molecular orbitals was introduced in 1928 by B. N. Finklestein and G. E. Horowitz, while the LCAO method for solids was developed by Felix Bloch, as part of his doctoral dissertation in 1928, concurrently with and independent of the LCAO-MO approach. A much simpler interpolation scheme for approximating the electronic band structure, especially for the d-bands of transition metals, is the parameterized tight-binding method conceived in 1954 by John Clarke Slater and George Fred Koster, sometimes referred to as the SK tight-binding method. With the SK tight-binding method, electronic band structure calculations on a solid need not be carried out with full rigor as in the original Bloch's theorem but, rather, first-principles calculations are carried out only at high-symmetry points and the band structure is interpolated over the remainder of the Brillouin zone between these points.

In this approach, interactions between different atomic sites are considered as perturbations. There exist several kinds of interactions we must consider. The crystal Hamiltonian is only approximately a sum of atomic Hamiltonians located at different sites and atomic wave functions overlap adjacent atomic sites in the crystal, and so are not accurate representations of the exact wave function. There are further explanations in the next section with some mathematical expressions.

In the recent research about strongly correlated material the tight binding approach is basic approximation because highly localized electrons like 3-d transition metal electrons sometimes display strongly correlated behaviors. In this case, the role of electron-electron interaction must be considered using the many-body physics description.

The tight-binding model is typically used for calculations of electronic band structure and band gaps in the static regime. However, in combination with other methods such as the random phase approximation (RPA) model, the dynamic response of systems may also be studied.

Mathematical Formulation

We introduce the atomic orbitals $\varphi_m(\boldsymbol{r})$, which are eigenfunctions of the Hamiltonian H_{at} of a single isolated atom. When the atom is placed in a crystal, this atomic wave function overlaps adjacent atomic sites, and so are not true eigenfunctions of the crystal Hamiltonian. The overlap is less when electrons are tightly bound, which is the source of the descriptor "tight-binding". Any corrections to the atomic potential ΔUr required to obtain the true Hamiltonian H of the system, are assumed small:

$$H(\mathbf{r}) = \sum_{\mathbf{R_n}} H_{\mathrm{at}}(\mathbf{r} - \mathbf{R_n}) + \Delta U(\mathbf{r}).$$

A solution ψ_m to the time-independent single electron Schrödinger equation is then approximated as a linear combination of atomic orbitals $\varphi_m(\mathbf{r} - \mathbf{R})$:

$$\psi_m(\boldsymbol{r}) = \sum_{m,\boldsymbol{R}_n} b_m(\boldsymbol{R}_n)\ \varphi_m(\boldsymbol{r} - \boldsymbol{R}_n),$$

where m refers to the m-th atomic energy level and R_n locates an atomic site in the crystal lattice.

Translational Symmetry and Normalization

The Bloch theorem states that the wave function in crystal can change under translation only by a phase factor:

$$\psi(\boldsymbol{r} + \boldsymbol{R}_\ell) = e^{ik\cdot\boldsymbol{R}_\ell}\psi(\boldsymbol{r}),$$

where $\mathbf{k}$ is the wave vector of the wave function. Consequently, the coefficients satisfy,

$$\sum_{m,\boldsymbol{R}_n} b_m(\boldsymbol{R}_n)\ \varphi_m(\boldsymbol{r} - \boldsymbol{R}_n + \boldsymbol{R}_\ell) = e^{ik\cdot\boldsymbol{R}_\ell} \sum_{m,\boldsymbol{R}_n} b_m(\boldsymbol{R}_n)\ \varphi_m(\boldsymbol{r} - \boldsymbol{R}_n).$$

By substituting $\boldsymbol{R}_p = \boldsymbol{R}_n - \boldsymbol{R}_\ell$ we find,

$b_m(\mathbf{R_p} + \mathbf{R}_\ell) = e^{i\mathbf{k}\cdot\mathbf{R}_\ell} b_m(\mathbf{R_p})$, (where in RHS we have replaced the dummy index $\mathbf{R_n}$ with $\boldsymbol{R}_p$)

or ,

$$b_m(\boldsymbol{R}_l) = e^{ik\cdot\boldsymbol{R}_l} b_m(\mathbf{0}).$$

Normalizing the wave function to unity:

$$\int d^3r\ \psi_m^*(\mathbf{r})\psi_m(\mathbf{r}) = 1$$

$$= \sum_{\mathbf{R_n}} b_m^*(\mathbf{R_n}) \sum_{\mathbf{R}_\ell} b_m(\mathbf{R}_\ell) \int d^3r\ \varphi_m^*(\mathbf{r} - \mathbf{R_n})\varphi_m(\mathbf{r} - \mathbf{R}_\ell)$$

$$= b_m^*(0)b_m(0)\sum_{\mathbf{R_n}} e^{-i\mathbf{k}\cdot\mathbf{R_n}} \sum_{\mathbf{R}_\ell} e^{i\mathbf{k}\cdot\mathbf{R}_\ell} \int d^3r\, \varphi_m^*(\mathbf{r}-\mathbf{R_n})\varphi_m(\mathbf{r}-\mathbf{R}_\ell)$$

$$= Nb_m^*(0)b_m(0)\sum_{\mathbf{R_p}} e^{-i\mathbf{k}\cdot\mathbf{R_p}} \int d^3r\, \varphi_m^*(\mathbf{r}-\mathbf{R_p})\varphi_m(\mathbf{r})$$

$$= Nb_m^*(0)b_m(0)\sum_{\mathbf{R_p}} e^{i\mathbf{k}\cdot\mathbf{R_p}} \int d^3r\, \varphi_m^*(\mathbf{r})\varphi_m(\mathbf{r}-\mathbf{R_p}),$$

so the normalization sets $b(o)$ as,

$$\mathrm{b}_\mathrm{m}^*(0)\mathrm{b}_\mathrm{m}(0) = \frac{1}{\mathrm{N}} \cdot \frac{1}{1+\sum_{\mathbf{R_p}\neq 0} \mathrm{e}^{\mathrm{i}\mathbf{k}\cdot\mathbf{R_p}} \alpha_\mathrm{m}(\mathbf{R_p})},$$

where $\alpha\,(R_\mathrm{p})$ are the atomic overlap integrals, which frequently are neglected resulting in,

$$b_m(0) \approx \frac{1}{\sqrt{N}},$$

and,

$$\psi_m(\mathbf{r}) \approx \frac{1}{\sqrt{N}} \sum_{\mathbf{R_n}} e^{i\mathbf{k}\cdot\mathbf{R_n}}\, \varphi_m(\mathbf{r}-\mathbf{R_n}).$$

The Tight Binding Hamiltonian

Using the tight binding form for the wave function, and assuming only the *m-th* atomic energy level is important for the *m-th* energy band, the Bloch energies ε_m are of the form,

$$\varepsilon_m = \int d^3r\, \psi^*(\mathbf{r})H(\mathbf{r})\psi(\mathbf{r})$$

$$= \sum_{\mathbf{R_n}} b^*(\mathbf{R_n}) \int d^3r\, \varphi^*(\mathbf{r}-\mathbf{R_n})H(\mathbf{r})\psi(\mathbf{r})$$

$$= \sum_{\mathbf{R}_\ell} \sum_{\mathbf{R_n}} b^*(\mathbf{R_n}) \int d^3r\, \varphi^*(\mathbf{r}-\mathbf{R_n})H_{\mathrm{at}}(\mathbf{r}-\mathbf{R}_\ell)\psi(\mathbf{r}) + \sum_{\mathbf{R_n}} b^*(\mathbf{R_n}) \int d^3r\, \varphi^*(\mathbf{r}-\mathbf{R_n})\Delta U(\mathbf{r})\psi(\mathbf{r}).$$

$$\approx E_m + b^*(0)\sum_{\mathbf{R_n}} e^{-i\mathbf{k}\cdot\mathbf{R_n}} \int d^3r\, \varphi^*(\mathbf{r}-\mathbf{R_n})\Delta U(\mathbf{r})\psi(\mathbf{r}).$$

Here terms involving the atomic Hamiltonian at sites other than where it is centered are neglected. The energy then becomes,

$$\varepsilon_m(\mathbf{k}) = E_m - N\,|\,b(0)\,|^2 \left(\beta_m + \sum_{\mathbf{R_n}\neq 0} \sum_{l} \gamma_{m,l}(\mathbf{R_n}) e^{i\mathbf{k}\cdot\mathbf{R_n}} \right),$$

$$= E_m - \frac{\beta_m + \sum_{\mathbf{R_n}\neq 0}\sum_l e^{i\mathbf{k}\cdot\mathbf{R_n}}\gamma_{m,l}(\mathbf{R_n})}{1+\sum_{\mathbf{R_n}\neq 0}\sum_l e^{i\mathbf{k}\cdot\mathbf{R_n}}\alpha_{m,l}(\mathbf{R_n})},$$

where E_m is the energy of the m-th atomic level, and $\alpha_{m,l}$, β_m and $\gamma_{m,l}$ are the tight binding matrix elements.

The Tight Binding Matrix Elements

The element:

$$\beta_m = -\int \varphi_m^*(\mathbf{r})\Delta U(\mathbf{r})\varphi_m(\mathbf{r})d^3r,$$

is the atomic energy shift due to the potential on neighboring atoms. This term is relatively small in most cases. If it is large it means that potentials on neighboring atoms have a large influence on the energy of the central atom.

The next term:

$$\gamma_{m,l}(\mathbf{R_n}) = -\int \varphi_m^*(\mathbf{r})\Delta U(\mathbf{r})\varphi_l(\mathbf{r}-\mathbf{R_n})d^3r,$$

is the inter atomic matrix element between the atomic orbitals m and l on adjacent atoms. It is also called the bond energy or two center integral and it is the most important element in the tight binding model.

The last terms:

$$\alpha_{m,l}(\boldsymbol{R_n}) = \int \varphi_m^*(\boldsymbol{r})\varphi_l(\boldsymbol{r}-\boldsymbol{R_n})d^3r,$$

denote the overlap integrals between the atomic orbitals m and l on adjacent atoms.

Evaluation of the Matrix Elements

As mentioned before the values of the β_m-matrix elements are not so large in comparison with the ionization energy because the potentials of neighboring atoms on the central atom are limited. If β_m is not relatively small it means that the potential of the neighboring atom on the central atom is not small either. In that case it is an indication that the tight binding model is not a very good model for the description of the band structure for some reason. The inter atomic distances can be too small or the charges on the atoms or ions in the lattice is wrong for example.

The inter atomic matrix elements $\gamma_{m,l}$ can be calculated directly if the atomic wave functions and the potentials are known in detail. Most often this is not the case. There are numerous ways to get parameters for these matrix elements. Parameters can be obtained from chemical bond energy data. Energies and eigenstates on some high symmetry points in the Brillouin zone can be evaluated and values integrals in the matrix elements can be matched with band structure data from other sources.

The inter atomic overlap matrix elements $\alpha_{m,l}$ should be rather small or neglectable. If they are large it is again an indication that the tight binding model is of limited value for some purposes. Large overlap is an indication for too short inter atomic distance for example. In metals and transition metals the broad s-band or sp-band can be fitted better to an existing band structure calculation by the introduction of next-nearest-neighbor matrix elements and overlap integrals but fits like that don't yield a very useful model for the electronic wave function of a metal. Broad bands in dense materials are better described by a nearly free electron model.

The tight binding model works particularly well in cases where the band width is small and the electrons are strongly localized, like in the case of d-bands and f-bands. The model also gives good results in the case of open crystal structures, like diamond or silicon, where the number of neighbors is small. The model can easily be combined with a nearly free electron model in a hybrid NFE-TB model.

Connection to Wannier Functions

Bloch wave functions describe the electronic states in a periodic crystal lattice. Bloch functions can be represented as a Fourier series:

$$\psi_m(\mathbf{k},\mathbf{r}) = \frac{1}{\sqrt{N}} \sum_n a_m(\mathbf{R}_\mathbf{n},\mathbf{r}) e^{i\mathbf{k}\cdot\mathbf{R}_\mathbf{n}},$$

where R_n denotes an atomic site in a periodic crystal lattice, k is the wave vector of the Bloch wave, r is the electron position, m is the band index, and the sum is over all N atomic sites. The Bloch wave is an exact eigensolution for the wave function of an electron in a periodic crystal potential corresponding to an energy $E_m(k)$, and is spread over the entire crystal volume.

Using the Fourier transform analysis, a spatially localized wave function for the m-th energy band can be constructed from multiple Bloch waves:

$$a_m(\mathbf{R}_\mathbf{n},\mathbf{r}) = \frac{1}{\sqrt{N}} \sum_\mathbf{k} e^{-i\mathbf{k}\cdot\mathbf{R}_\mathbf{n}} \psi_m(\mathbf{k},\mathbf{r}) = \frac{1}{\sqrt{N}} \sum_\mathbf{k} e^{i\mathbf{k}\cdot(\mathbf{r}-\mathbf{R}_\mathbf{n})} u_m(\mathbf{k},\mathbf{r}).$$

These real space wave functions $a_m(\mathbf{R}_\mathbf{n},\mathbf{r})$ are called Wannier functions, and are fairly closely localized to the atomic site R_n. Of course, if we have exact Wannier functions, the exact Bloch functions can be derived using the inverse Fourier transform.

However it is not easy to calculate directly either Bloch functions or Wannier functions. An approximate approach is necessary in the calculation of electronic structures of solids. If we consider the extreme case of isolated atoms, the Wannier function would become an isolated atomic orbital. That limit suggests the choice of an atomic wave function as an approximate form for the Wannier function, the so-called tight binding approximation.

Second Quantization

Modern explanations of electronic structure like t-J model and Hubbard model are based on tight binding model. Tight binding can be understood by working under a second quantization formalism.

Using the atomic orbital as a basis state, the second quantization Hamiltonian operator in the tight binding framework can be written as:

$$H = -t \sum_{\langle i,j \rangle,\sigma} (c_{i,\sigma}^{\dagger} c_{j,\sigma} + h.c.),$$

$c_{i\sigma}^{\dagger}, c_{j\sigma}$ - creation and annihilation operators

σ - spin polarization

t - hopping integral

$\langle i, j \rangle$ -nearest neighbor index

Here, hopping integral t corresponds to the transfer integral γ in tight binding model. Considering extreme cases of $t \to 0$, it is impossible for an electron to hop into neighboring sites. This case is the isolated atomic system. If the hopping term is turned on ($t > 0$) electrons can stay in both sites lowering their kinetic energy.

In the strongly correlated electron system, it is necessary to consider the electron-electron interaction. This term can be written in,

$$H_{ee} = \frac{1}{2} \sum_{n,m,\sigma} \langle n_1 m_1, n_2 m_2 | \frac{e^2}{|r_1 - r_2|} | n_3 m_3, n_4 m_4 \rangle c_{n_1 m_1 \sigma_1}^{\dagger} c_{n_2 m_2 \sigma_2}^{\dagger} c_{n_4 m_4 \sigma_2} c_{n_3 m_3 \sigma_1}$$

This interaction Hamiltonian includes direct Coulomb interaction energy and exchange interaction energy between electrons. There are several novel physics induced from this electron-electron interaction energy, such as metal-insulator transitions (MIT), high-temperature superconductivity, and several quantum phase transitions.

Example: one-dimensional s-band

Here the tight binding model is illustrated with a s-band model for a string of atoms with a single s-orbital in a straight line with spacing a and σ bonds between atomic sites.

To find approximate eigenstates of the Hamiltonian, we can use a linear combination of the atomic orbitals:

$$| k = \frac{1}{\sqrt{N}} \sum_{n=1}^{N} e^{inka} | n \rangle$$

where N = total number of sites and k is a real parameter with $-\frac{\pi}{a} \leq k \leq \frac{\pi}{a}$. (This wave function is normalized to unity by the leading factor 1/√N provided overlap of atomic wave functions is ignored.) Assuming only nearest neighbor overlap, the only non-zero matrix elements of the Hamiltonian can be expressed as,

$$\langle n | H | n \rangle = E_0 = E_i - U .$$

$$\langle n \pm 1 | H | n \rangle = -\Delta$$

$$\langle n | n \rangle = 1; \quad \langle n \pm 1 | n \rangle = S.$$

The energy E_i is the ionization energy corresponding to the chosen atomic orbital and U is the energy shift of the orbital as a result of the potential of neighboring atoms. The $\langle n \pm 1 | H | n \rangle = -\Delta$ elements, which are the Slater and Koster interatomic matrix elements, are the bond energies $E_{i,j}$. In this one dimensional s-band model we only have σ-bonds between the s-orbitals with bond energy $E_{s,s} = V_{ss\sigma}$. The overlap between states on neighboring atoms is S. We can derive the energy of the state $| k$ using the above equation:

$$H | k \rangle = \frac{1}{\sqrt{N}} \sum_n e^{inka} H | n \rangle$$

$$\langle k | H | k \rangle = \frac{1}{N} \sum_{n,m} e^{i(n-m)ka} \langle m | H | n \rangle$$

$$= \frac{1}{N} \sum_n \langle n | H | n \rangle + \frac{1}{N} \sum_n \langle n-1 | H | n \rangle e^{+ika} + \frac{1}{N} \sum_n \langle n+1 | H | n \rangle e^{-ika}$$

$$= E_0 - 2\Delta \cos(ka),$$

where, for example,

$$\frac{1}{N} \sum_n \langle n | H | n \rangle = E_0 \frac{1}{N} \sum_n 1 = E_0,$$

and,

$$\frac{1}{N} \sum_n \langle n-1 | H | n \rangle e^{+ika} = -\Delta e^{ika} \frac{1}{N} \sum_n 1 = -\Delta e^{ika}.$$

$$\frac{1}{N} \sum_n \langle n-1 | n \rangle e^{+ika} = S e^{ika} \frac{1}{N} \sum_n 1 = S e^{ika}.$$

Thus the energy of this state $| k$ can be represented in the familiar form of the energy dispersion:

$$E(k) = \frac{E_0 - 2\Delta \cos(ka)}{1 + 2S \cos(ka)}.$$

- For $k = 0$ the energy is $E = (E_0 - 2\Delta)/(1 + 2S)$ and the state consists of a sum of all atomic orbitals. This state can be viewed as a chain of bonding orbitals.
- For $k = \pi/(2a)$ the energy is $E = E_0$ and the state consists of a sum of atomic orbitals which are a factor $e^{i\pi/2}$ out of phase. This state can be viewed as a chain of non-bonding orbitals.

- Finally for $k = \pi / a$ the energy is $E = E_0$ and the state consists of an alternating sum of atomic orbitals. This state can be viewed as a chain of anti-bonding orbitals.

This example is readily extended to three dimensions, for example, to a body-centered cubic or face-centered cubic lattice by introducing the nearest neighbor vector locations in place of simply $n\,a$. Likewise, the method can be extended to multiple bands using multiple different atomic orbitals at each site. The general formulation above shows how these extensions can be accomplished.

Fermi Liquid Theory

Fermi liquid theory (also known as Landau–Fermi liquid theory) is a theoretical model of interacting fermions that describes the normal state of most metals at sufficiently low temperatures. The interaction between the particles of the many-body system does not need to be small. The phenomenological theory of Fermi liquids was introduced by the Soviet physicist Lev Davidovich Landau in 1956, and later developed by Alexei Abrikosov and Isaak Khalatnikov using diagrammatic perturbation theory. The theory explains why some of the properties of an interacting fermion system are very similar to those of the Fermi gas (i.e. non-interacting fermions), and why other properties differ.

Important examples of where Fermi liquid theory has been successfully applied are most notably electrons in most metals and Liquid He-3. Liquid He-3 is a Fermi liquid at low temperatures (but not low enough to be in its superfluid phase.) He-3 is an isotope of helium, with 2 protons, 1 neutron and 2 electrons per atom. Because there is an odd number of fermions inside the atom, the atom itself is also a fermion. The electrons in a normal (non-superconducting) metal also form a Fermi liquid, as do the nucleons (protons and neutrons) in an atomic nucleus. Strontium ruthenate displays some key properties of Fermi liquids, despite being a strongly correlated material, and is compared with high temperature superconductors like cuprates.

Description

The key ideas behind Landau's theory are the notion of *adiabaticity* and the exclusion principle. Consider a non-interacting fermion system (a Fermi gas), and suppose we "turn on" the interaction slowly. Landau argued that in this situation, the ground state of the Fermi gas would adiabatically transform into the ground state of the interacting system.

By Pauli's exclusion principle, the ground state Ψ_0 of a Fermi gas consists of fermions occupying all momentum states corresponding to momentum $p < p_F$ with all higher momentum states unoccupied. As interaction is turned on, the spin, charge and momentum of the fermions corresponding to the occupied states remain unchanged, while their dynamical properties, such as their mass, magnetic moment etc. are *renormalized* to new values. Thus, there is a one-to-one correspondence between the elementary excitations of a Fermi gas system and a Fermi liquid system. In the context of Fermi liquids, these excitations are called "quasi-particles".

Landau quasiparticles are long-lived excitations with a lifetime τ that satisfies $\frac{\hbar}{\tau} \ll \epsilon_p$ where ϵ_p is the quasiparticle energy (measured from the Fermi energy.) At finite temperature, ϵ_p is on the order of the thermal energy $k_B T$, and the condition for Landau quasiparticles can be reformulated as $\frac{\hbar}{\tau} k_B T$..

For this system, the Green's function can be written (near its poles) in the form,

$$G(\omega, p) \approx \frac{Z}{\omega + \mu - \epsilon(p)}$$

where μ is the chemical potential and $\epsilon(p)$ is the energy corresponding to the given momentum state.

The value Z is called the *quasiparticle residue* and is very characteristic of Fermi liquid theory. The spectral function for the system can be directly observed via ARPES experiment, and can be written (in the limit of low-lying excitations) in the form:

$$A(\vec{k}, \omega) = Z\delta(\omega - v_F k_{\|})$$

where v_F is the Fermi velocity.

Physically, we can say that a propagating fermion interacts with its surrounding in such a way that the net effect of the interactions is to make the fermion behave as a "dressed" fermion, altering its effective mass and other dynamical properties. These "dressed" fermions are what we think of as "quasiparticles".

Another important property of Fermi liquids is related to the scattering cross section for electrons. Suppose we have an electron with energy ϵ_1 above the Fermi surface, and suppose it scatters with a particle in the Fermi sea with energy ϵ_2. By Pauli's exclusion principle, both the particles after scattering have to lie above the Fermi surface, with energies $\epsilon_3, \epsilon_4 > \epsilon_F$ Now, suppose the initial electron has energy very close to the Fermi surface $\epsilon \approx \epsilon_F$ Then, we have that $\epsilon_2, \epsilon_3, \epsilon_4$ also have to be very close to the Fermi surface. This reduces the phase space volume of the possible states after scattering, and hence, by Fermi's golden rule, the scattering cross section goes to zero. Thus we can say that the lifetime of particles at the Fermi surface goes to infinity.

Similarities to Fermi Gas

The Fermi liquid is qualitatively analogous to the non-interacting Fermi gas, in the following sense: The system's dynamics and thermodynamics at low excitation energies and temperatures may be described by substituting the non-interacting fermions with interacting quasiparticles, each of which carries the same spin, charge and momentum as the original particles. Physically these may be thought of as being particles whose motion is disturbed by the surrounding particles and which themselves perturb the particles in their vicinity. Each many-particle excited state of the interacting system may be described by listing all occupied momentum states, just as in the non-interacting system. As a consequence, quantities such as the heat capacity of the Fermi liquid behave qualitatively in the same way as in the Fermi gas (e.g. the heat capacity rises linearly with temperature).

Differences from Fermi gas

The following differences to the non-interacting Fermi gas arise:

Energy

The energy of a many-particle state is not simply a sum of the single-particle energies of all occupied states. Instead, the change in energy for a given change δn_k in occupation of states k contains terms both linear and quadratic in δn_k (for the Fermi gas, it would only be linear, $\delta n_k \grave{o}_k$, where $\grave{o}_k$ denotes the single-particle energies). The linear contribution corresponds to renormalized single-particle energies, which involve, e.g., a change in the effective mass of particles. The quadratic terms correspond to a sort of "mean-field" interaction between quasiparticles, which is parameterized by so-called Landau Fermi liquid parameters and determines the behaviour of density oscillations (and spin-density oscillations) in the Fermi liquid. Still, these mean-field interactions do not lead to a scattering of quasi-particles with a transfer of particles between different momentum states.

The renormalization of the mass of a fluid of interacting fermions can be calculated from first principles using many-body computational techniques. For the two-dimensional homogeneous electron gas, *GW* calculations and quantum Monte Carlo methods have been used to calculate renormalized quasiparticle effective masses.

Specific heat and compressibility

Specific heat, compressibility, spin-susceptibility and other quantities show the same qualitative behaviour (e.g. dependence on temperature) as in the Fermi gas, but the magnitude is (sometimes strongly) changed.

Interactions

In addition to the mean-field interactions, some weak interactions between quasiparticles remain, which lead to scattering of quasiparticles off each other. Therefore, quasiparticles acquire a finite lifetime. However, at low enough energies above the Fermi surface, this lifetime becomes very long, such that the product of excitation energy (expressed in frequency) and lifetime is much larger than one. In this sense, the quasiparticle energy is still well-defined (in the opposite limit, Heisenberg's uncertainty relation would prevent an accurate definition of the energy).

Structure

The structure of the "bare" particle's (as opposed to quasiparticle) Green's function is similar to that in the Fermi gas (where, for a given momentum, the Green's function in frequency space is a delta peak at the respective single-particle energy). The delta peak in the density-of-states is broadened (with a width given by the quasiparticle lifetime). In addition (and in contrast to the quasiparticle Green's function), its weight (integral over frequency) is suppressed by a quasiparticle weight factor $0 < Z < 1$. The remainder of the total weight is in a broad "incoherent background", corresponding to the strong effects of interactions on the fermions at short time-scales.

Distribution

The distribution of particles (as opposed to quasiparticles) over momentum states at zero temperature still shows a discontinuous jump at the Fermi surface (as in the Fermi gas), but it does not drop from 1 to 0: the step is only of size Z.

Electrical Resistivity

In a metal the resistivity at low temperatures is dominated by electron-electron scattering in combination with umklapp scattering. For a Fermi liquid, the resistivity from this mechanism varies as T^2, which is often taken as an experimental check for Fermi liquid behaviour (in addition to the linear temperature-dependence of the specific heat), although it only arises in combination with the lattice. In certain cases, umklapp scattering is not required. For example, the resistivity of compensated semimetals scales as T^2 because of mutual scattering of electron and hole. This is known as the Baber mechanism.

Optical Response

Fermi liquid theory predicts that the scattering rate, which governs the optical response of metals, not only depends quadratically on temperature (thus causing the T^2 dependence of the dc resistance), but it also depends quadratically on frequency. This is in contrast to the Drude prediction for non-interacting metallic electrons, where the scattering rate is a constant as a function of frequency.One material in which optical Fermi liquid behavior was experimentally observed is the low-temperature metallic phase of Sr_2RuO_4.

Instabilities of the Fermi Liquid

The experimental observation of exotic phases in strongly correlated systems has triggered an enormous effort from the theoretical community to try to understand their microscopic origin. One possible route to detect instabilities of a Fermi liquid is precisely the analysis done by Pomeranchuk. Due to that, the Pomeranchuk instability has been studied by several authors with different techniques in the last few years and in particular, the instability of the Fermi liquid towards the nematic phase was investigated for several models.

Non-fermi Liquids

The term non-Fermi liquid, also known as "strange metal", is used to describe a system which displays breakdown of Fermi-liquid behaviour. The simplest example of such a system is the system of interacting fermions in one-dimension, called *Luttinger liquid*. Although Luttinger liquids are physically similar to Fermi liquids, the restriction to one dimension gives rise to several qualitative differences such as the absence of a *quasiparticle peak* in the momentum dependent spectral function, spin-charge separation, and the presence of spin density waves. One cannot ignore the existence of interactions in one-dimension and has to describe the problem with a non-Fermi theory, where Luttinger liquid is one of them. At small finite spin-temperatures in one-dimension the ground-state of the system is described by spin-incoherent Luttinger liquid (SILL).

Another example of such behaviour is observed at quantum critical points of certain second-order phase transitions, such as Heavy fermion criticality, Mott criticality and high-T_c cuprate phase transitions. The ground state of such transitions is characterized by the presence of a sharp Fermi surface, although there may not be well-defined quasiparticles. That is, on approaching the critical point, it is observed that the quasiparticle residue $Z \to 0$.

On the other hand, specific quantum critical point represented by fermion condensation quantum phase transition (see e.g. strongly correlated quantum spin liquid) supports quasiparticles with the finite quasiparticle residue $Z > 0$. This phase transition does preserve the Pomeranchuk stability conditions, and proffers a new way to violate the stability of Fermi liquid. These unique properties of the phase transition allow one to explain both the scaling and the non-Fermi liquid behaviour observed in heavy fermion compounds.

Understanding the behaviour of non-Fermi liquids is an important problem in condensed matter physics. Approaches towards explaining these phenomena include the treatment of *marginal Fermi liquids*; attempts to understand critical points and derive scaling relations; and descriptions using *emergent* gauge theories with techniques of holographic gauge/gravity duality.

Fermi Level

The Fermi level is the total chemical potential for electrons (or electrochemical potential for electrons) and is usually denoted by μ or E_F. The Fermi level of a body is a thermodynamic quantity, and its significance is the thermodynamic work required to add one electron to the body (not counting the work required to remove the electron from wherever it came from). A precise understanding of the Fermi level—how it relates to electronic band structure in determining electronic properties, how it relates to the voltage and flow of charge in an electronic circuit—is essential to an understanding of solid-state physics.

In a band structure picture, the Fermi level can be considered to be a hypothetical energy level of an electron, such that at thermodynamic equilibrium this energy level would have a *50% probability of being occupied at any given time*. The position of the Fermi level with the relation to the band energy levels is a crucial factor in determining electrical properties. The Fermi level does not necessarily correspond to an actual energy level (in an insulator the Fermi level lies in the band gap), nor does it require the existence of a band structure. Nonetheless, the Fermi level is a precisely defined thermodynamic quantity, and differences in Fermi level can be measured simply with a voltmeter.

The Fermi Level and Voltage

Sometimes it is said that electric currents are driven by differences in electrostatic potential (Galvani potential), but this is not exactly true. As a counterexample, multi-material devices such as p–n junctions contain internal electrostatic potential differences at equilibrium, yet without any accompanying net current; if a voltmeter is attached to the junction, one simply measures zero volts. Clearly, the electrostatic potential is not the only factor influencing the flow of charge in a material—Pauli repulsion, carrier concentration gradients, electromagnetic induction, and thermal effects also play an important role.

A voltmeter measures differences in Fermi level divided by electron charge.

In fact, the quantity called *voltage* as measured in an electronic circuit has a simple relationship to the chemical potential for electrons (Fermi level). When the leads of a voltmeter are attached to two points in a circuit, the displayed voltage is a measure of the *total* work transferred when a unit charge is allowed to move from one point to the other. If a simple wire is connected between two points of differing voltage (forming a short circuit), current will flow from positive to negative voltage, converting the available work into heat.

The Fermi level of a body expresses the work required to add an electron to it, or equally the work obtained by removing an electron. Therefore, $V_A - V_B$, the observed difference in voltage between two points, *A* and *B*, in an electronic circuit is exactly related to the corresponding chemical potential difference, $\mu_A - \mu_B$, in Fermi level by the formula,

$$V_A - V_B = \frac{\mu_A - \mu_B}{-e}$$

where $-e$ is the electron charge.

From the above discussion it can be seen that electrons will move from a body of high μ (low voltage) to low μ (high voltage) if a simple path is provided. This flow of electrons will cause the lower μ to increase (due to charging or other repulsion effects) and likewise cause the higher μ to decrease. Eventually, μ will settle down to the same value in both bodies. This leads to an important fact regarding the equilibrium (off) state of an electronic circuit:

> *An electronic circuit in thermodynamic equilibrium will have a constant Fermi level throughout its connected parts.*

This also means that the voltage (measured with a voltmeter) between any two points will be zero, at equilibrium. Note that thermodynamic equilibrium here requires that the circuit be internally connected and not contain any batteries or other power sources, nor any variations in temperature.

The Fermi Level and band Structure

In the band theory of solids, electrons are considered to occupy a series of bands composed of single-particle energy eigenstates each labelled by ϵ. Although this single particle picture is an approximation, it greatly simplifies the understanding of electronic behaviour and it generally provides correct results when applied correctly.

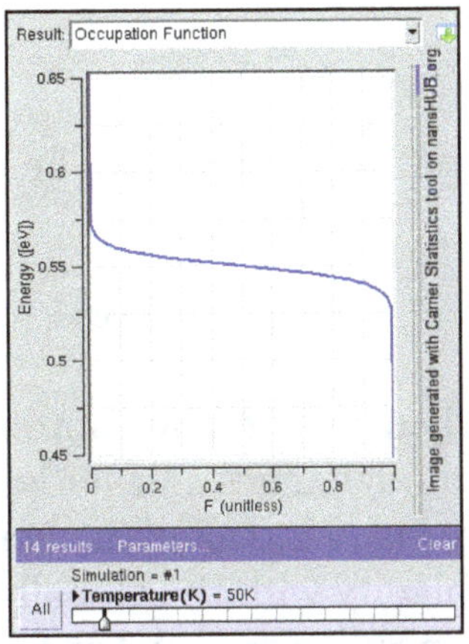

Fermi function F(Ȯ) vs. energy Ȯ , with μ = 0.55 eV and for various temperatures in the range 50K ≤ T ≤ 375K.

The Fermi–Dirac distribution, $f(\epsilon)$, , gives the probability that (at thermodynamic equilibrium) a state having energy ∈ is occupied by an electron:

$$f(\epsilon)=\frac{1}{e^{(\epsilon-\mu)/kT}+1}$$

Here, T is the absolute temperature and k is Boltzmann's constant. If there is a state at the Fermi level ($\in = \mu$), then this state will have a 50% chance of being occupied. The distribution is plotted in the left figure. The closer f is to 1, the higher chance this state is occupied. The closer f is to 0, the higher chance this state is empty.

The location of μ within a material's band structure is important in determining the electrical behaviour of the material.

- In an insulator, μ lies within a large band gap, far away from any states that are able to carry current.
- In a metal, semimetal or degenerate semiconductor, μ lies within a delocalized band. A large number of states nearby μ are thermally active and readily carry current.
- In an intrinsic or lightly doped semiconductor, μ is close enough to a band edge that there are a dilute number of thermally excited carriers residing near that band edge.

In semiconductors and semimetals the position of μ relative to the band structure can usually be controlled to a significant degree by doping or gating. These controls do not change μ which is fixed by the electrodes, but rather they cause the entire band structure to shift up and down (sometimes also changing the band structure's shape). For further information about the Fermi levels of semiconductors, see (for example) Sze.

Local Conduction Band Referencing, Internal Chemical Potential and the Parameter ζ

If the symbol ε is used to denote an electron energy level measured relative to the energy of the edge of its enclosing band, $\in_C$, then in general we have $\varepsilon = \in - \in_C$. We can define a parameter ζ that references the Fermi level with respect to the band edge ,

$$\zeta = \mu - \epsilon_C.$$

It follows that the Fermi–Dirac distribution function can be written as,

$$f(\mathcal{E}) = \frac{1}{e^{(\mathcal{E}-\zeta)/kT} + 1}.$$

The band theory of metals was initially developed by Sommerfeld, from 1927 onwards, who paid great attention to the underlying thermodynamics and statistical mechanics. Confusingly, in some contexts the band-referenced quantity ζ may be called the *Fermi level*, *chemical potential*, or *electrochemical potential*, leading to ambiguity with the globally-referenced Fermi level. In this article, the terms *conduction-band referenced Fermi level* or *internal chemical potential* are used to refer to ζ.

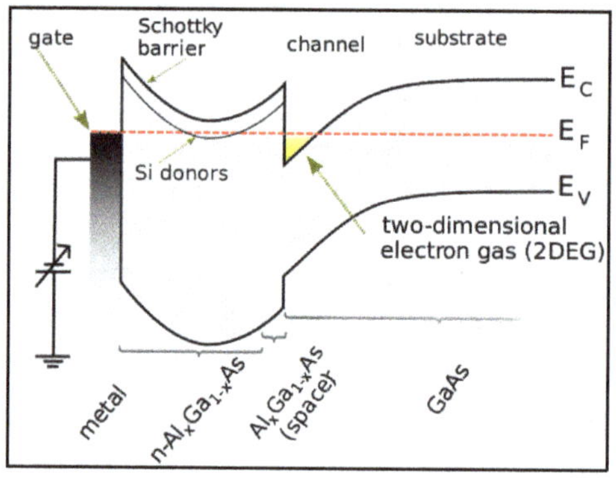

Example of variations in conduction band edge E_C in a band diagram of GaAs/AlGaAs heterojunction-based high-electron-mobility transistor.

ζ is directly related to the number of active charge carriers as well as their typical kinetic energy, and hence it is directly involved in determining the local properties of the material (such as electrical conductivity). For this reason it is common to focus on the value of ζ when concentrating on the properties of electrons in a single, homogeneous conductive material. By analogy to the energy states of a free electron, the E of a state is the kinetic energy of that state and ϵ_C is its potential energy. With this in mind, the parameter, ζ, could also be labelled the *Fermi kinetic energy*.

Unlike μ, the parameter, ζ, is not a constant at equilibrium, but rather varies from location to location in a material due to variations in ϵ_C, which is determined by factors such as material quality and impurities/dopants. Near the surface of a semiconductor or semimetal, ζ can be strongly controlled by externally applied electric fields, as is done in a field effect transistor. In a multi-band material, ζ may even take on multiple values in a single location. For example, in a piece of aluminum metal there are two conduction bands crossing the Fermi level (even more bands in other materials); each band has a different edge energy, ϵ_C, and a different ζ.

The value of ζ at zero temperature is widely known as the Fermi energy, sometimes written ζ_0. Confusingly (again), the name *Fermi energy* sometimes is used to refer to ζ at non-zero temperature.

The Fermi Level and Temperature out of Equilibrium

The Fermi level, μ, and temperature, T, are well defined constants for a solid-state device in thermodynamic equilibrium situation, such as when it is sitting on the shelf doing nothing. When the device is brought out of equilibrium and put into use, then strictly speaking the Fermi level and temperature are no longer well defined. Fortunately, it is often possible to define a quasi-Fermi level and quasi-temperature for a given location, that accurately describe the occupation of states in terms of a thermal distribution. The device is said to be in *quasi-equilibrium* when and where such a description is possible.

The quasi-equilibrium approach allows one to build a simple picture of some non-equilibrium effects as the electrical conductivity of a piece of metal (as resulting from a gradient of μ) or its thermal conductivity (as resulting from a gradient in T). The quasi-μ and quasi-T can vary (or not exist at all) in any non-equilibrium situation, such as:

- If the system contains a chemical imbalance (as in a battery).
- If the system is exposed to changing electromagnetic fields. (as in capacitors, inductors, and transformers).
- Under illumination from a light-source with a different temperature, such as the sun (as in solar cells).
- When the temperature is not constant within the device (as in thermocouples).
- When the device has been altered, but has not had enough time to re-equilibrate (as in piezoelectric or pyroelectric substances).

In some situations, such as immediately after a material experiences a high-energy laser pulse, the electron distribution cannot be described by any thermal distribution. One cannot define the quasi-Fermi level or quasi-temperature in this case; the electrons are simply said to be *non-thermalized.* In less dramatic situations, such as in a solar cell under constant illumination, a quasi-equilibrium description may be possible but requiring the assignment of distinct values of μ and T to different bands (conduction band vs. valence band). Even then, the values of μ and T may jump discontinuously across a material interface (e.g., p–n junction) when a current is being driven, and be ill-defined at the interface itself.

Technicalities

Terminology Problems

The term, *Fermi level,* is mainly used in discussing the solid state physics of electrons in semiconductors, and a precise usage of this term is necessary to describe band diagrams in devices comprising different materials with different levels of doping. In these contexts, however, one may also see Fermi level used imprecisely to refer to the *band-referenced Fermi level,* $\mu - \epsilon_C$, called ζ above. It is common to see scientists and engineers refer to "controlling", "pinning", or "tuning" the Fermi level inside a conductor, when they are in fact describing changes in ϵ_C due to doping or the field effect. In fact, thermodynamic equilibrium guarantees that the Fermi level in a conductor is *always* fixed to be exactly equal to the Fermi level of the electrodes; only the band structure (not the Fermi level) can be changed by doping or the field effect. A similar ambiguity exists between the terms, *chemical potential* and *electrochemical potential.*

It is also important to note that Fermi *level* is not necessarily the same thing as Fermi *energy*. In the wider context of quantum mechanics, the term Fermi energy usually refers to *the maximum kinetic energy of a fermion in an idealized non-interacting, disorder free, zero temperature Fermi gas*. This concept is very theoretical (there is no such thing as a non-interacting Fermi gas, and zero temperature is impossible to achieve). However, it finds some use in approximately describing white dwarfs, neutron stars, atomic nuclei, and electrons in a metal. On the other hand, in the fields of semiconductor physics and engineering, *Fermi energy* often is used to refer to the Fermi level described in this article.

Fermi Level Referencing and the Location of zero Fermi Level

Much like the choice of origin in a coordinate system, the zero point of energy can be defined arbitrarily. Observable phenomena only depend on energy differences. When comparing distinct bodies, however, it is important that they all be consistent in their choice of the location of zero energy, or else nonsensical results will be obtained. It can therefore be helpful to explicitly name a common point to ensure that different components are in agreement. On the other hand, if a reference point is inherently ambiguous (such as "the vacuum") it will instead cause more problems.

A practical and well-justified choice of common point is a bulky, physical conductor, such as the electrical ground or earth. Such a conductor can be considered to be in a good thermodynamic equilibrium and so its μ is well defined. It provides a reservoir of charge, so that large numbers of electrons may be added or removed without incurring charging effects. It also has the advantage of being accessible, so that the Fermi level of any other object can be measured simply with a voltmeter.

Why it is not Advisable to use "the Energy in Vacuum" as a Reference Zero

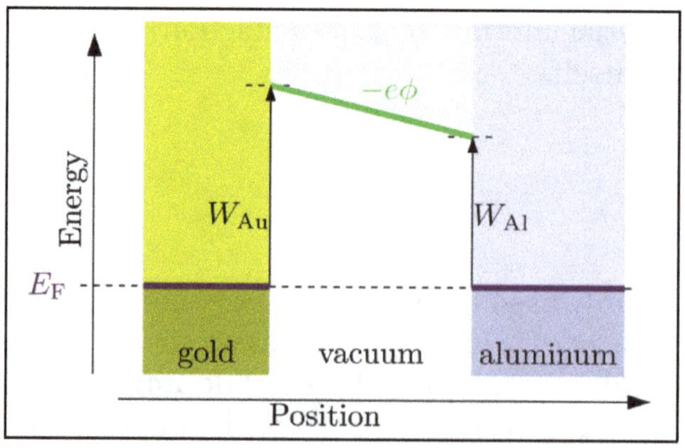

When the two metals depicted here are in thermodynamic equilibrium as shown (equal Fermi levels E_F), the vacuum electrostatic potential ▪ is not flat due to a difference in work function.

In principle, one might consider using the state of a stationary electron in the vacuum as a reference point for energies. This approach is not advisable unless one is careful to define exactly where *the vacuum* is. The problem is that not all points in the vacuum are equivalent.

At thermodynamic equilibrium, it is typical for electrical potential differences of order 1 V to exist in the vacuum (Volta potentials). The source of this vacuum potential variation is the variation in work function between the different conducting materials exposed to vacuum. Just outside a conductor, the electrostatic potential depends sensitively on the material, as well as which surface is selected (its crystal orientation, contamination, and other details).

The parameter that gives the best approximation to universality is the Earth-referenced Fermi level suggested above. This also has the advantage that it can be measured with a voltmeter.

Discrete charging effects in small systems

In cases where the "charging effects" due to a single electron are non-negligible, the above definitions should be clarified. For example, consider a capacitor made of two identical parallel-plates. If the capacitor is uncharged, the Fermi level is the same on both sides, so one might think that it should take no energy to move an electron from one plate to the other. But when the electron has been moved, the capacitor has become (slightly) charged, so this does take a slight amount of energy. In a normal capacitor, this is negligible, but in a nano-scale capacitor it can be more important.

In this case one must be precise about the thermodynamic definition of the chemical potential as well as the state of the device: is it electrically isolated, or is it connected to an electrode?

- When the body is able to exchange electrons and energy with an electrode (reservoir), it is described by the grand canonical ensemble. The value of chemical potential μ can be said to be fixed by the electrode, and the number of electrons N on the body may fluctuate. In this case, the chemical potential of a body is the infinitesimal amount of work needed to increase the *average* number of electrons by an infinitesimal amount (even though the number of electrons at any time is an integer, the average number varies continuously.):

$$\mu(\langle N\rangle, T) = \left(\frac{\partial F}{\partial\langle N\rangle}\right)_T,$$

  where $F(N, T)$ is the free energy function of the grand canonical ensemble.

- If the number of electrons in the body is fixed (but the body is still thermally connected to a heat bath), then it is in the canonical ensemble. We can define a "chemical potential" in this case literally as the work required to add one electron to a body that already has exactly N electrons,

$$\mu'(N,T) = F(N+1,T) - F(N,T),$$

  where $F(N, T)$ is the free energy function of the canonical ensemble, or alternatively as the work obtained by removing an electron from that body,

$$\mu''(N,T) = F(N,T) - F(N-1,T) = \mu'(N-1,T).$$

These chemical potentials are not equivalent, $\mu \neq \mu' \neq \mu''$, except in the thermodynamic limit. The distinction is important in small systems such as those showing Coulomb blockade. The parameter, μ, (i.e., in the case where the number of electrons is allowed to fluctuate) remains exactly related to the voltmeter voltage, even in small systems. To be precise, then, the Fermi level is defined not by a deterministic charging event by one electron charge, but rather a statistical charging event by an infinitesimal fraction of an electron.

Empty Lattice Approximation

The empty lattice approximation is a theoretical electronic band structure model in which the potential is *periodic* and *weak* (close to constant). One may also consider an empty irregular lattice, in which the potential is not even periodic. The empty lattice approximation describes a number of properties of energy dispersion relations of non-interacting free electrons that move through a crystal lattice. The energy of the electrons in the "empty lattice" is the same as the energy of free electrons. The model is useful because it clearly illustrates a number of the sometimes very complex features of energy dispersion relations in solids which are fundamental to all electronic band structures.

Scattering and Periodicity

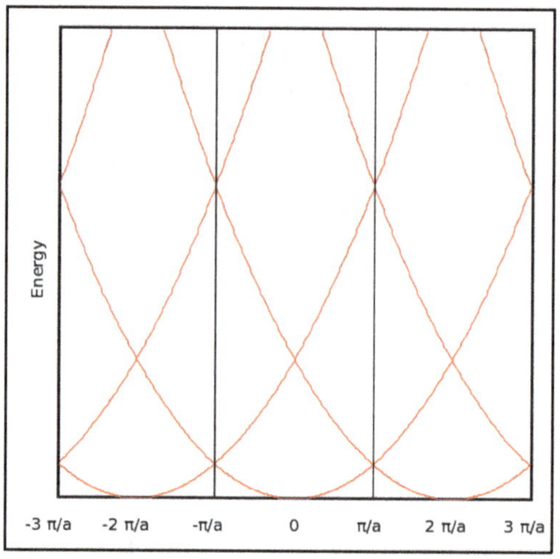

Free electron bands in a one dimensional lattice.

The periodic potential of the lattice in this free electron model must be weak because otherwise the electrons wouldn't be free. The strength of the scattering mainly depends on the geometry and topology of the system. Topologically defined parameters, like scattering cross sections, depend on the magnitude of the potential and the size of the potential well. For 1-, 2- and 3-dimensional spaces potential wells do always scatter waves, no matter how small their potentials are, what their signs are or how limited their sizes are. For a particle in a one-dimensional lattice, like the Kronig-Penney model, it is possible to calculate the band structure analytically by substituting the values for the potential, the lattice spacing and the size of potential well. For two and three-dimensional problems it is more difficult to calculate a band structure based on a similar model with a few parameters accurately. Nevertheless, the properties of the band structure can easily be approximated in most regions by perturbation methods.

In theory the lattice is infinitely large, so a weak periodic scattering potential will eventually be strong enough to reflect the wave. The scattering process results in the well known Bragg reflections of electrons by the periodic potential of the crystal structure. This is the origin of the periodicity of the dispersion relation and the division of k-space in Brillouin zones. The periodic energy dispersion relation is expressed as:

$$E_n(\mathbf{k}) = \frac{\hbar^2(\mathbf{k}+\mathbf{G}_n)^2}{2m}$$

The $\mathbf{G}_n$ are the reciprocal lattice vectors to which the bands $E_n(\mathbf{k})$ belong.

The figure on the right shows the dispersion relation for three periods in reciprocal space of a one-dimensional lattice with lattice cells of length a.

The Energy Bands and the Density of States

In a one-dimensional lattice the number of reciprocal lattice vectors $\mathbf{G}_n$ that determine the bands in an energy interval is limited to two when the energy rises. In two and three dimensional lattices the number of reciprocal lattice vectors that determine the free electron bands $E_n(\mathbf{k})$ increases more rapidly when the length of the wave vector increases and the energy rises. This is because the number of reciprocal lattice vectors $\mathbf{G}_n$ that lie in an interval $[\mathbf{k},\mathbf{k}+d\mathbf{k}]$ increases. The density of states in an energy interval $[E, E+dE]$ depends on the number of states in an interval $[\mathbf{k},\mathbf{k}+d\mathbf{k}]$ in reciprocal space and the slope of the dispersion relation $E_n(\mathbf{k})$.

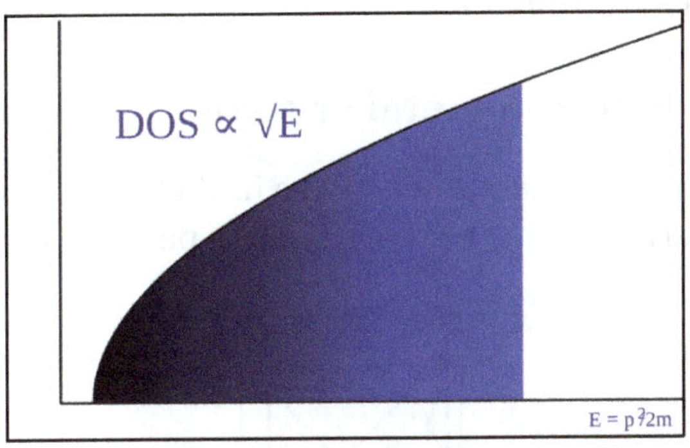

Figure 3: Free-electron DOS in 3-dimensional k-space.

Though the lattice cells are not spherically symmetric, the dispersion relation still has spherical symmetry from the point of view of a fixed central point in a reciprocal lattice cell if the dispersion relation is extended outside the central Brillouin zone. The density of states in a three-dimensional lattice will be the same as in the case of the absence of a lattice. For the three-dimensional case the density of states $D_3(E)$ is;

$$D_3(E) = 2\pi\sqrt{\frac{E-E_0}{c_k^3}}\,.$$

In three-dimensional space the Brillouin zone boundaries are planes. The dispersion relations show conics of the free-electron energy dispersion parabolas for all possible reciprocal lattice vectors. This results in a very complicated set intersecting of curves when the dispersion relations are calculated because there is a large number of possible angles between evaluation trajectories, first and higher order Brillouin zone boundaries and dispersion parabola intersection cones.

Second, Third and Higher Brillouin Zones

"Free electrons" that move through the lattice of a solid with wave vectors **k** far outside the first Brillouin zone are still reflected back into the first Brillouin zone.

The Nearly Free Electron Model

In most simple metals, like aluminium, the screening effect strongly reduces the electric field of the ions in the solid. The electrostatic potential is expressed as,

$$V(r) = \frac{Ze}{r} e^{-qr}$$

where Z is the atomic number, e is the elementary unit charge, r is the distance to the nucleus of the embedded ion and q is a screening parameter that determines the range of the potential. The Fourier transform, $U_{\mathbf{G}}$, of the lattice potential, $V(\mathbf{r})$, is expressed as,

$$U_{\mathbf{G}} = \frac{4\pi Ze}{q^2 + \mathbf{G}^2}$$

When the values of the off-diagonal elements $U_{\mathbf{G}}$ between the reciprocal lattice vectors in the Hamiltonian almost go to zero. As a result, the magnitude of the band gap $2|U_{\mathbf{G}}|$ collapses and the empty lattice approximation is obtained.

The electron bands of common metal crystals

Apart from a few exotic exceptions, metals crystallize in three kinds of crystal structures: the BCC and FCC cubic crystal structures and the hexagonal close-packed HCP crystal structure.

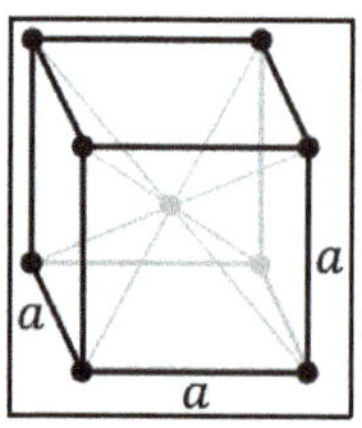

Body-centered cubic (I).

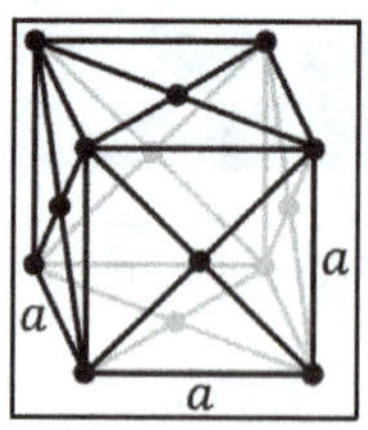

Face-centered cubic (F).

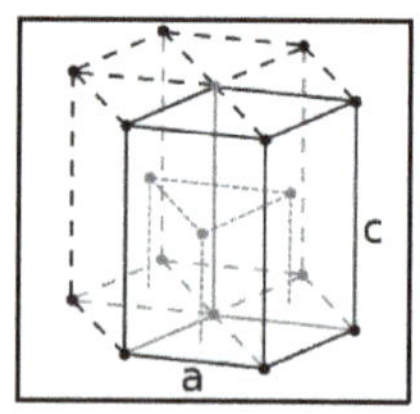

Hexagonal close-packed.

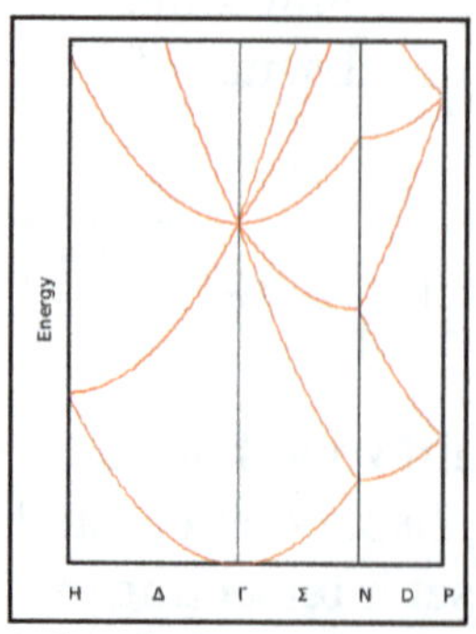

Free electron bands in a BCC crystal structure.

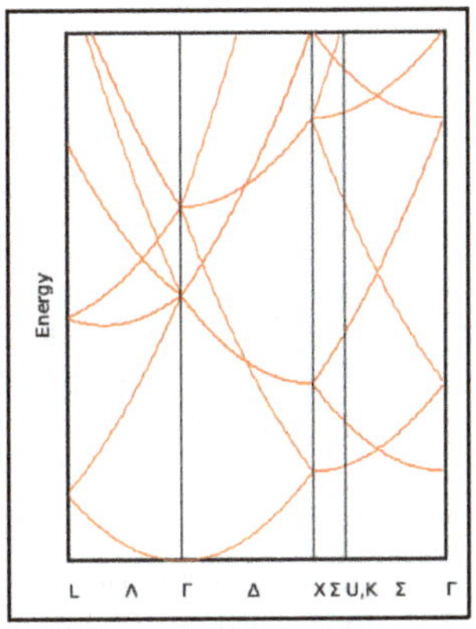

Free electron bands in a FCC crystal structure.

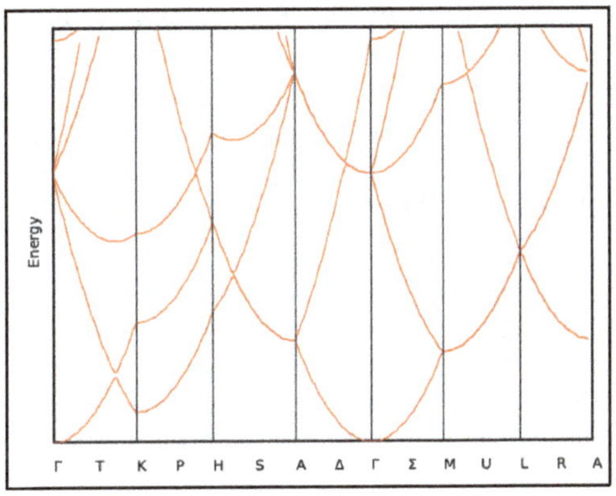

Free electron bands in a HCP crystal structure.

References

- Streetman, Ben G.; Sanjay Banerjee (2000). Solid State electronic Devices (5th ed.). New Jersey: Prentice Hall. p. 524. ISBN 0-13-025538-6
- Amusia, M.; Popov, K.; Shaginyan, V.; Stephanovich, V. (2014). "Theory of Heavy-Fermion Compounds - Theory of Strongly Correlated Fermi-Systems". Springer. ISBN 978-3-319-10825-4
- For example: D. Chattopadhyay (2006). Electronics (fundamentals And Applications). ISBN 978-81-224-1780-7. and Balkanski and Wallis (2000-09-01). Semiconductor Physics and Applications. ISBN 978-0-19-851740-5
- Alexander Altland and Ben Simons (2006). "Interaction effects in the tight-binding system". Condensed Matter Field Theory. Cambridge University Press. pp. 58 ff. ISBN 978-0-521-84508-3
- Ong, edited by N. Phuan; Bhatt, Ravin N. (2001). More is different : fifty years of condensed matter physics. Princeton (N.J.): Princeton university press. p. 65. ISBN 0691088667. Retrieved 2 February 2020
- Wysokiński, Carol; et al. (2003). "Spin triplet superconductivity in Sr2RuO4" (PDF). Physica Status Solidi. 236 (2). arXiv:cond-mat/0211199 . Bibcode:2003PSSBR.236..325W. doi:10.1002/pssb.200301672. Retrieved 8 April 2020

Permissions

All chapters in this book are published with permission under the Creative Commons Attribution Share Alike License or equivalent. Every chapter published in this book has been scrutinized by our experts. Their significance has been extensively debated. The topics covered herein carry significant information for a comprehensive understanding. They may even be implemented as practical applications or may be referred to as a beginning point for further studies.

We would like to thank the editorial team for lending their expertise to make the book truly unique. They have played a crucial role in the development of this book. Without their invaluable contributions this book wouldn't have been possible. They have made vital efforts to compile up to date information on the varied aspects of this subject to make this book a valuable addition to the collection of many professionals and students.

This book was conceptualized with the vision of imparting up-to-date and integrated information in this field. To ensure the same, a matchless editorial board was set up. Every individual on the board went through rigorous rounds of assessment to prove their worth. After which they invested a large part of their time researching and compiling the most relevant data for our readers.

The editorial board has been involved in producing this book since its inception. They have spent rigorous hours researching and exploring the diverse topics which have resulted in the successful publishing of this book. They have passed on their knowledge of decades through this book. To expedite this challenging task, the publisher supported the team at every step. A small team of assistant editors was also appointed to further simplify the editing procedure and attain best results for the readers.

Apart from the editorial board, the designing team has also invested a significant amount of their time in understanding the subject and creating the most relevant covers. They scrutinized every image to scout for the most suitable representation of the subject and create an appropriate cover for the book.

The publishing team has been an ardent support to the editorial, designing and production team. Their endless efforts to recruit the best for this project, has resulted in the accomplishment of this book. They are a veteran in the field of academics and their pool of knowledge is as vast as their experience in printing. Their expertise and guidance has proved useful at every step. Their uncompromising quality standards have made this book an exceptional effort. Their encouragement from time to time has been an inspiration for everyone.

The publisher and the editorial board hope that this book will prove to be a valuable piece of knowledge for students, practitioners and scholars across the globe.

Index

www.ingramcontent.com/pod-product-compliance
Lightning Source LLC
LaVergne TN
LVHW080734210926
848485LV00003B/554